AF841635

392571

SOAS Studies on South Asia
Understandings and Perspectives series

Meanings of Agriculture

SOAS Studies on South Asia

Terence J. Byres, ed., *The State and Planning in India*
Nigel Crook, *India's Industrial Cities. Essays in Economy and Demography*
Dagmar Engels, *Beyond Purdah? Women in Bengal, 1890-1939*
Sudipta Kaviraj, *The Unhappy Consciousness. Bankimchandra Chattopadhyay and the Formation of National Discourse in India*
Peter Robb, ed., *Rural India, Land, Power and Society under British Rule*
Peter Robb, ed., *Society and Ideology. Essays in South Asian History presented to Professor K. A. Ballhatchet*
Peter Robb, ed., *Dalit Movements and the Meanings of Labour in India*

South Asia: Understandings and Perspectives

Nigel Crook, ed., *The Transmission of Knowledge in South Asia. Essays on Education, Religion, History, and Politics*
C. J. Fuller, ed., *Caste Today*
Peter Robb, ed., *The Concept of Race in South Asia*
Burton Stein and Sanjay Subrahmanyam, eds., *Institutions and Economic Change in South Asia (forthcoming)*

SOAS Studies on South Asia
Understandings and Perspectives Series

Meanings of Agriculture

Essays in South Asian History and Economics

Edited by

Peter Robb

DELHI
OXFORD UNIVERSITY PRESS
BOMBAY CALCUTTA MADRAS
1996

Oxford University Press, Walton Street, Oxford OX2 6DP

Oxford New York
Athens Auckland Bangkok Bombay
Calcutta Cape Town Dar es Salaam Delhi
Florence Hong Kong Istanbul Karachi
Kuala Lumpur Madras Madrid Melbourne
Mexico City Nairobi Paris Singapore
Taipei Tokyo Toronto

and associates in
Berlin Ibadan

© P. G. Robb and contributors, 1996

ISBN 0 19 563981 2

Camera-ready copy prepared in the Centre of South Asian Studies
School of Oriental and African Studies, University of London
Printed at Pauls Press, Okhla, New Delhi 110020
and published by Manzar Khan, Oxford University Press
YMCA Library Building, Jai Singh Road, New Delhi 110001

SOUTH ASIA: UNDERSTANDINGS AND PERSPECTIVES
GENERAL EDITOR'S INTRODUCTION

The present volume continues a project sponsored by the School of Oriental and African Studies, University of London, and initiated during my tenure as chairman of the SOAS Centre of South Asian Studies. It involves many members and associates of the Centre. The intention is to take a number of important terms or concepts, and to discuss each one's different meanings. In each case a volume will be prepared comprising a critical introduction to sum up and reflect upon the state of our knowledge of this concept, plus a number of illustrative essays containing their own analysis, which may or may not agree with that of the introduction. The essays are also intended to be fairly detailed and empirical in emphasis, so as to stand in regard to the introduction in something of the relationship of evidence to interpretation. The project is directed both at specific problems and at a number of fundamental debates on the nature of discourse; and yet it is not intended primarily to generate new theory but rather to make its contribution by approaching questions from a new direction.

Part of the dissatisfaction which lies behind the project is with Eurocentric terminology. This is not because we deny the possibility of there being any universal terms, nor because we think all knowledge produced by Europeans essentially the same and equally corrupted by power. It is because we are impressed by the need to avoid all essentialism, and by the importance, both intellectually and in practical situations, of an appreciation of difference. It is because we are uncertain how larger categories may properly be constructed. Similar concerns are expressed in various ways in many disciplines, and constitute a crisis of interpretation.

We want to start with detailed and concrete studies. Part of our question is: what is it that constitutes 'South Asian' perceptions? We do not think there is—that is, that anyone can provide—an adequate answer to this question as a whole. Moreover we do not intend our work to be merely comparative, especially not between South Asia and Europe, as if the latter were a yardstick. Though we will investigate the question, we are not at all sure that there *are* peculiarly South Asian perceptions, as a type, and we are as interested in differences within the region and over time as in discrepancies between the terminology of our disciplines and that of its subjects. On this basis, we think it worthwhile, with regard to concepts in frequent and often uncritical use, to examine the different meanings they have, and then reflect upon the differences. The exercise can be substantive rather than terminological. We question whether the 'virtual reality' of our studies bears any close relation to the complexities of its subjects. We think we can get closer, provided we work with small components. We do so not to be restricted to them out of context, but to use them to focus on larger issues.

Peter Robb

PREFACE

Other volumes in this series consider such questions as race, caste and gender, transmissions of knowledge, economic units and legal rights. This volume thus adds to a list of issues crucial to the understanding of South Asia. It presents a selection of papers from a conference on the 'Meanings and Purposes of Agriculture in South Asia' held in SOAS in July 1992, plus several papers commissioned later.

Methods of citation and other details have deliberately not been standardised beyond a certain minimum: the conventions of different authors and disciplines have been retained.

I am grateful to all the contributors for co-operation and patience in about equal measure. I am grateful also to other paper-givers, discussants, participants and assistants (especially Janet Marks) at our meetings in July 1992, and to all who were involved in two other workshops held in tandem with them. The first concerned the nature of economic units, and was organised by Burton Stein and Sanjay Subrahmanyam. The second workshop presented Japanese and comparative insights into Indian local economies, and was organised by me, Kaoru Sugihara and Haruka Yanagisawa. Revised proceedings of both of these conferences are being published.

Generous financial support was given to this work by the School of Oriental and African Studies, University of London. It included a temporary fellowship in the SOAS Centre of South Asian Studies for Utsa Patnaik. She gave me invaluable assistance with the design and execution of the conference and this volume.

Very warm thanks are due to Janet Marks and Joyce Hutchinson for the preparation of the camera-ready copy (their help will be grievously missed in future), to Barbara Lazoi who completed the final stages, and to the support and efficiency of Rukun Advani and the Oxford University Press in New Delhi.

Peter Robb
London, 1996

NOTE ON CONTRIBUTORS

B. B. Chaudhuri	Department of History, University of Calcutta
Rajat Datta	Centre for Historical Studies, Jawaharlal Nehru University
Sumit Guha	Fellow, Nehru Memorial Museum and Library
Benedicte Hjejle	Independent scholar, Copenhagen
Karin Kapadia	Formerly London School of Economics
David Ludden	Department of History, University of Pennsylvania
Eric Meyer	Centre d'Etudes de l'Inde et de l'Asie du Sud, Paris
Utsa Patnaik	Centre for Economic Studies and Planning, Jawaharlal Nehru University
Peter Robb	Department of History, School of Oriental and African Studies
Abhijit Sen	Centre for Economic Studies and Planning, Jawaharlal Nehru University
Chitra Sivakumar	Department of Econometrics, University of Madras
S. S. Sivakumar	Department of Econometrics, University of Madras
Ravi S. Srivastava	Department of Economics, University of Allahabad

CONTENTS

Chapter 1

VARIANT MEANINGS OF AGRICULTURE: IMPLICATIONS FOR RESEARCH AND POLICY

Peter Robb

As *social* sciences, economics and hence economic history are fundamentally about exchange. As social *sciences*, they provide explanations of processes. Even microeconomics (to adapt the definition of a well-known textbook) analyses 'how *society* decides what, how, and for whom to produce':[1] in positive economics, theory is predictive of how *systems* work. This tends to mean that particular forms are generalised. Certain kinds of and motives for production, and in particular those involving markets, inevitably occupy the central or normal place in economic explanations. It is evident that this produces problems and generates heterodox interpretations in almost all specific studies, sometimes because of the unavoidable 'intrusion' of non-economic factors, but perhaps most of all when such 'modern' methods and theory are being applied to earlier or other situations, as in histories relating to 'pre-capitalist' or non-European societies. This book considers some of these difficulties, in regard to the understanding of agriculture in South Asia.

In a recent work on India, the finding of K.N. Raj was that new seed strains, improved technology, and necessarily concentrated irrigation have produced dualism and not transformation in agriculture, and hence that very high increases in productivity are either illusory or due to specific favourable conditions.[2] The argument implies a 'modern', technologically-advanced agriculture on one hand, and a 'traditional' agriculture on the other. One is at least 'formally free' (in Jon Elster's sense) and the other is bound or limited by its environment. Raj argued against the expectation that technology alone can provide increased productivity; in particular, institutional changes may be needed because

[1] David Begg, Stanley Fisher and Rudiger Dornbusch, *Economics* (2nd edn., Maidenhead 1987), p.2; emphasis added.

[2] K.N. Raj *et al.*, eds., *Organizational Issues in Indian Agriculture* (Delhi 1990), pp.10-21.

of, say, tenurial conditions, size of production units, or utilisation of labour. He wrote: '...there are risks associated not only with agricultural production but with institutional arrangements like leasing of land and hiring of labour. There are also differences in the capacity to bear risks and in the willingness to bear them. All this can weight the scales in favour of alternatives different from those one might expect under conditions of certainty.'[3]

The argument is a curiously exact replay of disputes among colonial agronomists and officials in late nineteenth-century India, as I and others have shown.[4] Raj's conclusions seem transparently correct. How do we make use of them? Clearly there is a need to go beyond single-order assumptions, such as the universal value and appropriateness of technology, and to take up instead concrete questions about production. But how many descriptions do we have of the details of these multiple, different conditions? Do they imply, in effect, not a traditional or a modern but an infinite number of agricultures? What agreed means of generalisation do we have, to group and make practical use of these differences?[5]

We know that agriculture is undertaken for different purposes—for food, for profit, for status, and so on. Its variety is obvious, and yet we tend to regard it as a single universal category. Two kinds of question arise from the variant meanings of agriculture. The first considers what features are really general, and whether we should be constructing any basic principles and procedures, any laws of agricultural production.

[3] Ibid., pp.44-5.

[4] See P. Robb, 'Bihar, the colonial state and agricultural development in India, 1880-1920', *Indian Economic and Social History Review* xxv, 2 (1988), pp.205-35.

[5] The implication is that analyses of production details are relatively rare in regard to the history of South Asian agriculture, and that general theories have not tended to take South Asian conditions into account. For exceptions from different perspectives, see the concern with the socio-political context of peasant production in a hierarchical society, in Jacques Pouchepadass, *Paysans de la plaine du Gange. Le district de Champaran 1860-1950* (Paris 1989), for example pp.xxvii-xxxi and 594-615, or the study of production calendars in Shahid Amin, *Sugar and Sugarcane in Gorakhpur* (Delhi 1984), or Ian Stone, *Canal Irrigation in British India* (Cambridge 1984). For examples of what is needed, one might note the methodology of theoretical outline plus details in, say, Jules N. Pretty, 'Sustainable agriculture in the Middle Ages: the English manor', *Agricultural History Review* 38 (1990), and its parallels with the Sivakumars' chapter below.

Can we map out the many differences of purpose and meaning? How important are they, both in practice and for understanding? Can we explain why certain crops are grown or certain methods used? How different are the answers for different persons, fields, crops or seasons? Are there distinctive sets of answers which fit particular periods or areas? Are there, for that matter, even wholly 'subsistence' and 'market' *strategies* let alone economies? The second question asks how we should respond to the diversity in particular cases—here with regard to South Asia. Are some of the differences of purpose and meaning peculiar consequences of the forms and conditions of agriculture to be found in our region?

Political economists have long debated these questions of course, for example during the Indian 'mode of production' debate. When introducing a collection of essays from that debate, Utsa Patnaik (who also contributes to this book) wrote: 'the process of commoditisation under colonial conditions was...quite distinct from commercialisation of agriculture in capitalistically developing countries'.[6] The key to the difference was, she says (as again in this volume), the 'unilateral transfer' of resources and profits away from India to Britain. In other words, there is a distinctive context in India which has significantly shaped the nature of its agriculture; there is after all an Asiatic mode, though in an altogether different sense from Marx's. The basis of Patnaik's argument —that India is different, and not a mere appendage of the colonial and capitalist world—is one very sympathetic to the approach proposed in this volume. But the equation of macroeconomic with colonial and capitalist forces may seem to leave India an appendage after all (as in one of Hamza Alavi's articles included in Patnaik's anthology); and, more seriously, we still have very great problems in linking the specifics of macro- and micro-levels by means of what is after all, at least as an initial hypothesis, another single-factor explanation. Utsa Patnaik herself has stated that the concept of a mode of production is not descriptive but analytical, that it 'could not be treated as one might an elastic glove, stretching it here and there to fit varying empirical reality',[7] reality in which there must be as many modes as there are social formations.

[6] Utsa Patnaik, ed., *Agrarian Relations and Accumulation* (Bombay 1990), p.6.

[7] Ibid., p.4.

Most historians will experience difficulty in applying concepts which do not answer to empirical examples. Of course theory is necessary in social as in other sciences to deal with that order of things, the most interesting, which cannot be wholly described in empirical terms. But this does not free theory from the need to confirm as far as possible with what *can* be known, and constantly to be adjusting to new discoveries or applications. Otherwise it would stand in relation to reality rather in the fashion of God to the universe in the teleological argument: a connection has to be assumed, as a matter of faith, by analogy with man's own ability to build tools and mechanisms. The problem with such reasoning of course, as science indicates with growing certitude in recent centuries, is that the natural world does not arise from design but from processes of a different kind, from physical and evolutionary laws which strictly are without cause, purposeless. To explain the meanings of particular practices in Indian agriculture, therefore, we cannot merely refer to the colonial legacy (as a monolith) but must demonstrate specific connections between observed conditions and effects. An accumulation of these will embolden us to suggest a hypothesis which is more general than our examples.

Some such demonstrations have been provided in current theories of Indian agriculture, but more detailed work is still needed. Moreover, such notions as 'village community', 'moral economy', 'paternalism', 'capitalism', 'mode of production' and so on, properly describe not systems but aspects, elements which exist in all economies to varying degrees. We may well be able to discover dominant features, such as colonial impact on India, but only within strict limits. And if there are movements over time from, say, so-called pre-modern to modern characteristics, then they are engendered by powerful general forces, and not by any law of inexorable economic motion. In this way, studies of agriculture in South Asia now find themselves in the position reached some time ago by historians of medieval Europe when they abandoned the idea of the progressive breakdown of feudalism, and began to emphasise fluctuations between serfdom and wage labour as *strategies*. More controversially, it may be that the end of such tactical fluctuations was capitalist production and wage labour only when (and so far as) larger conditions and perceptions required. Parallel points may be made about modes of agricultural production more generally.

M.S.S. Pandian notes that the Indian mode-of-production debate, with a few exceptions, paid too little critical attention to concrete data on Indian agriculture, and that this was especially problematic in regard to assessments of change.[8] One might add that it was most disadvantageous of all with regard to the early history of change and continuity. Pandian concluded that it was not enough to describe the macro-economic factors as being exclusively and monolithically 'colonial'. It was necessary to study (as some scholars had) the indigenous context, the political economy of production relations in terms of landownership, labour deployment and so on. Then, in addition, it was necessary to relate such contexts and parameters 'holistically' to each other and particularly to productive forces, which had received too little attention. We need to take a closer look at the details of all aspects of production. In particular we need disaggregation, at an early stage in our inquiries, asking, for example, who controlled what, what influenced which decision? By no means need this imply a re-run of the earlier debates. Rather, we should be responding to the extent to which they reached an impasse, and to which as a result there is now no consensus about which general theories to apply. At issue is more than peasant *mentalité*; this we may well be unable to recover or reproduce. It may often be possible, however, to arrive at the subjective factors in decision-making by examining objective limits and outcomes; and these after all are as important a part of the story (though not more so) than the consciousness which reflects them.[9]

Thus this book advocates a fundamental re-thinking of the connections between the different forms and functions of agriculture in practice, and the theoretical analysis of agriculture-related issues, especially economic change and development. The intention is to cast doubt on current procedures and debates by reflecting on the frequent discrepancies between empirical findings on the conditions of agricultural production, and the assumptions and generalised explanations underlying many assessments of agrarian change or policy. Instead of treating agricultural production only as a stage in a process of exchange, we ask

[8] M.S.S. Pandian, *The Political Economy of Agrarian Change* (New Delhi 1990).

[9] But for example see the discussion of J.C. Scott and the supposed peasant subsistence ethic in Michael Adas, '"Moral economy" or "contest state"? Elite demands and the origin of peasant protest in Southeast Asia', *Journal of Social History* 13 (Summer 1980), pp.523-30.

about its different meanings and purposes. The underlying issue here is firstly the appropriateness to South Asia of available theories, and the applicability of analytical and empirical techniques. We seek to identify which are universal and which specific features, mainly in order to understand what is distinctive about the conditions and potential of agriculture in South Asia. A more concrete set of questions concerns the nature and direction of any changes in agricultural purposes, for example (as is most commonly assumed to be the case) *from* the reproduction of communities *to* the maximising of individual profit. How do longterm trends in agriculture's context—in state policies, landed-property rights, the access to markets and terms of trade—affect the nature of agricultural production? Or are agricultural strategies always specific, multiple, flexible and contingent?

At the workshop which gave rise to this volume, the issues uniting the papers seemed to be threefold. First, we identified three kinds of *knowledge*: localised, fragmented and universalist. More generally there were questions of how knowledge is used, and how it changes over time and space: for example, in the formulation of 'archaic' knowledge (as here by Ludden), there is an implicit opposition with newer or more 'developed' knowledges. We also considered agriculture as an 'object' of study and science, rather than as practice (or vice versa). Secondly, we discussed obvious differences in *kinds of agriculture*—that is, in agricultural regimes from slash-and-burn to plantations. In general these are the kinds of distinction which form the basis of analysis—that is, we categorise according to the organisation or relations of production. Agriculture also relates differently to physical environments. Its role and condition are different in marginal areas and in areas with various forms of artificial irrigation. It appears differently in villages, small-holdings and estates. It has different imperatives and rhythms according to whether the main crop is wheat or rice, sugarcane, or coco palms, or groundnuts, or cotton, and so on. Hence different practices are associated with the demands of particular crops. But there are also standardisations arising from beliefs and customs, social or gender roles, or indeed in accordance with broader forces derived from law, terms of trade and so on—factors which might extend across different agricultural regimes. Several papers implicitly or explicitly re-examined these principles of categorisation. We also considered several further deconstructions of factors of production: namely of labour, credit, prices; these examples form a significant part of the illustrative material provided in the papers in this book.

Thirdly, we considered different *purposes of agriculture.* Cultivation has different meanings for those involved in it in different roles—for labourers, share-croppers, peasant farmers, village elites, landlords, providers of credit, processors, buyers, transporters, and tax-collectors; and also for botanists, agronomists, engineers, economists, and historians. It is also said to have different social implications in its various forms and modes—for slash-and-burn cultivation, for market-gardening, for cash-cropping, or for subsistence production. In addition to its roles of providing food, employment and commodities, agriculture has its various sociologies. It may fulfil ritual obligations, express the *dharma* or self-image of groups, indicate masculinity and feminity or family and kinship roles, define bondage or superiority or social status more generally, denote affinity with particular locations, create rights of property (such as a legal presumption of ownership through use), and generate group cohesion or resistance and avoidance in reaction to external authority.

This volume gathers some of the papers in the workshop, along with several newly commissioned. The method adopted is to illustrate the different meanings of agriculture—for cultivators, for labour, and in regard to markets or socio-political structures—with regard firstly to different types and levels of agricultural decision-making, and secondly to inappropriate theory, which in itself provides one of the contexts for agricultural production, for example by influencing states and international agreements. Taken as a whole, the work is thus partly a comparative analysis of concepts.

II

In the first essay, David Ludden stresses the multiplicity, conditionality and temporal specificity of meanings in agricultural terminology. This was especially true of India, he suggests. He offers functional and socio-political explanations of the agricultural terms current among elites and more broadly in populations. For example, in medieval Europe, a discourse of tenure was appropriate to a feudal order which allocated land without controlling the details of production, but about the same time in China agrarian techniques and conditions were rendered more visible to external elites in order to serve a system of state intervention and improvement. For India one might add Mughal manuals which, for equally 'appropriate' reasons, related to gardens, special crops such as out-of-season mangoes, and revenue assessment and collections.[10] Ludden is anxious not to ring-fence, as 'agricultural', understandings which arose from social and political conditions which did not treat agriculture as a distinct branch of behaviour and knowledge. But chiefly, in India, he argues, there were local distinctions between agricultural vocabularies and perceptions that were largely environmentally determined, for want of centralised control over agricultural production—as in land tenure, measurements, currency, taxation and so on. A desire for external control later generated stronger generalising and 'scientific' influences upon terminology. Ludden then employs the differences in terms and perceptions in different regions of south India in order to uncover reality. He assumes that variations in the incidence of proverbs and sayings in nineteenth-century collections reflect actual variations in their incidence, and thus in agricultural knowledge, in different regions.

He also uses the terminology to define changes over time. The biggest change was that a discourse of science replaced heterogeneity, a revolution which produced the current terminology and understandings of agriculture. Ludden shows that the 'scientific' norms are inadequate to explain the south Indian past. But it is also his contention that earlier, localised forms of knowledge became 'archaic'—that is, they were superseded by new generalised knowledge attuned to commodity production, labour mobility, education, science, and the standardisations of the state and capitalism. Ludden does not address this question very fully. That it is true of the prevailing discourse of agriculture need not be doubted. But whether or not that discourse is wholly appropriate to India is one of the major subjects of this volume. In the colonial period it can be argued that a quasi-feudal emphasis on tenurial issues was still at least as influential as the scientific and capitalist assumptions. Even in more recent times, a major issue is whether Indian agriculture is organised according to the rationale and functions attributed to it in outsiders' analysis.

[10] See I.G. Khan, 'Changing patterns of authority and control in post-Mughal north India: c.1740-1830' in J.P. Neelson, ed., *Gender, Caste and Power in South Asia: social status and mobility in a transitional society* (New Delhi 1991).

Two important aspects of this question are the manner of commercialisation, and the status and role of agricultural labour. In his essay below on commercialisation in eastern India, Binay Chaudhuri first provides a summary of some of the views against which this volume may be judged. His concern then is to ask how far commercial agriculture was 'forced', especially through the manipulation of harvest loans by those who sought to buy the crop. He finds that such advances, though prominent in particular circumstances, were not generally needed for the great expansion of jute cultivation in East Bengal where alternative sources of credit and alternative cash crops (namely rice) precluded the dependency that developed for some other crops and areas. Even in those cases, Chaudhuri draws attention to apparent resistance by cultivators, including withdrawal from the cultivation of the crops in question—the implication being that elements of consent and self-interest may be deduced where the cultivation continued. Where this is leading, as the example of sugar makes plain, is towards profit motives in aspects of Bengal agriculture. Elsewhere I have suggested an additional and rather different conclusion, that the forms assumed by forced commercialisation, such as advances, also reflected difficulties experienced by outside capitalists in their attempts to penetrate the Indian countryside.[11]

Rajat Datta's concern in his paper is to extend his earlier discussions of the expansion of the rice market in eighteenth-century Bengal, particularly with regard to share-cropping. Though he dissents from aspects of my earlier discussion of agricultural commercialisation,[12] in several ways his paper complements my arguments and others in this volume. My main point has not been the imperfection of markets in colonial India—of course virtually all markets are imperfect in some respect. My concern is with the different levels and types of choices in agricultural production. Though Datta spends time, convincingly, in demonstrating that Bengali agriculture was capable of responding to demand, it is also plain from his examples that, as he says, the price-responsive increases in rice production were engineered largely by intermediaries,

[11] See Peter Robb, 'Peasants choices? Indian agriculture and the limits of commercialization in nineteenth-century Bihar', *Economic History Review* xlv, 1 (1992). This essay was provided as the starting-point for the workshop from which this volume developed. Hence several further references will be made to it in this introduction.

[12] Ibid.

for example through land reclamation and harvest loans. It is not clear that his agriculturists enjoyed the same range and potential of decision-making as, say, their Japanese counterparts. Many in the eighteenth-century Bengali countryside certainly seem to have participated in markets—in some sense even smallholders could do so on portions of their holdings, as did share-croppers at least in the proportion of their output claimed by their principals. But, except in a relatively few substantial and thus independent households, the cultivators' motivations and opportunities seem often to have been limited by their immediate needs for cash or repayments. It was local conditions, including to some degree the entrenched position and expectations of the cultivators, which helped perpetuate the relations of production; yet, because of them, it was often the controllers alone who were motivated by and responsive to the market, and only they who could calculate the risks of new ventures. The 'village', as Datta puts it, 'was at a relative disadvantage' when it came to market participation. He postulates a functional relationship between share-croppers' labour-power and their principals' investment capital, but of course, as he shows, there was also a division of interest and responsibility. This amounts to another significant disjunction between those responding to the market and those controlling agricultural methods.

Some other papers in this book attend to the details of production, in two main ways—by considering land-use in relation to different social levels, and by assessing the standing of, and the range of decisions available to, different kinds of agricultural labour. First, Benedicte Hjejle studies the details of land use and management in several villages in south India on which extremely detailed information is available for the period of transition to East India Company rule. Her account provides a salutary lesson, through statistics, on the resilience of localities, its main point being the importance of more marginal land and of local managers of cultivation to the overall agrarian economy. Hjejle shows the variety of different and co-existing strategies of land use, and not only that use was subject to ecological conditions and seasonal adjustment, as well as to local socio-political considerations, but also that the figures collected by officials (and hence their perceptions) were susceptible to local manipulation and distortion.[13] On

[13] The argument that Indian colonial statistics are unreliable is thus endorsed, but also the idea that they nonetheless may be used; for such a discussion of the

the other hand, Hjejle uses the same information to argue for a transformation of the rural economy after the intervention of the Company's government. She suggests that the agrarian controllers and intermediaries (whose function and character she analyses) were affected fundamentally. Certainly it appears that different roles became instrumental in the revenue and marketing systems, and in particular that power shifted to groups uninvolved in actual processes of agriculture. This caused, according to Hjejle, social division, decline of production and increased vulnerability. Later the nature of such change is taken up here more fully (though with less statistical detail) in the complementary paper by S.S. and Chitra Sivakumar.

The different purposes of different lands within a single 'village' and their disruption under and since colonial rule are also graphically illustrated by Eric Meyer in his discussion of *chena* (temporary slash-and-burn) agriculture in Sri Lanka. His analysis of its role is founded on a characterisation of cultivators according to the range of decisions they were able to take, and a consequential consideration of the size of decision-making units; like Hjejle, he shows that the individual farm is 'not...entirely irrelevant...but...is encompassed within a larger unit'. Similarly different types of cultivation—irrigated paddy, market gardening, *chena*—commonly formed a single system of production, providing a distinctive definition of agriculture in many parts of Sri Lanka. The conflicts between this and other understandings can then be explained. Meyer shows how it is necessary to comprehend whole systems and their peculiarities before deciding on optimal tactics: it was partly a misunderstanding of the functions of *chena* which produced the ill-effects of state agrarian policies in Sri Lanka, before and after independence. For colonial administrators, with their stereotype of settled cultivation as economically superior and more virtuous, the proper mode of agriculture was expressed in the plantation: this led to dispossession and disruption for many Sri Lankan farmers. For nationalist politicians, purporting to resurrect a golden age of the peasant, agriculture came to mean the supremacy of paddy: this led to non-peasant development and the exacerbation of tensions, including ethnic conflict,

---

problems of Indian statistics of a later period, in the context of the vexed question of agricultural yields, see Sumit Guha, ed., *Growth, Stagnation or Decline? Agricultural productivity in British India* (Oxford in India Readings, Delhi 1994), pp.1-48.

in the north and eastern parts of Sri Lanka. Thus outcomes, the different strategies at all levels, were influenced by understandings.

Similarly, in detailed case studies of recent conditions in north India, Ravi Srivastava illustrates some of the particular characteristics of Indian agriculture at the local level. He turns our attention from the land to the cultivators. His unit of decision-making is not the world or the state or even the village community. It is the set of owners, employers, creditors, share-croppers and workers focused on each agricultural plot, a set whose parameters differ according to the kind of decison being made. The reason for this specificity is in large measure the resilience of historical relations of production. He takes two contrasting villages. In the first, agriculture is a possibly alternate occupation for workers, as well as a basis for peasant production. Employers have sought to meet challenges to their authority and profits by increasingly paying for casual labour and by piece-work rather than daily wages, by letting out land to poor families on a seasonal basis, by raising the landlord's share of output and of the costs of production, and by imposing a closer control over production decisions. Debt is a major means of control over the labour-force, virtually replacing more 'traditional' ties between employers and employees.

In the second village, economic change has been less rapid, but there too the coercion of workers has increased. The landlords and richer tenants are not intervening quite so much in production decisions, though they are tending to become more involved. A similar earlier stage is visible in regard to changes in the patterns of share-cropping. But, again, though service ties still prevail, debt and other forms of instrumental social control are being added to these 'traditional' terms for labour employment. The characteristics of these relations of production, Srivastava argues, are influenced by local conditions, including problems of supervision, availability of alternatives, possibilities for resistance or desertion, and so on. Obviously ecological and economic factors would also play their part.

Srivastava's examples reflect interestingly on the persistence of interlinked forms of coercion (already mentioned);[14] as Chaudhuri outlined, the subject is familiar from many accounts by economists of India. Here the coercion was being strengthened, both in the course of

[14] See Peter Robb, 'Peasants choices?' Ensuing quotations and references are to this paper also.

radical change (in Srivastava's first village), and where there was greater continuity (his second village). Agriculture in this context had meaning as an arena for economic dominance on the one hand, and of labour subordination on the other. Though Srivastava does not use such terms, it seems that his villages continued what I have called a 'strong tendency for economic functions to be subdivided'—even though this had now largely ceased to be expressed in interdependence across a local community. The meaning of agriculture was clearly quite different for different classes among those involved in agricultural production; it was fragmented, a fact which rendered it unlikely that there could be any common response to opportunities or pressures. Though the tendency towards greater intervention by landlords went some way towards constructing a unified space for decision-making and practice in agriculture, the process was by no means complete in Srivastava's villages, even in the more advanced case. Landlords increasingly decided cropping patterns and material inputs, but were much less frequently involved in labour deployment or other decisions about methods of production. This view accords, interestingly, with some nineteenth-century observations. In terms of decision-making, such sets of agrarian actors are a long way from the optimising and supposedly independent farmers so often assumed to exist by economic policy-makers.

By considering our second main question, agricultural labour, Guha too provides further room for disaggregation, using the opposite strategy to that of Chaudhuri—that is, ignoring the broader socio-political context of agriculture. For the purposes of the analysis, this is assumed to impinge more or less equally on local conditions, which implies a consideration of the means rather than the ends of cultivation. Guha nonetheless implicitly introduces different modes of agricultural work (from labour supervision to crop-watching) and hence different purposes for different persons—distinguishing not only between employers and dependants, but among the labourers themselves, from agrestic slaves and household servants to yearly-hired and piece-workers. Guha's purpose is partly to consider why different forms existed; thus household labour, for example, either reduced the need for or eased the problems of supervision. But Guha's ultimate question is how labour input may be measured. Partly using proverbs, like Ludden, and partly from observers' accounts of agricultural practice early in the colonial era, he demonstrates the difficulty of applying one current, standardised concept, that of agricultural work. Such difficulties of categorisation exemplify a point which was central to Ludden's argument. In a concrete instance, Guha once more demonstrates that the different meanings and purposes of agriculture matter very greatly for its assessment and analysis.

His contribution casts into interesting relief a second paper on labour, by Karin Kapadia, studying the recent history of women's involvement in labour gangs in south India. Echoing a major concern of this volume, her starting-point is the inadequacy of theoretical models for the particular conditions she describes. By the workers she studied, agriculture was experienced in terms of wage-gathering. Distinct types of work were recognised, such as for piece work with its higher rates of remuneration and greater effort, and for daily wages, a lower rate for less efficient labour. Kapadia associates piece-rate contracts with 'modernity' and exploitation, in contrast with the 'customary rights' of 'traditional' labour relations. These terms already imply models. However, through Guha's and her paper in particular we are invited to start at a point before the theories to which Kapadia objects; we are able to consider the three main questions which may be asked about all labour, answers to which define its character: how is it attracted and retained, how is it controlled and supervised, and how is it paid? Not all theoretically-possible answers to these questions are ever available to any real employer of labour; the possibilities are variously constrained, for individuals and societies. As they change so do the conditions of labour—in our case the meanings of agriculture for cultivators and rural workers. We will return below to the evolution of labour conditions, and to the context of agriculture, as issues raised by all these papers.

These examples illustrate the different perspectives upon agriculture achievable by adjustments of the arena of analysis. Further papers, by economists, draw a possible moral for policy-makers from these differences. They address the meanings of agriculture in the broad context of international policy and economic theory, and also of course reflect on change. Utsa Patnaik's is a vigorous statement of the view, frequently heard, that agricultural decisions are not made freely, and that in India they are particularly limited by the extent to which producers are in thrall, through 'forced commercialisation', not only to the locally dominant but to international commerce and a worldwide economic system. The substitution of subsistence by cash-crop production (a necessary part of per capita economic growth) is here read in the context of two analytical constructs, the national economy and the peasant society, and as indicating that long-distance demands and not cultivators are deciding cropping patterns. In such circumstances agriculture means at best a bare livelihood, but above all subordination. This picture is a potent one, and reinforces the arguments for different meanings in agriculture; but it is also one which (as in Chaudhuri's paper) this volume helps assess. For Patnaik, and for very many other commentators, the origins of this common condition of Indian agriculture are to be found in the colonial period. As already noted, Patnaik blames what she calls colonial transfers (as opposed to trade), and their equivalents in the post-colonial world. In the present paper too she describes the annual transfers of agricultural surplus to Britain in the colonial period, and, without quantifying them, regards them as a net depressor of domestic demand. More than this, she argues that, under colonialism *and* in the modern system, state and other investment is restricted deliberately in tropical countries in order to depress domestic demand, and thus (in an argument which looks only to the supply side internationally) the export prices of tropical products. In these producing countries, domestic demand is further reduced by the increase in domestic prices to international levels for exported crops, and to some degree for other subsistence crops by the squeeze on total output caused by loss of land to cash-cropping for export. Patnaik concludes that food-consumption and wealth were and are being diverted from developing to developed countries.

This leads to a more startling proposition that developed countries' wealth was not only built upon but still depends on cheap tropical imports. Patnaik claims, rather paradoxically, that it is only these low prices which conceal the dependance of the developed countries upon Third World products, or, more precisely, that there is a causal relation between the wealth of the first world and its cheap imports from the third. Presumably low prices reflect excessive supply, but Patnaik's argument is that that excess is produced artificially from geopolitical rather than economic factors. She implies too that prices are depressed to levels well below those at which existing levels of consumption could be maintained (though, as said, she objects to the excessive external consumption); and also, a different point, that the prices are inequitable—depressed to levels below those at which producers would be fairly rewarded, a moral as well as an economic argument. To explain this situation, she considers the role of debt in the enthralling of cultivators, but ultimately attributes their vulnerability to global forces. Such arguments have the effect of reinforcing the definition of Indian agriculture as a arena for producers coerced to provide raw materials for consumers in developed countries. Indications of transfers of wealth are taken to demonstrate the mechanism for forced commercialisation.

We shall consider that matter further in the light of other papers, but first it is also relevant to assess the explanation offered by Patnaik. The 1970s oil crisis showed the damage that could be inflicted (not only on developed countries) by changing the price and availability of one indispensable import. Many international cartels have sought, more or less ineffectually, to dictate prices for other commodities (which would seem one possible policy conclusion from Patnaik's analysis). But from the oil example in the longer run, and given the present composition of the international trade of developed countries, one might rather think that the fundamental problem for tropical countries was the *declining* significance of their produce and markets for developed economies, and hence their lack of bargaining power. A common assertion is that tropical and agricultural economies are damaged by the import substitution (such as vegetable oils and beet-sugar) and the subsidised exports of food and raw materials by developed countries. As that complaint indicates, the present problem derives from the price for most Third World commodities—and especially the low real returns to producers. Ironically, it is the relatively low returns—in the form of wages and other social costs, though not in the returns to multinationals or to foreign and local capital—that are nowadays regarded in many developed countries as a mark of the success and the competitive threat of industrialising Asian nations.

Abhijit Sen's arguments further develop these international meanings of agriculture. He shows that the World Bank anticipates increases in food production through new technology and the workings of the market (that is, private investment for profit), and that the Bank objects to government interventions on the argument that they distort the market and discourage this productive investment. It follows that, from this perspective, agriculture is defined as an aspect of world trade. Its purpose is to produce goods for export not only from the countryside to the town, but also from one country to another. The unit within which production is determined is the world economy, and the chief instruments of decision-making are prices, working upon individual, profit-maximising farmers. Sen disagrees that this 'World Bank' meaning of agriculture should be applied to India. He shares with Patnaik an appreciation of the malign impact of international trade upon Indian cultivators, and generally of the importance of external forces. For him agricultural decisions depend also on state interventions, with government contributions determining the scale of private investment, the prosperity of producers, and the possibilities for growth. Reinforcing this, he notes in India the increasing commercialisation of inputs, notably fertilisers, even for parts of the output which are not intended for the market, and he attributes recent improvements in Indian agriculture to increases in yield (with a consequent danger of over-cropping) rather than to the intensification of land-use or the extension of cultivation. For Sen, the price mechanism is insufficient to increase production equitably, and hence his suspicion of the prescriptions of the World Bank and GATT. Instead (as Patnaik also argues) another outside element, state policy, has a crucial role in defining the conditions in which agriculture is carried on. Accordingly Sen fears the effects of unbridled price mechanisms, as he does those of a reduction in fertiliser subsidies in the absence of a countervailing increase in other public investment. In short, for Sen, agriculture is an element of a national economy and indeed polity, its decision-making contained within the sphere of each responsible state. His criticism of the present Indian government is that it is ignoring its responsibility to improve the terms for agriculture.

Underlying his arguments against liberalisation and the end of subsidies to agriculture, because of what will be at best uneven benefits for the agricultural sector in India, Sen thus offers a particular appreciation of the function and mechanisms of agricultural production, one which differs from that of the World Bank and free-market economists. He anticipates political as well as economic stresses from these contradictions. However, his analysis is not ahistorical. For example, he notes changes occurring in the meaning of agriculture within the Indian economy, how its importance is declining due to a weakening of inter-sectoral linkages. The tragedy of India, as Sen shows, is that this change is not being furthered by a reduction in the dependence of the population upon agriculture. Here we return (as a consequence of this anomaly) to another function and arena for agriculture: its meaning solely as a source of employment and subsistence. In more advanced economies the primary purpose of agriculture is to produce a source of profit for producers, on similar terms to those prevailing for the producers of any other commodity or service. In India, as Richard Jones pointed out in his critique of classical rent theory, agricultural production often does not operate on those terms. For many areas and large sections of the population, agriculture is subsistence oriented.

The final paper by S.S. and Chitra Sivakumar begins where the opening paragraph of this introduction left off, and attempts to provide an alternative paradigm which encompasses the variant and specific aspects displayed in this book. It applies the kind of insight discernible from Patnaik and Sen, to illustrate a trajectory of change similar to that proposed by Ludden and Hjejle. Its hypothesis depends on the contexts in which agriculture was carried on, a subject pursued through a study of agrarian society in Tondaimandalam over two hundred years. In particular the Sivakumars give elaborate and convincing explanations of the nature and origin of the overlapping character of local entitlements, leading to intermixed or overarching control: they provide a full account of the interdependence between employer and worker inherent in such systems, and of their relationship to the nature of agricultural production, to historical evolution, to caste and belief systems, and so on. Employing an approach which has been fairly commonplace among historians, though recently hailed by some economists as a new institutional economics, these studies postulate that, in the colonial era, a disjunction developed between local and external environments of production, and that this had serious consequences for the productivity of Indian agriculture and the distribution of wealth.

By this account, in pre-British days the village was the locus of transaction costs and political power. In colonial times these costs were incurred increasingly at the demand of the state and its bureaucracy. Such costs increased further during the post-Independence period of centralised planning. The different regimes affected agricultural performance. The period saw a long struggle by these external forces against the existing social order which had been promoted by communal irrigation and paddy cultivation: collective caste and kin institutions were opposed or transformed by governments which sought to manage the agrarian order centrally. The incompatibility between the imperatives of indigenous agriculture and those of politics and the state was eventually resolved by the victory of centralisation and of individualistic irrrigation and production. It was a revolution that contributed to the present destitution of the agricultural labourers. The costs of centralisation, moreover, help explain the sluggish economic performance of the colonial and planning eras. A similar disjunction and consequences were apparent in Ludden's and Hjejle's accounts, also on south India, and they appear repeatedly in the other papers as well.

III

We can go a long way with attention to the detail. But we also seek the answers to broader questions of historical change and possibilities for economic development. These imply general questions about the construction of theories, and specific comments about aspects revealed in this volume. In the end the issue must be how far there are general rules which explain why one kind of agricultural regime is more efficient than another. Moreover, though the arrangement of this volume is thematic and not chronological, all the papers offer some explanations of change over time. But some of the papers operate at a level of higher generalisation than others. The Sivakumars certainly seek for general rules, for an understanding of trends, such as the nature of emerging 'world systems', and hence for prescriptions in economic policy. Another attempt to construct such an analysis is represented by Patnaik's indictment of colonialism. Generally in these explanations, as said, the writers exemplify the economists' tendency to look outside the field of agriculture itself. But a contextual emphasis invites comparison with the focus upon agricultural production in other papers—the volume is divided between the two approaches, as in the age-old opposition between theoretical and mimetic knowledge.

To take the instance of colonial impact (it may stand for other generalisations), we find inevitably that the empirical examples raise difficulties. For example, the power and the distinctive European character of the early East India Company, as portrayed by the Sivakumars, may seem exaggerated to modern historians. In their account, however, it is virtually a metaphor for the engine of the economic and political change which they believe to have been ushered in during the nineteenth century. Similarly, for Patnaik's analysis the main outlines are familiar (not least from her own earlier work), but, because the details are slippery, the conclusions are also contentious,[15] and there are some particular puzzles about aspects of the interpretation. For example, it is unlikely that India's increased consumption of imports occurred in contexts of wholly flat levels of domestic demand so that (as Patnaik argues) the imports always displaced indigenous production. The colonial era displays the oddity of rural poverty apparently worsening as agricultural prices rise, increases of rent are legally restricted, and land taxation is reduced in real terms. This was a time when India generally had a favourable trade balance, and received increased prices for vastly increased exports. This does not reduce the undoubted importance of India to Britain's trade balances, and the transfers of various kinds from India to Britain, but the fact remains that India was the major importer of bullion over much of that period. Famine may be taken to prove impoverishment, but, even leaving aside the muddy waters of population and G.D.P., there are uncertainties about famines—whether they were specially frequent or severe in the colonial period; what part in the death-toll was played by epidemic diseases; how far and why famines were less frequent after 1901; and so on. Moreover, some of colonial India's external transfers were payments for works of irrigation, infrastructure and administration. The usual argument is that these works did not increase productivity or at least the well-being of the Indian majority; but later Patnaik considers such investment crucial to economic development. Similarly she suspects the role of markets but wants to see Indian farmers able to produce larger surpluses.

The obvious answer to conflicts of interpretation is not to assert the supremacy either of theory or of empirical detail—a false dichotomy anyway—but rather to try to construct more accommodating theories. Certainly we cannot do without them. The analyses of the past help

[15] For a recent balanced discussion, see B.R. Tomlinson, *The Economy of Modern India 1860-1970* (New Cambridge History of India, III.3; Cambridge 1993). For a debate over the applicability to South Asia of a comparably embracing theory, that of Immanuel Wallerstein, see Sugata Bose, ed., *South Asian and World Capitalism* (Delhi 1990). Though the collection is uneven, David Washbrook's emphasis on 'intermediation', in his 'South Asia, the world system and world capitalism', will be found to be relevant, in a general way, to some of the arguments to be made below; and note also C.A. Bayly's insistence on the confounding of external influences with endogenous changes (and incidentally, relevant to the series in which this volume appears, his call for a new history of knowledge), in his 'Beating the boundaries: South Asian history, *c.* 1700-1850'.

determine the prescriptions for the future. It is interesting that others have argued that, however it is explained (and colonial rule is a prime candidate), the key problem for modern India has been its failure to maintain its share of world trade—[16] the consequent depression in G.D.P. plus inequitable income distribution restricted demand and discouraged investment. It is that vicious circle which, some may argue, Manmohan Singh's reforms currently seek to redress. Similarly, earlier and until recently, rather than a more diffused xenophobia or even socialist ideals, it was the emphasis on colonial *exploitation* (the malign influence of outsiders and of export trade, as in the 'drain' theory or in the explanation of famine in terms of reduced food availability per capita, repeated here by Patnaik) which provided the main intellectual and emotional bulwark of the closed, regulated economy which India enjoyed. By the same token, it is difficult to see what other, new policy recommendation could flow from Patnaik's diagnosis.

What of that other line which she introduced for an inquiry into *general* conditions of agriculture, namely the systematic disadvantage of agricultural in comparison with industrial production, especially important in an age of technological innovation, improvements of productivity, and product diversification? The disadvantage may be mitigated by investment but it remains, Patnaik implies, both relative, in terms of the average returns from a given investment, and absolute, because of the finite capacity of land to produce and, in the case of food crops, of the consumer to consume. We might conclude that this is one reason for developed countries' subsidies of their agriculture, their barriers against agricultural imports, and the consequentially high prices they pay for food; and also that most tropical agriculture suffers because comparable subsidies are not being transferred to it, aid budgets providing only a puny contribution. As colonial Malthusians also argued even in the nineteenth century, the answer for developing countries must be investment to enable them too not only to cultivate efficiently but to diversify away from agriculture. This was recognised in India by the planning of the Nehruvian era; and, by different methods, it is the strategy of India's rulers in the 1990s. Thus *the* question, of any alternative prescription to open-market policies, is whether it can create as much investment and as much competitiveness and efficiency. Patnaik's own

[16] See, for example, Colin Simmons, 'Deindustrialization and the Indian economy, c.1850-1947', *Modern Asian Studies* 19, 3 (1985).

gloomy survey of the international predominance of free-market rhetoric suggests that at least a show of concurrence will be needed to satisfy the financiers and multinationals (given that their preferences can neither be transformed nor ignored, by smallish players in the real world). It may be also that domestic demand is almost bound to be insufficient, especially today, to support fledgling industries, in that large increases in mass consumption may be expected only after industrialisation and a rise in living standards: hence the advantage to industrialising economies of initially selling internationally, to existing elites. The comparative inherent weakness of agriculture, identified by Patnaik and the Sivakumars, is thus writ large: it implies that the gap between local and general conditions will remain difficult and painful.

In short, general theories are at once desirable, instrumental and dangerous. How then are differences of interpretation to be resolved, and the priorities of macroeconomic policy decided? What we have demonstrated (as if the lesson needed repeating) was that detail undermines generalisation. Thus (*pace* Gandhi in *Hind Swaraj* and the Sivakumars here) there are assumptions current in the literature which make pre-colonial societies resemble a glorious amalgam of the golden age, a moral economy, the village community and Indian spirituality. But closer examination often reveals a more complex and contradictory picture, with greater contestation and continuities from the past into the present. More detailed historical study will often reveal too that some innovation, a favourite associated with a supposedly emerging or changing system, in fact has a long history. Our problem then becomes one of deciding whether the function or effect are as unchanged as the form.

Sometimes one can place one detail against another in order to qualify an interpretation derived from theory. For example we may decide that, as a description of change in Indian labour conditions, Kapadia's verdict does not work at an absolute level. It may be that, in a particular time and place, piece rates are, as she says, more exploitative than some other forms of payment. The Sivakumars, however, stress instead a shift from annual to casual labour contracts (a frequent characterisation, for example by Jan Breman).[17] Guha's account shows

[17] See also the more nuanced conclusions of V.K. Ramachandran, *Wage Labour and Unfreedom in Agriculture* (Oxford 1990). For convenient summaries on casualisation (and labour migration) see Jan Breman, *Of Peasants, Migrants and Paupers. Rural labour circulation and capitalist production in West*

that piece rates have long been one of the available means of labour control and remuneration in India, one with obvious advantages for the means of supervision. Kapadia's view therefore has to work quantitatively and contextually: in her case she presents powerful evidence of a decline, in agrculital production and employment, of collective arrangements of a vertical or mixed kind, in favour of individualistic or market relations, in which a limited degree of horizontal solidarity in the workforce does not disguise its basic division and vulnerability. Yet, as Ludden's account reminds us, very much this same decline has been reported for generations. Kapadia also shows therefore that it takes *particular* forms. Perhaps the wider use of piece rates represents a means of increasing exploitation and control in conditions of changing bargaining power and gender-composition for the labour force. Kapadia notes that it has coincided not only with the increasing employment of women, but with their winning of higher wages. In her examples, socially agriculture provided an arena for rivalry as well as solidarity; and, for at least one labour-gang leader, it provided an opportunity for enterprise beyond the expected role of gender and caste.

Sometimes generalisation can also mount an attack on detail; that too is its function. For example, we may decide that in his favourable view of *chena*, Meyer is at one with those economists who regard the so-called informal sector as specially dynamic, even democratic.[18] But *chena*, like the informal sector, may also have been appropriate only for certain economic conditions and goals. For all its value as safety-valve or in allowing upward mobility, it may have reflected poverty of opportunity: lack of capital, barriers to enterprise (in this case royal or state and other controls over labour and irrigated land), leisure-choices, low productivity, and weak economic demand. It may thus have facilitated low-level subsistence while helping to perpetuate an inequitable system overall.

The persistence of different forms of social and economic rationale, within the prevailing norms of colonial and then international commerce, makes it important that general prescriptions should not be carelessly applied, whether in scholarship or policy. We see clearly then

---

India (Oxford 1985) and *Labour Migration and Rural Transformation in Colonial Asia* (Comparative Asian Studies 5, Amsterdam 1990).

[18] This comparison was suggested to me by a paper by Nalini Vittal, 'An Indian view of the informal sector', read at SOAS, 18 January 1995.

that our real task is to identify the precise configuration in which external and internal contexts impinge, to shape the conditions for agriculture. This was what is implied by Meyer's paper and by M.S.S. Pandian's insistence on 'holistic' explanations. The implication of this discussion is not that there may be no general theory, but that distinct local circumstances must be allowed continually to modify it. Equally, comparisons help refine empirical interpretations. From this perspective we can begin to construct some common ground from the varying interpretations of these papers.

Part of the difference between them is merely a matter of priorities or vantage point. Patnaik, for example, is not much concerned with the way the global context is translated to the cultivator, the local form of the coercion he suffers (as discussed in other papers). But part of her argument depends on the skewed enjoyment of the benefits of higher prices and increased production, not only as between developed and developing countries, but within the developing countries themselves. In India, it is supposed, the export trade (indeed the market economy generally) produced a relatively small class of consumers, rather than a widespread increase in demand. Why should this have been so? Comparisons will help. Sharp extremes of social differentiation, depressing mass consumption, are not peculiar to poor, tropical countries. Britain too, not least during its industrial revolution, and increasingly again in recent years, is a society with grotesquely skewed incomes and unequal opportunities (as in education). Like its business corporations, British society is often regarded by people of power and influence (especially when justifying their own inflated incomes) as comprised of individual effort and competition, rather than as a common enterprise needing to draw on all talents, or in operation as a team in which different roles are necessary and deserving of reward. No doubt such attitudes were expressed and spread by colonialism. But before British rule and since, India has had its own extremes of wealth and power. It also has mechanisms of redistribution, not so much through the state as in Britain, as through institutions and customs, especially those relating to work. The mix is peculiar to India, and it is in that particular context (as well as in theoretical and global norms) that Indian agriculture is situated.

What of the possible controls over labour, as invited in Guha's paper, and to some extent by Kapadia, in comparison with those evident from Meyer, Hjejle and the Sivakumars? As is well-known, labour came to be rewarded and organised differently in eighteenth-century Britain, with enclosures, scientific agriculture and industrialisation. A regimen of work-place, specialisation of task, time-keeping and supervision was needed to replace labour patterns that were less bound to place and time, and less sub-divided between workers, and management that was looser, less supervisory, and often profit-sharing (including piece-work payments); the new regimen was needed, in part, in order to restrict the returns to labour at a time of increasing capital investment. With these changes and a rising population, poor laws and other provisions ceased to be (so far as they had been) instruments of charity and became rather a means of disciplining the workforce, of ensuring a 'deserving' (that is, a labouring) poor. Philanthropic ameliorations of working conditions were then needed through further, public regulation—factory acts and the like—but present-day social service provisions still depend upon recipients being 'available for work', even though patterns of employment are otherwise changing once again. For some occupations a comparable change was achieved by professional bodies, training, knowledge, career structure, job descriptions and rules of conduct. Under colonial rule some South Asian production was certainly subject to such changes—as was evidenced by the steady streams of aspirant Indian barristers, doctors or officials at London crammers, or by the lines of workers in Assam tea plantations. But other production was not, even much of that which seemed capitalist or industrial. It was subject to a different logic.[19]

Therefore was the labour casualisation, arguably to be observed in India, merely an imposition by external capital and administration, as the Sivakumars insist? They hint at demographic factors, but otherwise attribute the problem to a decisive break with pre-colonial society, occurring both before and after independence. (Implicitly they endorse the common attack, made until recently mainly from the political left, on post-colonial Congress governments in India as a kind of continuation of the British raj in social basis and other features.) Though the general trend in labour and production must be undeniable, the sharpness and completeness of the change may be exaggerated. Differential impact (admitted by the Sivakumars in their characterisation of Tondaimandalam, with one fifth of all the irrigation pumpsets in India),

[19] See Peter Robb, ed., *Dalit Movements and the Meanings of Labour in India* (New Delhi 1994).

implies that the Indian countryside in general was not wholly re-made from an imported blueprint. Accordingly, as the Sivakumars argue, the colonial administration had had constantly to make an accommodation with Indian conditions, in detail and locally.

It is true nonetheless that colonial policy, from famine codes and prison regimes to property and labour laws, was frequently based on schemes imported from Britain and generally on a belief in quantification and averages—of conditions, wages and prices—derived from the European and indeed post-industrial definitions of labour and production. There were at least three elements, therefore, in colonial policy. Two of them were such pragmatic accommodations and imported principles. The third reflected continuing indigenous features which themselves fitted with the foreign influences. For example, there had been motives for both divisions of and control over labour in pre-colonial times, despite or rather because of the availability of land. The implication is that change and conditions must be explained with at least as much complexity as colonial policy. There may well be rules or connections to be discerned, but they will not be single-stranded.

Part of the problem with the supposed norms of economic evolution, and the assumed opposition of modes of production, has been with the distinction between subsistence and market orientation. It is interesting that Datta quotes with approval Wharton's definition of 'subsistence' as individual or group production involving little if any exchange.[20] An elastic meaning for 'group' and for 'production' would seem necessary to make this fit the range of non-market motivations and practices apparent in Indian agriculture: in other senses there certainly seem to have been forms of effectively subsistence production that involved exchange, not least in share-cropping and harvest loans.[21] The issue has been addressed more sensitively by economic historians of some other regions, for example by making a distinction between 'subsistence-oriented' and 'market-oriented' trade, with a further subdivision of the latter according to whether or not there was 'an economy integrated by the market principle'.[22] These arguments take us some way towards

[20] See below, and C. Wharton, ed., *Subsistence Agriculture and Economic Development* (London 1970), p.133.

[21] Perhaps it is only in this confusion of terminology that my interpretation differs from that of Datta.

[22] See Richard Gray and David Birmingham, *Pre-colonial African Trade. Essays on trade in central and eastern Africa before 1900* (London 1970), pp.1-

understanding the complexities, even though they will not quite suffice for the Indian case, and perhaps not elsewhere as well, in that the expanding involvement of small producers in markets—for land, labour and food—need not necessarily remove their involvement in or attachment to the social organisation and belief-systems supportive of subsistence economies, or require the producers' own direct attachment to the profit motives usually associated with the growing of crops for the market. On the one hand they may adopt maximising or avoidance alongside risk-averting, co-operative and subsistence strategies. On the other hand they may sell in the market while remaining the creatures of others.

That last point takes us back to differentiation, which may be explained by the needs of pre-colonial supra-local elites to manage localities, but which also seems, more generally, to be a common theme in the economic history of countries under colonialism—that is, there is evidence of intrusions by state and market (to varying degrees) into more localised communities *by means of local intermediaries* who benefit socially, politically and economically, usually on the basis of a pre-existing position of relative strength, even though on occasion, when disappointed, they also lead resistance or even rebellions. Many examples have been described in Vietnam and Indonesia and other parts of South East Asia, and in most parts of Africa, as well as in colonial India. A distinction should be made between this situation, where the colonial intrusion extends an existing differentiation (which may have been produced in part by earlier states and by pre-colonial trade), from another circumstance, when regions are being incorporated into settled government and agriculture, and their inhabitants resist as a whole under existing leaders. In both cases solidarities are sought, often through religion, and in economic terms both may represent a stronger

23; and Ken Swindell, *Farm Labour* (African Society Today; Cambridge 1985), esp. pp.55-90. See also Michael Adas, '"Moral economy" or "contest state"?', pp.521-46; it is worth noting that, while E.P. Thompson's moral economy depended upon shared ethics, access to land and the absence of industrial work regimes, that of J.C. Scott depended upon a peasant ethos focused on subsistence priorities in a local community. On the linear evolutionary implications of colonial and capitalist impact, compare the arguments for an alternation rather than a progression between more or less 'closed' local economies, depending largely upon external political conditions, in G. William Skinner, 'Chinese peasants and the closed community: an open and shut case', *Comparative Studies in Society and History* 13 (1971), pp.270-81.

involvement in international trade; but the social contexts are different in that they represent different stages or processes of economic change.

Chaudhuri's essay seeks to restate the importance of the colonialist dislocation by arguing against those who have explained conditions in terms of internalised contexts and parameters of production, but also by insisting that some of the distortions introduced during colonial rule were the result of opportunities taken, more or less freely, by certain cultivators at the expense of others. These arguments, while differing in emphasis, nonetheless seem to fit quite well with the characterisations of the colonial period by Patnaik and the Sivakumars. But we should notice how, for various analyses, the independent producer is a sort of residual category: he is assumed to exist at a level below the lowest one at which distinctions are being made, in the analysis, between different types and interests. By contrast, a point which in large measure is the subject of this volume, is the need to distinguish between types and levels of decision-making in agriculture.

For their purposes here, as we noted, neither Chaudhuri nor Datta, in their papers here, give detailed consideration to the middle and lower levels of production relations. Chaudhuri does not make a particular effort to disaggregate the category of peasant cultivators, in the ways that Guha and Kapadia do. But he does report, for example, that technical innovations were made in sugar production, as with the Behea mill, under the influence of intermediaries, sugar factors and moneylenders; and that, conversely, the Opium Department was unable to exercise sufficient control over production so as to prevent the cultivators from engaging in double-cropping, or abandoning poppy in favour of more profitable crops, or even concealing and illegally selling opium. Thus it is plain from Chaudhuri's account first that market interventions might have relatively little impact on the methods of production in the fields, while methods could change rapidly where controllers of capital had direct interest and means of influence. It is plain secondly that resistance to particular production regimes would often depend on support for change on the part of intermediaries and village controllers. Chaudhuri's opium or indigo cultivators who avoided supplying their creditors, or successfully changed to another crop, or from one planter to another, often did so where the change was in the interest of local moneylenders or opium or indigo agents. A transfer from one planter to another, asserted by Chaudhuri to prove the peasants' exercise of will, may also reflect a more effective tyranny, or a temporary respite in a generally disadvantageous relationship. For the jute cultivators, similarly, Chaudhuri makes a distinction between crop-buyers' advances and general borrowing. The latter, it is implied, left the borrower free to sell on an open market. But such other sources of finance in some cases might have exercised a hold over the crop not unlike that sought by jute-buyers. Thus Chaudhuri's examples do not refute the assertion that the role of elite and intermediate people was crucial and distinctive over a long period in India. Rather (like the Sivakumars' evidence) they reinforce the argument that there were arenas of production and exchange where 'peasant' decisions were supreme, or in which myth and ritual were important, and other arenas in which such local interests were more or less powerless—and that parallel exclusions and capacities affected other actors (labourers, merchants, agents, landlords). The thesis also allows for changes resulting in different combinations of power and independence, either from internal shifts or from external affects which altered the value of a particular local asset.

There are some other differences of interpretation, which may be similarly resolved. Hjejle argues that colonial rule and international capital produced new and specially malign developments in these perennial systems. The effect seems to have been from ignorance and through various inadvertences, whereas Patnaik goes further and interprets the distortions as deliberate policy. The Sivakumars relate interlinked or, as they call them, 'overarching' credit markets (an important aspect of this intermediation) to fundamental characteristics of agriculture and the risks and uncertainties they induced. The precise changes in the roles of intermediaries will need to be examined case by case. The British in India provided trade, communications, taxation, measurements, and property and other laws which (as many of them believed) shifted the balance of power in the countryside; but they also repeatedly argued for market rather than coercive systems of production. Social differentiation, landlessness and harvest loans and other forms of tied or coerced labour were common before and after colonial rule, when India's economy was apparently less open. Hjejle posits a more equitable and socially efficient system of production when *racavar* agents commanded the resources, because (she believes) they understood agricultural priorities and were subject to social influence. However, in other cases, and perhaps also in this one, one may trace the perpetration of intermixed intermediary control to features of Indian local society and agriculture. Even the capitalists who benefited from the system in the colonial era often complained that the real costs of coercion—management and supervision, the loans themselves (many not rccovered), rents to landlords, and so on—at best ate away at its advantages (more secure supply of the crop, reduced prices to the producers). As a result the total purchase price for crops was very much higher than the rates paid to actual cultivators or workers. Others have noticed these anomalies; and at the time parallel points were made about the costs and wages of Indian factory labour. The explanation is that commercialisation was enforced through debt not only to reduce procurement prices, but also to secure supply in conditions of dispersed small-holdings, mixed cropping, entrenched local power, and potential competition from traders in the same or other produce. The effects of such a system could be (as Hjejle notes and as has been suggested in other instances here) that middlemen drew profits away from those best able to invest in productivity, and that moneylenders were attracted to deploy their capital in ways which might extend but were unlikely to intensify production. The existence of local elites illustrates the inequitable nature of pre-colonial society. But it may be that particular effects resulted during the colonial era from a combination—or rather an inconsistency—between such structures and the burgeoning market system. Such is the argument of the Sivakumars.

The effect of several papers in this volume is thus to draw attention to the role of intermediaries such as those who controlled Indian cultivators through debt, and to the 'putting-out' of functions of production and exchange—and also to entrenched and dispersed landholding, and to social as well as economic barriers, features which, for example, both invite and require undoubtedly oppressive practices of traders and landowners, those which are held to produce 'forced commercialisation'. India had long had devices for extending credit to cultivation, and for extracting agricultural surplus. The Sivakumars, in this volume, are particularly interested in the changing linkages between rural communities and the wider world. They set out a number of forces which helped create and define local managing elites. These include the complexity of irrigation, the need for management of disputes, the control of labour (and the evening out of payments and supply across the seasons), and so on. Broadly it is plain that the local society they describe is—not impermeable certainly—but resistant to outside influence, not least in regard to production, even as market rationales impinge. Resistance works, for example, interestingly, through entrenched attitudes to time and hence to production. More generally it does so by means of what the Sivakumars call 'politico-legal rules and norms', what is sometimes called custom. Caste in particular is seen as the chief among customs thus defining and defending localities.

Recent scholarship urges caution when placing such arguments in historical time; for example the origin of Vellalas, and of their equation with agriculturists and 'Tamils', outlined by the Sivakumars, has also been discussed by Dagmar Hellmann-Rajanayagam, who stresses the nineteenth-century elaborations of such ideas, in a context of socio-political change which the Sivakumars describe.[23] It should be remembered too that the pre-colonial moral order celebrated by the Sivakumars does not preclude poverty and oppression for those members of society who were regarded, by that order, as being of no consequence. There is clear evidence of institutions of charity and of social pressure for moral behaviour. But there is nothing in such means which determines that the ends should be egalitarian. What matter for the poor and weak are not the instruments of enforcement and redistribution, but the moral and political norms, the belief system, which determine the extent and beneficiaries of charity. In the case described by the Sivakumars, as they indicate, the seasonal nature of the demand for agricultural labour virtually required that some workers should be left at a low level of subsistence if land and surplus were not to be excessively redistributed. Of course free land is likely to contribute to social mobility, and yet, in the pre-colonial system described by the Sivakumars, the consent of local elites was necessary for new settlements. The ready availability of cultivable land might well demand coercive practices, if social differentiation and contributions to religious and other collective needs were to be ensured.

The broad principle of local resilience, in part through social custom and belief, is not altered by our debates and qualifications, except in so far as they indicate that there will be *some* interchange between prevailing socio-economic norms and the experience of localities—in this book Chaudhuri's and Datta's arguments and Meyer's examples all

[23] Her essay is published in Peter Robb, ed., *The Concept of Race in South Asia* (Delhi 1995).

suggest as much. An insistence upon the importance of intermediaries and local elites, whether as obstacles to or transmitters of such influence, is not a suggestion that conditions remained unchanging. For example, as has often been explained, changes in the bargaining power of either employer or labour could have serious effects, especially if employers were freed of the need for concern for the subsistence of the labour force between periods of peak demand. Just such a change seems likely to have affected the society described here by the Sivakumars and by Hjejle.

Two contrasting perspectives are on offer in this book, as generally in the literature. One stresses the local resilience of forms of agricultural production, the entrenched character of particular roles. The other believes in the hegemony of external forces, for good or ill. Some conclusions result from successful (or inappropriate) comparisons of detail. Others are achieved by single-minded attention to the minutiae of cultivation. One key to reconciling these differences is the middleman, a term which may denote any of a number of persons of *local* influence who are in touch with *external* power and wealth—Hjejle's *racavar* renters are a good example, but other village-level elites or traders' agents could perform similar roles. The resilience of the locality and of non-market rationality often helps explain the damage wrought by external forces; it will result from the disjunction between such practice of Indian agriculture and the logic of international commerce or economic theories. The middleman is the means whereby the effects are transmitted; he circumvents local conditions, and delivers suffering and poverty. But he is also an agent for local resistance.

This argument is not about the hegemony of the intermediary, as a mere replacement for those other overwhelming forces, the market, the state, and so on. That would be merely to ascribe to the middleman an instrumentality in securing the effect of those usually agreed means of generalisation. Nor is this argument necessarily a denial of the importance of overarching forces. We see these from Chaudhuri in a sense, in accordance with classic market theory, when he identifies some cultivators' profit motives independent of or below the level of the controlling intermediaries. We saw them more obviously when Patnaik and Sen attach significance to the influence of colonialism or capitalism or GATT. The point of stressing the middleman is rather to allow for different and even opposed meanings of agriculture within the one system of production and exchange. It allows for such distinctions both horizontally (say, in co-existing forms of land-use, or of labour) and vertically (as between cultivator and landlord or moneylender or trader). Then, when market-oriented or external explanations are read together with the detailed and disaggregated accounts of local conditions also offered in this book, then the combination offers a new generalisation: one based upon difference and discrepancy. The Sivakumars call this a 'transaction regime', and suggest that the modern economic history of India may be explained largely in terms of the ways in which that regime has changed. The Sivakumars posit a transformation in its size and locus and in the role and independence of its village-level managers and institutions. They identify their own version of the disjunction illustrated in other papers. But this change of personnel and focus does not remove the importance of brokerage and diversity in agricultural decision-making. A transaction regime is a single system composed of different (even, I would add, contradictory) parts and interests.

Awareness of inappropriate theory, of the gap between agricultural practice and economic theory, need not imply an unqualified acceptance of the trajectory identified by Ludden, as in the Sivakumars' broad brush marking an absolute essentialist contrast between Tamils and the East India Company. For, though the effects of colonialism or free trade may be contentious, we conclude that they are transmitted through indigenous institutions. It is also agreed that the conditions of Indian agriculture differ, and that these differences have to be understood if policy or analysis is to succeed. Thus some of the papers in this book provide micro- and some macro analysis, and this may lead to disagreements. But in the end what is shown from both perspectives is how the prescriptions of market economics do not relate to the actual experience of most Indian cultivators, so that to apply them is to run the risk of, at best, serious dislocation and suffering, whether or not they improve the Indian economy. Thus Ravi Srivastava's paper, or that of Eric Meyer, bears out those implications of Patnaik's and Sen's thesis empirically and in detail. And state interventions are indeed needed, as I believe colonial administrators also realised, to help break through specific limits of purpose and function, from which most individual producers and workers cannot escape. It is because of these limits too that there are convincing fears of distortions being exaggerated in India, between classes and regions, as a result of prevailing market rationale and prescriptions, World Bank-style. In short, the variant meanings of agriculture are crucial to the choice of methods of development, to the future as well as the past.

Chapter 2

ARCHAIC FORMATIONS OF AGRICULTURAL KNOWLEDGE IN SOUTH INDIA[1]

David Ludden

Historians search for the meaning of words in the past, but history intervenes. Time destroys data about past contexts of usage. Passing generations create new contexts of meaning by reinventing old ones. Texts employ methodologies they often hide and lift words out of the contexts of everyday usage. The words most likely to survive are alienated by their very durability from their speech environment; as we are alienated from them by a history that separates our reading from their composition. Trapped in our present, we understand relics of past discourse in our own terms, with their own historically acquired meanings. There is no escaping history's uncertainties and alienations as we describe past contexts, interpret past terms, and reconstruct past discourse. Our search for meaning is a disputation about plausible present interpretations of past usage (see Bernal 1987: 1-38). For, in the end, the ancestors speak to us only on our terms (Goldstein 1976).

Here I explore archaic formations of agricultural knowledge—configurations of usage and context—in south India, and especially in Tamil Nadu. To situate their archaism, I begin by considering the historic treatment of agriculture as an object of knowledge. In defining a semantic domain called 'agriculture' and making a bundle of terms cohere as 'agricultural terminology', ordinary farmers have very little to say, except to provide words for us. Interrogating farmers today, we see their expressions in a context of our creation, which shapes their meaning: we categorise them as farmers, use their categorical status to collect and authenticate our semantic data, and then reproduce a discourse in social science about agriculture. This discursive formation was born in Europe, as self-consciously urban, progress-oriented, scientific specialists assumed their modern power in the division of

[1] Research for this paper was funded by grants from Fulbright-Hays and the Social Science Research Council supported by the National Endowment for the Humanities. Comments from Peter Claus, Frank Perlin and Arjun Appadurai were helpful in revision.

labour during the cultural consolidation of industrial capitalism. In this context, science, management, marketing and government converse about the farmer as a social other who lives outside cities, in alien rural contexts of farm labour, including village India. The formations of knowledge that I explore became archaic as modern agricultural discourse conquered India.

The agricultural text

In English and other languages, the appearance of agriculture as a defined semantic domain emerges from its definition as a textual subject matter. My own study of relevant texts could be compared to 'examining flowers from horseback', as a Chinese saying puts it; but from horseback it seems that the textualisation of agriculture derives from managerial efforts by intellectual elites. Across two millennia before 1800, strenuous efforts were made to manage agrarian production at the extremities of Eurasia, as attested by texts in Greek, Latin, northern European languages, Chinese and Japanese. Despite their diversity of cultural environments, these agricultural calendars, manuals and technical compendia have many features in common that suggest similarities in the contexts of their composition. They emerge where literate elites have control not only of agricultural output but also of critical inputs, above all, it seems, labour. State power is crucial in China, where virtually all texts on agriculture were written by officials, many themselves landlords, and where the most influential texts were printed and distributed by the state. In China and the West, estate management for commercial production is the context for many texts, which disproportionately concern profitable agricultural commodities (Needham 1984: 47-85).

The textualisation of agriculture, as a managerial act, employs an instrumental attitude towards farming, in which a classification of production outputs defines a sphere of human life. Agriculture is defined by organic produce (Fussell 1983: 7-8); agricultural terminologies establish instrumental relations among elements of input and output. The contexts of this instrumentality were quite different in Europe and China, where agricultural texts proliferated after the tenth century. In Europe, texts concentrated more on tenurial relations; in China, more on technology. This contrast reflects a more general one: European texts document in more detail agrarian social conditions and individual farm life-situations; Chinese evidence reveals more about material conditions and government policy (Needham 1984: 85-93). We can imagine a basis for this contrast in the social relations of production. In China, authors of agricultural texts were landlords and officials, who used intimate observations of farm technique to further state-supported agricultural expansion. In Europe, feudal institutions were both mechanisms for labour control and obstacles to greater control of the work-force by the landed classes, so that land tenure was central for authors of agricultural texts.

As Michel Foucault would lead us to expect, when landowners broke through feudal barriers to gain new powers over production, Europeans produced a more domineering agricultural expertise. Capitalist relations of production used science to write the new agricultural text. Authors in Europe and China had for centuries recorded lore and observations about farming, to make peasant wisdom known to literati: for the old texts, farming practice determined agricultural knowledge. By the eighteenth century, Europe, but not China, moved towards scientific determinations of agricultural knowledge. Authors and publishers now disseminated progressive, advanced technical expertise, authorised, tested and stored at the top and distributed from the top down, by states and their allied economic and scientific experts. At the top, specialists defined the semantics of agricultural discourse and the nature of relevant data. Localistic, non-scientific data from farmers became folklore. Experts used research stations, statistics, and scientific techniques with claims to universal validity to forge world hegemony. A scientific formation of agricultural knowledge with global reach was established in Europe by 1830. Francesca Bray argues that it did not affect China before 1900 (ibid.). By then it had conquered India.

British colonial bureaucracy codified agriculture as it did other objects of knowledge, and in the 1840s modern formations of agricultural power/knowledge began to engage agrarian India. The world was opening up to European plantation enterprise, which needed to know about crops, labour and production techniques everywhere in uniform, economical, English terms. As cotton production in Egypt, America and India began to preoccupy London in the 1840s, railway, telegraph and steamship began to lower transport costs, as managerial and print technologies lowered information transaction and processing costs, to make the cities of the empire data-processing centres for world-wide labour and product markets. By 1880, a modern information apparatus was in place in British India and agricultural expertise developed rapidly, thriving on standardisation. Imperial bureaucrats worked under what has been called 'rule by the manual' (Smith 1985: 161; also Ludden 1985a: 115-29; 1988). The improvisation of surveyors and collectors gave way to the rule-driven, machine-like routines of department specialisation. The hand-written manuscript gave way to the printed report. The motley assortment of information in Company correspondence gave way to specialised, uniform data in census, wage, price, production and ethnographic series (Ludden 1990b).

By 1900, the colonial state's discursive domination of agrarian India was complete and its attitude towards farmers set in a modern mould. Bills and reports had found farmers helpless in the face of debt and drought. Farmers *out there* in the village were objects of official concern, scorn and paternalism. The instrumental powers of science nested happily inside the language of imperial domination (Cohn 1985). Lord Mayo founded the Agriculture Department and set its tone in 1869 (Manak 1979: 27):

> For generations to come the progress of India...must be directly dependent on her progress in agriculture.... There is perhaps no country in the world in which the State has so immediate and direct an interest in such questions.... Throughout the great part of India, every measure for the improvement of the land enhances the value of the property of the State. The duties which in England are performed by a good landlord fall in India, in a great measure, upon the government. Speaking generally, the only Indian landlord who can command the requisite knowledge is the State.

The Madras government produced an *Agricultural Class-book* in 1880 and held an Agricultural Exhibition in 1883 (Robertson 1880). The state no longer simply told Europeans about farming in India (Greenaway 1864; Royle 1851); it became *the* expert on Indian agriculture and in doing so defined the style and substance of modern agricultural expertise (see Schrottky 1876; Pogson 1883; Shah 1888; Moreland 1904, 1913; Randhawa 1966). The *Moral and Material Progress Report* for 1907-8 reported that 'thirty years of experience and record gathering since the institution of the Departments of Agriculture lead to the recognition that Government's primary task was to apply European scientific methods to Indian agriculture' (Manak 1979: 73). That application transformed agricultural discourse.

In India as in England, the collected oral wisdom of farmers did provide data for science. The Madras Agriculture Department compiled 'sayings and proverbs' on agriculture in Telugu and Tamil in the 1890s, to advance 'knowledge of the agriculture of the country'. Queries to Collectors generated lists of sayings 'more or less to the point'. Sayings selected for publication were those that 'really apply to agricultural matters', setting a stamp of modern codification on 'wisdom...of ages'.[2] C. Benson, who compiled the first 'sayings and proverbs' volume, had the idea while working in Kurnool, but compiled Telugu sayings for the whole presidency, without reference to their provenance. Agriculture as semantic domain, Telugu as container of a culture, and Madras Presidency as a unit for aggregations of empirical knowledge were thus codified simultaneously. Science, orientalism and empire imposed their classifications on farmers' words to produce scientific knowledge. When C.K. Subba Rao, Sub-Assistant Director of Agriculture, compiled a Tamil 'sayings and proverbs' volume, he carried Benson's logic further, in a nationalist direction. Like his contemporary, U.V. Swaminathaiyar, he compiled verbal pieces of Tamil tradition and produced a text to document the Tamil heritage.[3] Subba Rao adds to his collection of sayings 'relevant' quotes from the post-classical compendium, the *Tirukkural*, and from medieval texts like the *Kamba Ramayanam*, to assign by this juxtaposition more weight to rustic wisdom and impress readers with the progressiveness and continuity of Tamil thought on agriculture. It is unlikely that these 'sayings and proverbs' volumes did agricultural science much good.[4] But for us they preserve—as they extract, decontextualise and reassemble—bits of knowledge produced in old formations of usage and context that were banished from discourses of development.

[2] Quotes are from C. Benson (1898). Benson was a Collector of Kurnool and author of *An Account of Kurnool District* (Madras, 1889), during the completion of which he became aware of 'proverbs and sayings...as illustrative of agricultural life and practice there' (1898, Introduction).

[3] Dr. U.V. Swaminathaiyar omitted provenance from his editions of old Tamil texts, but included variant readings. He compiled poems into anthologies as prescribed by much later commentaries, to create a classical corpus that did not survive as such and perhaps never existed. Swaminathaiyar describes local settings from which he collected his texts in his autobiography, published in 1940-42, and reprinted as *En Sarittiram* (Madras, 1958).

[4] At best these volumes helped to educate extension workers. Bhattacharya gives many examples of the ambivalence that attends modern extension workers' use of such information (1976: 174-230, 238, 243).

Archaic formations

A compilation such as C.K. Subba Rao's could have been written long before his time, but was not. Given their scrutiny of agrarian conditions and interactive interdependence with literate landed men in the districts of Madras Presidency, we can suppose that Company men would have unearthed vernacular texts on agriculture had they existed. We could expect mention of such texts among the thousands cited and transcribed under British rule.[5] And Subba Rao would have cited any known Tamil text about agriculture. A textualisation of agriculture, such as occurred in Europe and China, seems not to have occurred in the Indian subcontinent before 1800. Conditions that promoted it elsewhere were absent in South Asia. The literati do not seem to have had sufficient control over farm inputs to sustain a managerial stance towards agriculture.[6]

The colonial regime textualised Indian agriculture and replicated Europe's own transformation of agricultural knowledge in the process, moving from eclectic, wide-ranging reports to a modern, scientific, codification of agricultural expertise. In doing so, the raj gathered a huge store of data. Countless word relics help us to reconstruct what became archaic formations of agricultural knowledge in the nineteenth century. These fragments are alienated from the contexts that once made them meaningful, in part by the very activity that preserved them, in part by the passage of time. But then again, few if any texts we use today to study South Asia's past escaped colonial influence during their modern textualisation: that is the price we pay for preservation.

We can begin to use our relics by locating three huge temporal strata of Tamil language and agrarian history, each with a distinct way of putting agricultural information into texts. Each epoch sits somewhat isolated from the others because intervening centuries left relatively few texts. Across the millennia, south Indian authors subsume, scatter and subordinate agricultural terms in texts designed according to what pre-modern Chinese and Europeans as well as modern scholars would

5 The 'Translation of a Sanskrit Text of Land Measurement' gives an indication of the energy devoted to collecting texts under the Company. I do not mean to imply here that agricultural texts could not have existed or could not survive today; but that they have not come to light and that until they do their absence seems to me significant. Any information that an interested reader could provide on this matter would be greatly appreciated.

6 The comparative research on which this speculative statement is based has begun (see Byres and Mukhia 1985).

call non-agricultural motifs. The implication is not that agriculture was unimportant for authors or in everyday speech; rather that authors do not conceive agriculture as a discrete object of consciousness, a cognitive category by which to sort information about social life. They do not, of course, state their case negatively; they posit textual formations of terminology in which agriculture must be situated to make sense.

The earliest epoch of Tamil documentation spans the five centuries before 400 A.D. Its poetry is our only textual evidence for the period and its poets depict five habitats—mountain, forest, lowland, seashore and wasteland—where configurations of flora, fauna, gods, people, scenes and events organise poetic accounts of love and war. In milieux ecological, aesthetic and social, agriculture provides many items for poetic description, but does not emerge as a subject with a distinct identity. Extended poetic statements on agrarian matters centre on royalty and on irrigation, suggesting the importance of irrigated paddy-farming in the classical landscape. *Paranantru* no.18 advises a king (in lines 13-30):

If you crave riches in the next world
crave to break the strong shoulders
  of kings who guard this world
  to become supreme yourself,
or crave to hear glorious praise,
listen, great king,
to learn what deeds deliver such rewards.
Those who give food give life to all living beings,
who cannot live without water.
Food is first for all bodies made of food,
and food is but soil and water mixed.
So those who unite water and soil
create living bodies and life in this world.
Even kings with vast domains
strive in vain when the land is dry,
when fields sown with seed
must look to the sky for rain.
So Pandyan king, who makes dreadful war,
do not disdain my words: expand quickly those watery places
built where water moves across the land!
For those who harness water
harness rewards for themselves,
and those who fail cannot endure.

Other poems make pointed statements about tank irrigation and paddy irrigation as the basis for prosperity; and verses from the fifth century didactic corpus, *Tirukkural*, advise rulers in very similar terms (Singaravelu 1966: 43-4). Many classical and post-classical texts praise farmers and exhort leaders to feed the people. But agriculture is subsumed within descriptions of poetic milieux and prescriptions for dutiful rulers (see Ramanujan 1966; Ludden and Pillai 1976).

Separated from the classical age by several centuries, medieval south India (c.800-1400) produced a much larger and diverse body of texts, in many thousand inscriptions and a vast literature which focus on the institutional complex around temple deities built in this epoch. The five classical milieux disappear. They are replaced, in the inscriptions, by cultivated tracts called *nadus*, dominated by people with material assets for transactions for gods, kings and Brahmans; and, in the poetry, by natural imagery invoked to portray devotion, valour, fate and theology. We can see nature, gods and kings involved in medieval agriculture; and kings above all, for royal grants of land for irrigated paddy-farming are prominent subjects in the inscriptions. We can see agricultural assets in enough detail to reconstruct medieval villages and expanding cultivation (Heitzman 1985, 1987). But medieval texts portray agriculture within a discourse about power, piety and, above all, ritual (Stein 1980; Ludden 1985a: 15-42).

A third, early-modern epoch begins in the sixteenth century and continues into the mid-nineteenth. The centuries that separate it from medieval times saw migrations and technological change transform linguistic, social and agricultural geography as much as political economy in south India. Now the medieval *nadus* have disappeared from our texts. As we will see, many of Subba Rao's proverbs on cattle, livestock, cotton, wells, tanks and dry cultivation, which would have made no sense in the medieval context, would have done so by the eighteenth century. In these early-modern centuries, many immigrant non-Tamils, first Telegus and Kannadigas, then Europeans and north Indian Muslims, had a pervasive influence on the Tamil country, as manufacturers, farmers, merchants and military commanders during the rise of new states with economic foundations in agrarian commercialism. In a process of language development similar to that which gave rise to the medieval context (when Sanskrit became central in Tamil history with the rise of Brahman power), the Tamil language changed in the early-modern period by absorbing terms from non-Tamil peoples who became key actors in agrarian society. Non-Indian authors also wrote texts to document Tamil history, first the Portuguese and Dutch (Subrahmanyam 1990), and then the French and English.

English records from the early 1800s evidence agrarian conditions in the last early-modern century, and despite their European interests and cultural background these texts embody a consistent early-modern view of agriculture as of concern only for state revenue. Royal courts—their interactions, intrigues and incomes—form the central motif for early-modern texts. Commercial cultivation expanded dramatically in early-modern centuries, and involved many crops that Ch'ing dynasty manuals consider in China: tobacco, cotton, dyes and vegetables, for example. But textual evidence of this expansion in south India is indirect and comes from texts on the terms of taxation (Ludden 1988).

We could conceivably proceed to use classical, medieval, and early-modern texts to devise a Tamil formation of agricultural knowledge, embedded in Tamil culture over the centuries. But this would only reproduce old illusions of closure and continuity in Tamil tradition. A dispute over one Tamil term shows how these illusions have crystallised and how to avoid them. The term is *anangu*, and the *Tamil Lexicon* lists fifteen definitions, each with citations from early Tamil texts: pain, affliction, suffering; disease; fear; lust; killing; deity; celestial damsel; demoness that takes away one's life by awakening lust or by other means; beautiful woman, as resembling a celestial damsel; devil; dancing under religious excitement, especially when possessed by Skanda; low-caste person; beauty, form; young offspring.

The authors of the cited texts do not, of course, speak in the *Lexicon*; it gives meaning to their words in its terms, which resemble those of the *Oxford English Dictionary*. But unlike the *O.E.D*, the *Lexicon* neither lists citations chronologically, nor labels archaic usage. In its practical construction, the *lexicon* more closely resembles the glossaries by H.H. Wilson, H. Yule and A.C. Burnell, and those in the Madras *Manual of Administration*. These glossaries were technologies of imperial unification (Smith 1985; Ludden 1985a: 115-29). By the 1930s, when the *Lexicon* was completed, imperial technology could serve national purposes, as provincial governments built Tamil cultural identity (Irschick 1969; Washbrook 1976). The *Lexicon* defines terms for that identity. The banishment of Sanskrit from Tamil that has proceeded rapidly since 1960 is visible in the *Lexicon*, where we cannot find many of the Sanskrit terms used by U.V. Swaminathaiyar in his

Tamil commentaries on Tamil classics. The *Lexicon* sees Tamil literature as a unified text, as a closed tradition, with internal diversity, but no history. Asserting word origins, it locates lexical worlds outside Tamil, to mark boundaries around Tamil. It privileges old texts, and cites relatively few composed after 1500. Its real Tamil is old Tamil, an essential Tamil of classical and medieval texts. It makes archaic usage seem to be current in a unified Tamil text/tradition, erasing the history of change in Tamil literary usage.

It is a short step to the view that all Tamil usage constitutes one cultural text. George Hart took this step and redefined *anangu*. Finding female taboos in classical poetry, he argued that '[it] may be inferred that women were thought to possess a latent sacred power which at certain times became dangerous so that they had to be tabooed'. With this inference, *anangu* becomes 'the sacred power...of women' and its 'chief source...a woman's chastity (*karpu*), a virtue as important then as it is now in South India' (1973: 236). 'Then' and 'now' not only connect the classical poetry with modern ethnography but also identify indigenous beliefs that survive millennia, because for Hart the Tamil classics document Tamil culture before northern influence (1973: 233). Understandably, Hart's argument soon escaped literary studies. Susan Wadley (1980: xii) cites Hart (1973) when she says, '...there is strong evidence that the attribution of sacred powers to women derives from the beliefs and practices of the indigenous peoples of India and images of sacred femaleness are prevalent in the earliest Tamil literature'.

These arguments assume the existence of a unified Tamil textual tradition that constitutes one context for literary usage and reflects a single Tamil culture. Orientalist textualisation was built on that assumption and made it seem increasingly plausible by burying evidence about the changing contexts of Tamil usage and composition. In this century, therefore, the *Lexicon*, literary scholars, nationalism and social science can presume Tamil cultural unity and write their texts accordingly, forging unity from literary fragments. But historical research has begun to shatter this tautological unity. When V.S. Rajam considers *anangu*, she takes each poem as a usage context in temporal sequence. She has found not fifteen but eighty-three significant usages of *anangu* in classical poetry, and, in addition, a pattern of semantic change over centuries of classical and post-classical composition. Usage and meaning changed over time; old meanings became archaic. *Anangu* shrivelled 'from a multidimensional concept of diverse significances to a single-dimensional concept of a celestial female' (Rajam 1986: 266). Change even produced the illusion of continuity, as medieval scholars thrust their shrivelled sense of *anangu* back onto classical usage in their learned commentaries.

Rajam identifies three errors for us to avoid. Hart projected onto classical poetry a category of 'sacred' that the poets do not recognise. Rajam accepts that power is implied by *anangu*, but, in the context of classical poetic conventions, she has good reason to prefer implications of passion or sexuality. We would commit a similar error of classification to call terms 'agricultural' when authors do not classify their own terms in this way. We are looking for formations of knowledge that pertain to agriculture and for us some set of their terms may be agricultural. But authors in archaic south Indian contexts did not specify a set of terms as 'agricultural terminology', any more than classical Tamil poets categorised 'the sacred'.

Hart also errs when he finds one meaning for *anangu* that is not a point within a range of variation—even if it may be a central point, as a median or mean in statistics—but the fundamental meaning. We must not assume similarly that agricultural terms have fundamental meanings as such, simply because they are the kind of terms that we think are fundamental in farming, or because they have the instrumental clarity and efficacy we find in a modern agricultural terminologies. Ritual and social conflict may be as meaningful in agriculture as soils and seeds. Terms with ritual meaning do not forfeit agricultural significance; and the rituals of farming are no less significant for being mundane.

Finally, Hart defines *anangu* in what seems to be a single semantic system simply because its elements are Tamil, however extraneous they might be to the classical poems. But Rajam shows that semantic change accompanies cultural reproduction and indicates cultural change in post-classical times. She shows that cultural reproduction involves not only profound change in language use, but also intellectual efforts to colonise the past and manage its meaning in the present. In our present context, scholars have begun to reconstruct archaic contexts of social life and to interpret the textual evidence accordingly.[7] The evidence

[7] T.K.T. Viraraghavacharya, for instance, has reconstructed specific local conditions and activities that produced a great medieval poetess and then goddess, Andal. In this kind of work, the break in Tamil tradition produced by

weighs heavily against the presumption that a bounded, unified and continuous Tamil culture is described in Tamil texts or ever existed within the boundaries of Tamil Nadu.

Here I treat the Tamil country merely as part of the peninsula and use text relics to suggest ways in which context and usage may be combined to reconstruct archaic formations of agricultural knowledge. For me, therefore, Subba Rao's proverbs are like classical poems in that they are textualised fragments of discourses we cannot reconstruct or date precisely. We can presume that proverbs collected in the 1890s came out of discursive routines with a past, but not necessarily a future. In fact, I would argue that like classical poetry, Subba Rao's sayings were relics when they were collected. We cannot know their history, but can plausibly argue that they were lifted out of formations that became archaic in Subba Rao's lifetime.

By 'archaic' I do not mean 'traditional'. Archaic formations lost their vitality and virtually died under colonialism. They left relics like Subba Rao's proverbs, which could survive even today, in formations alien to those that spawned them. Formations are thus not collections of words to be labelled 'traditional' or 'modern'; they are configurations of usage and context. I will not posit a historical sequence of specific formations, but merely suggest key formative features—cognitive, spatial and temporal—which fit evidence left behind by historic formations. My concern is agricultural praxis. I seek the logic of agricultural word usage in practical contexts of production. But I am also concerned incidentally in this paper with why south Indian formations were not textualised in the Chinese or European manner and with how they became archaic.

I see agricultural knowledge as working within productive power relations (see Ludden, 1985a, 1985b, 1985c). In archaic south Indian settings, states were influential but did not control agricultural inputs to any great extent, whereas colonialism centralised state control of irrigation investments, currency, land tenure, weights and measures, taxation, and many other features of agrarian life. A localism that characterised productive power relations in pre-colonial times became archaic under the raj, but shaped the archaic formations of knowledge with which I am concerned here, for farmers had first and foremost to comprehend

---

medieval conditions is becoming increasingly clear. See Ludden 1985a, pp.31-2, 43, and 46.

the complexity of local conditions, in which taxes, gods, kings and markets rivalled rainfall, soil and manure in agricultural significance. History transformed the agrarian landscape, but knowing how to farm always meant knowing locally specific combinations of the many elements that affected prosperity.

Ecology and power

Agriculture is defined by societies in history. The English words *culture* and *agriculture* became nouns to denote crop cultivation in the seventeenth century, as did *cultivate* (with its antecedent verb *cultive*) —'to bestow labour and attention upon (land) in order to [raise]... crops' (*O.E.D.*). Today, the word used for *agriculture* or *cultivation* in Tamil is *vivasayam*, from the Sanskrit term for 'labour' or 'exertion' (*Lexicon*: 3,718). It appears in Subba Rao's volume only with other Sanskritic terms, for instance, in the saying 'agriculture is alchemy' (no.31).[8] Its reference to labour and the all-India currency of Sanskrit terms (bolstered by the vast importance of Brahmans in all Indian sciences) recommend it for modern usage. Also, *vivasayam* has no social or historical connotation, now a feature of some importance. To illustrate, the authors of a 1960s' film script used it instead of a Tamil term despite their Tamil nationalism. The film, *Vivasayi*, starring M.G. Ramachandran, depicted a prosperous Tamil farmer awash in the Green Revolution. Other terms for 'farmer' in the title would have been impolitic. *Ryot* or *raiyat* (ultimately from Arabic and Persian) connotes Mughal and British imperialism. Tamil terms either denote ploughmen, which would be demeaning, or dominant castes, which would be devisive.

The archaic Tamil term for *agriculture* has two forms: *vellanmai* (which seems to have been more common in speech),[9] and *velanmai*, which is more literary and etymologically correct (*Lexicon*: 3,793). It is composed of a verbal noun (*anmai*) meaning *control* or *rule*; and *vel*, which as a verb means *sacrifice*, *marry*, *desire*, but is here a noun, which is used in literature to mean *marriage*, *desire*, *Kama*, *Skanda*, *ruler*, and *great man*, and in inscriptions as a title (*Lexicon*: 3,842-3).

[8] All references to Telugu and Tamil sayings and proverbs are to the Benson and Subba Rao volumes, respectively; citations are by their numbering of sayings and by page number, as appropriate.

[9] Subba Rao's introduction says, '...the spelling of the Tamil follows the actual pronunciation of the ryots rather than being literally accurate'.

Thus *velanmai* came to mean *beneficence*, *gift*, *bounty*, and *liberality* in literary usage before being used as *agriculture*, a meaning for which the *Lexicon* (3,843) gives no literary citation. Reflecting the richness of its connotations, Subba Rao translates *vellanmai* often as *bounty*, *abundance* or even *success*, for it is frequently used to denote positive results from agricultural activity (for example, no.644). The caste title 'Vellala', is a third person, verbal noun, *velalar*, meaning therefore 'one who controls abundance, sacrifice, etc.'; *velan*, *velalan*, and *velalappillai* are its medieval Chola inscriptional forms (Karashima 1978: III, 436-8); and this is as close as Tamil comes to saying 'farmer', as distinct from 'ploughman' (*uluvan*) or 'villager' (*kudi*).

We can impose a rough historical sequence on terms for farmers and farming in use today. *Raiyat* is early-modern: it came south with a bundle of Mughal administrative terms; Anglicised under the Company, it refers to a landowner paying taxes to the state. Though *vivasayam* is older as a southern term, the *Lexicon*, true to form, cites no literary usage; yet we can be sure it became prominent in modern agricultural discourse, where it represents agriculture as a technical practice. Tamil usage surrounding *velanmai*, from the fifth century onward, focuses on the farmer's powers, not his official status or technical farming practice.

The *Tirukkural* does mention ploughing and manuring in a vein similar to sayings current in 1896 (Subba Rao 1896: nos.7, 8, 350-72), but more typically it lauds ploughmen as 'linchpins of the world' (no.2), for 'everyone else eats what they give, travelling behind' (no.3).[10] Among 69 verses on this theme in the *Kamba Ramayana*, one quoted by Subba Rao (p.21) says, 'Even students of the Vedas...must wait at the door of the ploughman. The prosperity for subjects of powerful kings depends on the ploughshare.' Many Tamil sayings current in 1896 pursue the same themes (for example, nos.16-31).

That success in farming relies on power, which depends in turn on morality, is enshrined in the etymology of *valanmai* itself, and is clear in the *Tirukkural*, which warns, 'however deeply one ploughs and whatever skill one exerts, the earth bears fruit only for those who have earned religions merit (*punniyam*)' (Subba Rao 1896: no.11). Using the bounty of the earth for religious merit, by patronising Brahmans,

[10] Subba Rao gives G.U. Pope's translations for *Tirukkural* verses. I have modified some translations slightly throughout this essay.

became the linchpin of agrarian power in medieval times (Stein 1980). To endow temples, sponsor ceremonies, give alms, and pursue other forms of generosity remained critical for dominant caste-power, influence and bountiful harvests throughout the nineteenth century. Proverbial injunctions in Subba Rao's collection against cost-benefit accounting and parsimony (nos.38-40) reflect a moral technology that distinguishes household economy, where accounts are necessary (no.38), from the social process of *vallanmai*, where generosity is power.

In that social process a central axis of power ran from farmers, through local leaders, to kings and gods. The term *anmai* is attached over centuries not only to other forms of *velala*, most importantly, *karalar* ('controller of the monsoon [water]'), but also to *nattanmai* ('control of the nadu') and *nattanmaikkaran* ('headman'). These terms reside in records of entitlement that describe interdependence among kings, local leaders and farming; records that arose from rituals. Rituals enabled the kings to empower cultivators, first local leaders who became dominant-caste Vellalas, then Brahmans, and finally other castes who established realms of dominance (Ludden 1985a: 590-68). Rituals brought gods to bear on the prosperity of kings, and kings and cultivators together in pursuit of power over prosperity. Politics and ritual not only described inequalities of power, divisions of tribute and distributions of authority: they were technologies of agricultural production.

Crises, above all war, pestilence and famine, tested these social technologies. They demanded solidarity along the central axis of power and made political decisions especially potent. In archaic agrarian settings, expressions of solidarity between kings and farmers in temple rituals were means to re-establish control of the elements. Francis Zimmerman, in his work on Hindu medicine (1987), argues that in Ayurvedic traditions epidemic disease is a problem for medicine, politics and religion simultaneously. Thus the theme of *niti* (justice) pervades Sanskrit literature and peasant perceptions because the king is responsible for the rituals that promote purity and prosperity.

Editors surely weeded out sayings about ritual from Agriculture Department volumes in the 1890s, because, clearly, in the nineteenth century, agrarian commitments to ritual indicate its importance in agriculture.[11] David Arnold (1984: 67) argues that famines represented

[11] See Ludden 1985a: 34-41, 59-67, 164-200, for ritual dimensions of agrarian political economy in Tinnevelly. Bhattacharya (1976: 178) describes rituals

for peasants the cataclysm of Kali Yuga, a time of trial for kings, sages and farmers alike: '...famine was...charged with exceptional religious significance...'. Capricious rains were associated with capricious deities who needed attention and failure of rain with human failures to attend deities properly. Peasants in Hampi attributed the 1876-77 famine to their failure to drag the temple car its full circuit, and a missionary in Salem reported, 'there are various opinions prevalent, but...all agree in attributing [famine] to some kind of sin'. Special rituals were necessary, in which all village castes came together. In Bellary,

> women tied a live female frog to a winnowing-fan...and almost covered it with the sacred leaves of the margosa tree. The fan and frog were then taken in procession by women and children singing 'mother frog should have her bath.... Give water, O Rain God!'

Similar performances occurred elsewhere, 'organized by the villagers themselves, sometimes helped by donations from wealthier households'.

> But peasants also expected the *sarkar*...to arrange and pay for the holding of special ceremonies. During the early colonial period officials of the English East India company complied with these expectations.... The Collector of Tirunelveli paid 600 'chucrums' in December 1811 for rain-making ceremonies to be performed in the principal temples of the district 'with the view of inspiring the people with confidence and encouraging them to proceed with their preparations for cultivation...'.

The 'colonial state had long abandoned such responsibilities' when famine struck in 1877, and farmers complained that the raj neglected its ritual duty, according to the missionary at Salem (Arnold 1984: 68-74).

Archaic formations of agricultural knowledge embraced everything affecting prosperity. Knowing how to work effectively in society was critical; this meant being realistic about power relations among human beings as well as between people and gods. Few farmers do all the farm work themselves, and so many of Subba Rao's sayings stress personal supervision of ploughing, planting and harvest (no.52-60). One Sanskritic saying enjoins the (Brahman) landowner to 'visit your fields morning and evening' (no.58). Masters cheat workers in measuring grain wages, and workers cheat masters by loitering (no.69). The shrewd overseer will be all smiles at ploughing time but stern at the harvest (no.68). Scattered fields in more than one village are too diffi-

---

attending planting in Bengal today.

cult to manage (nos.61, 97-101), and land near village or road is convenient but easy picking for thieves (nos.105-8). Joint tenancy and ownership are no good, even with relatives (nos.76-7), and debt is necessary (no.52) but perilous (no.78).

A number of sayings spotlight sites and relations of power. The king's displeasure brings ruin (no.620); so does frequent inspection of the crop for tax assessment (nos.63-4); but then again, one saying advises that 'when measured, paddy will decrease but land increase' (no.647), insuring low official output figures and lowering farm taxes. Our recorded proverbial wisdom is dominant caste wisdom: in Tamil (nos.79-84) and Telugu (nos.37-43, 279, 288, 291), it boasts that only Vellalas and Reddis can be real cultivators. The Brahmans are incompetent, and Pariahs, Malas and Madigas too poor and ignorant. The sayings also depict male power. They refer repeatedly to demands by wives and concubines, compare selecting good land to finding a good woman, and use the birth of a first boy-child by the age of 24 as a measure of good fortune. To convey the value of low land that can be easily irrigated, one Tamil saying has it that a man who marries a flirt and one who farms high ground will be ruined alike (no.116). Sayings about weeds show that orderly control over unruly women and good fortune come together in farming: weeds are like deceitful hearts (no.475), a field overgrown with *korai* like a barren woman (no.477), fields infested by *hariali* like an unruly wife, and a garden full of *tadana* like a house with a harlot (no.480). But a field full of *hariali* weeds is also as likely to thrive as a man who opposes a king, to indicate that dominance at home requires obedience at court.

Subba Rao gathered rough-and-ready aphorisms that peasants used to teach their sons the ways of the world. But archaic formations include many other types of knowledge as well. Rulers from Maharajas to Nayaks and Poligars needed to know how to encourage and protect cultivation. The advice from the classical poet shows how royal self-interest and irrigation fit together notionally even in very distant times; and evidence from medieval inscriptions to Company records indicates that rulers learnt the lesson (Ludden 1979). Taxation could encourage farmers by lowering rates on newly-ploughed land; and in general rulers seem to have recognised that to keep farmers they had to adjust taxes to seasonal fluctuations. Often local histories even depict hands-on work by local leaders in rural development, not always successfully.

In northern Mysore, some time in the seventeenth century,

Doda Mastee and Chicka Mastee, two brothers with their families and cattle, came from the north and built two houses.... [T]hey cleared the jungle around and maintained themselves by cultivating the soil; they invited inhabitants from other parts *advancing them money to increase the cultivation;* the brothers next built a fort and gave their village the name Halla Goudennahully.... In time the brothers quarrelled because *the younger brother Chikka Mastee made a tank which endangered the village*. Doda Mastee being displeased with his brother's folly removed and built north of the first village another called also Goudennahully and all the inhabitants removed to the new village.... *Halla Goudennahully in the meantime was inundated by the water of the tank* (Mackenzie: 121; emphasis added).

Bringing their families and cattle to a suitable spot; building houses and forts; clearing the jungle; protecting people and animals from predators; farming and bringing others to expand settlement and cultivation; building a tank properly and seeing disaster on the horizon: all of these made Doda Mastee a man to follow, a wise and able leader. Manliness, power and knowledge connect the leader to his people's farmland in many local histories. One common motif from Mysore appears in the story of Heeroor: the land was jungle, when a raja went hunting and hares chased his hounds, a sign that this was 'male soil'. So the raja built a fort, encouraged people to come in, and founded a lineage (Mackenzie: 111).

More complex formations of knowledge than we could find in Subba Rao's sayings made agriculture work. They seem organised on lines we could call ecological. Classical Sanskrit and Tamil texts both depict forms of ecological knowledge. Tamil poets portray five milieux, and Sanskrit authors concentrate on a contrast between dry (*jangala*) and wet (*anupa*) to integrate medicine, government and morality. Both hold that the soil, water, flora and fauna determine the character (Tamil: *kunam*) of a place, also its human conditions and behaviour. Sanskrit authors show how soils infuse their qualities into the plants that grow in them, thus into animals eating plants, and humans eating animals and plants. A theory of substantialism emerges like that of Bachelard and Hippocrates.[12] In it, landscape gives Ayurvedic doctors diagnostic data, for health requires adaptation to land, compensation for imbalances in humours, and intervention to manage equality among them.

[12] Zimmerman (1987) quotes Bachelard's *La formation de l'esprit scientifique* and Hippocrates' *On Airs, Waters and Places*, IX.

Sanskrit substantialism, as described by Zimmerman, resonates with theories about the role of coded substances in Indian culture as well as with environmental determinism.[13] Perhaps people in south India did think in substantial terms, believing that farmers knew how to farm naturally, that *velanmai* was less the result of good decisions than a sign that the farmers' nature (*kunam*) was one with the land. In this view, land would belong to a family or a dominant caste not as a possession so much as a constituent of its character; land would be seen not as an instrument of subsistence so much as the material of existence. Land would be neither to farm nor to rule (Neale 1969), but to know as oneself. E. Valentine Daniel argues that this kind of logic can be found in Tamil Nadu today (1984: especially 182-225).

But our evidence does not suggest that an integrated science of the sort Zimmerman finds in Sanskrit texts has informed agricultural production in south India. Many of the Company's informants were Brahman landowners, and Subba Rao, a Brahman, quotes a number of Sanskritic aphorisms. But we find neither humoral references nor Ayurvedic homologies. There is not much to support a substantialist reconstruction either, even where we would most expect to find it, in arguments about land. Instead of claims about *kunam* we find stories of conquest, migration, genealogy, entitlement and dominion. Though these preoccupations do not preclude a substantialist underpinning (Daniel 1984: 95-104), they neither require nor imply it.

It seems that the Ayurvedic doctrine guided medical practice in relative isolation from south Indian agriculture,[14] and that archaic formations were more loosely ecological than strictly substantialist. What we have is bodies of knowledge composed of many aphoristic bits and pieces, not organised as a set of texts, certainly not as a doctrine or coherent theory. It was not designed for specialists like the men who passed down Ayurvedic science, so, like Tamil classics and Subba Rao's proverbs, agricultural knowledge was neither theoretically nor textually integrated. It developed inside agricultural praxis as an array of descriptions, prescriptive assertions, and observations of regularities, associations, and combinations of phenomena. It was concerned with

[13] Townsend (1791, 87ff.) explains the character of peoples by their environment, as do some early orientalists. Said (1978), Bernal (1987), and Jordan (1968:3-43) connect environmental determinism and racism.

[14] Zimmerman's argument pertains only to Ayurvedic ecology.

events and action all along the central axis of power running from the farm to the heavens. Each of its components is practical, instrumental; the ensemble is ecological in positing uncontrollable forces that humans must read in order to succeed.

A textualised representation of this knowledge can be imagined in the motley jumble of Sanskrit verses compiled as Varahamihira's *Brhat Samhita*,[15] which enjoins the astrologer to know 'indications of the approach of the monsoon...signs of immediate rainfall, prognostication through the growth of flowers and creepers...[celestial influences on the] fluctuation in the prices of commodities [and] growth of crops... treatment and fertilising of trees, water-divination [etc.]' (16). Because deities enjoy trees and water (537), the astrologer needs to know signs on the earth that indicate water sources below (499-561). He needs to know portents of famine: sunspots are a dire signal,[16] but also certain rainbows (29), shapes on the moon's face (36-8), eclipses (58), dust storms (67), appearances of Venus (105), and comets (146-51). In addition, famines are more likely in some planetary years than others (224).

Portents of rain are critical and introduced in the *Brhat Samhita* in such phrases as the poet's advice to the Pandyan king: 'As food forms the very life of living beings, and as food is dependent on the monsoon, it should be investigated carefully' (230). Seven chapters consider rain signs and parallel exactly the predictive wisdom of Tamil proverbs (Subba Rao 1896: nos.138-349), using not only configurations of the planets but signs such as rainbows, cloud shapes, insect and animal behaviour, the sounds and shapes of thunder and lightning, and rainfall in divisions of solar and lunar calendars. Many of Subba Rao's proverbs may come from astrologers (nos.186, 254, 258, 264-74, 275-349) who provided guidance for farmers. Archaic measurements of rainfall in south India, using implements that Benjamin Heyne described for Mysore (Heyne 1802), were for predicting rainfall, so that almanacs could presage 'the quantity of rain allotted to each country'; and the *Brhat Samhita* shows astrologers how to make rainfall measurements accurately (245-6).

[15] Thanks to Francis Zimmerman for pointing me to this text.

[16] There is a very convincing description of famine caused by evil dark shafts on p.26, which, reminiscent of the classical Tamil poet's advice to the Pandyan king, says 'Famished people...will condemn their king's conduct...'.

Accurate observation would have produced proverbial wisdom for predicting rain and storm, in India as in the West (Lee 1976), so that the transmission of astrological aphorisms need not be invoked to explain all similarities between the *Brhat Samhita* and Subba Rao's proverbs. But some similarities indicate transmission and diffusion. Their ways of predicting rain are the same. They agree in their concern for and methods of timing agricultural activities, above all planting. Both use the calendar to predict rainfall and to determine auspicious days; and precise timing by the calendar is the subject of a great many Tamil proverbs (nos.482-508). Both assert that the character of the soil can be told by what grows on it, and remedied by various means, though Tamil proverbs depict only soils of the south (nos.123-37).

They both theorise, implicitly, that the visible world is replete with signals of future events. They differ on what signals signify. For instance, the astrologer's text says that prospects for crops (332-6) and commodity prices (340-3) are foretold by constellations; whereas the Tamil proverbs selected by Subba Rao do not. But, just as the astrologer must know the signals in all varieties of plants and animals (chs.61-8, 86, 91-6), and, for instance, that 'the value of elephants is determined by the amount of ivory they have and [the value] of cows, horses, and the like...by their hair' (267), Tamil farmers paid close attention to plant and animal signs and particularly to swirling hairmarks (*suli*) on horses, cows and bulls, which foretold not only the value of the animals on the farm but the future prospects for the farmer as well (Holmes 1900; Littlewood 1936: 229-32).

Farmers, astrologers, doctors, poets and kings produced archaic formations of knowledge in which hares chasing hounds, rain filling poets, frogs tied to winnowing fans, deities dragged around temples, constellations, sunspots, and hair on a bull's back had agricultural meaning. Such significations have not vanished. I was told in 1968 that landing men on the moon caused drought. But I was also told that India's rural poverty was caused by farmers' failure to adopt birth control and high-yielding varieties. Both assertions depict a modern formation: global media put the moon landing into my informants' heads; and science would cite those about causes of poverty for the whole Third World. Archaic formations were not subject to science. They lacked the chasm between farmers and experts that opened up in the 1800s. Integrated from top to bottom, they employed a logic that was ecological in that it predicated prosperity on interactions among many natural forces, and worked towards harmony with forces outside human control. But archaic formations also lacked modern media, textuality and scale. They were not integrated across the peninsula, despite widespread and long-term diffusion of Brahmanic intelligence. They were ecological, therefore, in a second sense: their substance at any particular time and place was a localisation, defined by a particular configuration of all the countless forces pertaining then and therefore to agricultural prosperity. Hence the difficulty and indeed irrelevance of textualisation.

Place and time

To place this ecological formation on the ground, we can create a good geographical unit for an agrarian history of south India by cutting a map of the peninsula across at the latitude of the eighteenth parallel north.[17] The south (except Anaimalai and Cardomon peaks) lies below 1,350 metres, its relief dominated by the Western Ghats, mountain spurs (the Balaghat Range), and the high Deccan plateau, which slopes down from ghats at between 300 and 600 metres, to raise the southern interior, before sloping like the whole peninsula down to a flat eastern plain at less than 150metres. The south slopes from west to east behind the Western Ghats, on the north-west-south-east bias of the Krishna-Godavari system, its gradual slope interrupted by the high plateau and Eastern Ghat peaks.[18] This is a land of rain shadow. From the summer (western) monsoon, annual rainfall on the west coast and slopes is from less than 2,000 to more than 4,000 millimetres. Total rainfall declines rapidly around ghats north (around Surat) and east of the ghats to less than 800 millimetres, with pockets of less than 500. Rain from winter (south-east) monsoons raises total rainfall in the south-eastern area marked by Bangalore, Kanya Kumari and Vizagapatnam to a range from below 800 to over 1,000 millimetres, and even higher on the south-east coast.[19] But rainfall reliability declines in the south-east from the uncertainty of winter rain.

Behind the Western Ghats, a north-west south-east gradient is

[17] The northern boundary of south India would thus be a line running from Bankot on the west to Konsadu on the east just north of Godavari peninsula, through Phaton in Maharashtra and Warangal in Andhra.

[18] The exception is the bas relief around the Bastar rim.

[19] Again the lower Godavari is an exception, influenced by rain on the Bastar rim.

inscribed across semi-arid south India and its agrarian history by an increasing availability of drainage for irrigation with pre-modern technology.[20] In the Maratha Deccan, before modern dams, wells were almost the only irrigation (Taylor 1856) and still predominate.[21] But on the Karnataka plateau and north-east to Hyderabad and Warangal, tank irrigation was and is important, more so still to the south-east.[22] Pre-modern dam or channel irrigation works multiplied along the same gradient; because of greater total rainfall and irrigation groundwater recharge, farmland watered by wells would also have done so. The gradual increase from north-west to south-east in drainage availability implies parallel increases in irrigated acreage, multiple cropping and population density; but a major hole in this overlapping set of gradations lies in the very dry Andhra-Maratha interior and Pennar-Krishna interfluve (Anantapur, Bellary, Kurnool, Adoni, Raichur, Bijapur), where numerous tanks have long supported meagre irrigation and low population densities.[23] Agriculture expanded over centuries into three forest types that distinguish the south from the steppe land of thorny shrubs in Punjab, Rajasthan and Gujarat.[24] Dry tropical forest of deciduous trees covered the flatlands, most now reduced to more or less tree-covered savannah. Monsoon forest, losing leaves in the dry season, and full of teak, covered the slopes of the high plateau and Eastern Ghats. Evergreen rain forest covered the Western Ghats. Into each forest type, farming populations pushed agricultural frontiers, with technologies and knowledge developed in distinct environments; and overall, the peninsula's north-west south-east gradient organised the geographical diversity of agro-technological milieux. Pastoralism and long-fallow millet cultivation dominated the driest parts, especially north and west. Shortening fallow and well irrigation enabled more intense dry farming to take over where rainfall, technology and water-table allowed. Rainfall and drainage made wet paddy cultivation more

[20] This north-west to south-east gradient is shaped by three physical attributes—decreasing prevalence of black soil (which absorbs run-off), increasing total rainfall, and increasing localised relief.

[21] There was less than one per cent tank irrigation in 1965-6.

[22] In 1965-66, tanks irrigated less than 30 per cent irrigated net sown acreage in much of the Tamil Nadu-Karnaka interior, rarely less than 15 per cent from there towards the coast, in 1965-66.

[23] The two paragraphs above draw on *Census of India* and Singh (1974).

[24] See Zimmerman (1987) for more description and sources.

prominent in the south-east, and in western tropical wetlands where forest produce also fed populations.

Specialised formations of agricultural knowledge developed in agro-technological regimes that intersected and succeeded one another, but can be abstracted as in the table below (p.67). Depending on time and place, localities were dominated by one of these regimes and developed knowledge according to its specialisations. The forests and geology also helped to spatialise formations of knowledge, for instance, about carpentry near eastern teak forests, about granite carving in the Deccan, and about gold-mining near eastern hills, where in Kolar landless labourers are reported in 1802 to have predicted gold deposits by soil type and by flights of peacocks (Mackenzie 1802: 174).

Local knowledge aggregates into regional formations. Livestock-breeding, for instance, became a specialty of dry-farming regions of the high plateau and Krishna plain, where herds were fed fodder crops and sold to the more irrigated and worse breeding regions to the west, east and south.[25] This helps explain why collected Telugu sayings of the 1890s are more insightful on cattle-rearing than their Tamil counterparts, treating matters like bedding and signs of calving (nos.482, 507-8), which Tamil proverbs ignore. Two Tamil proverbs treat fodder, one to show that paddy straw yields little but sweet milk (no.851), and the other to praise *cholam* fodder (no.818), favoured by Telugu farmers in the Tamil country; whereas Telugu proverbs treat feed-timing and various fodders as well (nos.480, 481, 485). Tamil sayings treat cows and cattle only as objects of purchase and control (nos.819-867), comparing them to wives and children (nos.801, 802, 844).

Differences among regions arose from their histories as agrarian environments. Pastoralism had dominated much of the eastern coastal plain in medieval times (Heitzman 1985) and did pervade oral histories of the early-modern period from the Karnataka and Andhra uplands, but long before 1800 seems to have been reduced to insignificance along the eastern lowlands (Ludden 1985a: 46-59). Long-fallow dry farming remained important in the interior, but, by the eighteenth century, the eastern lowlands were dominated by the mixed wet-dry regime and the alluvial coast by irrigated agriculture. Andhra's long history of pastoralism and dry-farming explains why Telugu sayings speak poorly of irri-

[25] Specific breeds of cattle were thus developed in particular breeding regions. See Littlewood (1936).

gation compared to their Tamil counterparts, stressing that wet farming is risky and unmanly (nos.45-52), while dry farming is both noble and secure (pp.194-5, nos.275-94). Tamil sayings instead praise wet farming and treat its operations in more detail, with injunctions on locating (nos.116-22) and manuring (nos.135-7) paddy fields. Both Tamil and Telugu proverbs depict farming practices in contrasts between wet and dry, but turning dry land into wet was more a Tamil problem: several Tamil (but no Telugu) sayings refer to greater demands for labour control and taxes that attend the higher yields of paddy farming (nos.113-5). Archaic formations of knowledge in Tamil seem thus to have been more dedicated to expanding irrigated agriculture than those in Telugu.

We can speculate further on these lines to suggest historical successions. Medieval Tamil inscriptions depict an agrarian context in which pastoral activity is much more important than it would be in later centuries of Tamil history, and many medieval Tamil chiefdoms depended on pastoral economies. These political settings and their formations of knowledge were wiped out in post-medieval centuries of agricultural expansion, causing striking discontinuities between medieval and early-modern states in the Tamil country. The political power of pastoralists diminished to nothing in post-medieval centuries, when pastoral Maravas and Kallars were defeated by armed dry farmers, many led by Telugu Nayakas. Dry farmers conquered pastoralists, cleared scrub forests, dug wells, built forts, planted fields, and transformed the economy. Interactions among wet and dry farmers became the central motif in agrarian politics, after 1500, as the mixed wet-dry agricultural regime conquered Tamil Nadu. Therefore, by 1800, Tamil language formations of agricultural knowledge differed from both their medieval antecedents and from their Malayali, Telugu, Kannada and Marathi contemporaries, not only because of linguistic change, but also because Tamils ignored the tropical regime of the west, destroyed the pastoral regime that survived in the northern peninsula, and were bent on destroying the long-fallow dry regime that survived on the Tamil plain and dominated much of the peninsular uplands.

Two very long periods of agrarian history thus lie behind Subba Rao's proverbs and Company records. Irrigated rice farming dominated early Tamil formations of agricultural knowledge, but not the Tamil landscape, even as domination by irrigation interests was achieved in medieval times within kingdoms and *nadus*, described by Burton Stein

(1980). After 1300, migrations from the north brought dynamic, innovative, and ever more powerful short-fallow dry farmers. They combined with the lords of the wet land to conquer the landscape, drive out long-fallow farming and pastoralism, and dominate agriculture, a process that continued even more recently, as described by Christopher Baker (1984). Long-term change in the agricultural environment thus reproduced the powers of those farmers who controlled irrigated land but also transformed the meaning of dry land. History altered the relations of wet and dry, because short-fallow dry farming—with a heavy use of bulls to plough deeply and lift deep well-water for garden crops—steadily increased the relative value of dry land. In addition, many of the best dry-land farmers were immigrant Telugus, allied with the Nayakas who typified the political landscape after 1500 (Ludden 1985a: 183-96).

The increasing value of dry land in the Tamil agrarian landscape appears in changing uses of the term *kadu.* In classical poems, it is used to mean 'burning ground' (*sudukadu*). A poetic theme, *kaduvalttu*, is defined in the grammar *Tolkkappiyam* as praise of the burning ground to represent 'the transitoriness of earthly things' (*Lexicon*: 855-6). Thus *kadu* was a negation of comfortable social life. Though today *kadu* means 'forest' in everyday speech (*Kalakat tamil akarati*: 314), the world of the classical poets was so fully forested that they needed more than one word to capture the meanings of forest. Three of their five poetic milieux are forested: *kurinji* is tropical mountain forest; *mullai* is deciduous woods along hillsides, where animals threaten travellers; and *palai* is dry flatlands, thick with prickly scrub and cactus, and with robbers (*maravas*, hunter-pastoralists) in search of wealth, who threaten travellers. Only one milieu represents comfortable domesticity: a tract of irrigated villages (*maruytam*), watery lowlands, where forest all round means wildness and danger, where burning grounds represent life's impermanence.

By medieval times, the landscape was simplified by a dichotomy between *nadu* and *kadu*, where *kadu* implies untamed, rugged, forested habitation. Now *nadu* territories—not isolated lowland villages—are centres of agrarian civilisation. A series of identifications is thus established. The *kadu* is wild, inhabited by unruly folk who need to be brought into the orbit of royal authority by alliance and conquest. The *nadu* is civilised, controlled by Brahmans and Vellalas with trading towns, irrigation, temples, gods, and ceremonial order. Now *kadu* could mean the border of cultivated lands, a dry tract, or a small settlement (in the dry tract), and is often used as a place-name suffix. Kings and chiefs sought to incorporate the land and people of the *kadu* in their domains. Wars of conquest and incorporation brought labourers from the *kadu* into the expanding irrigation economy of the *nadu*, people of the *kadu* being hunters, pastoralists, and long-fallow farmers.[26]

After 1300, probably after 1500, *kadu* took on the meaning of dry land, whether cultivated, fallow or waste. *Nadu* lost its specific territorial meaning to become merely a term for 'country', a usage with classical precedent now enshrined in 'Tamil Nadu'. By 1800, Tamil analysis of agriculture rested on the contrast between dry land (*kadarambam* = *kadu*, dry land, plus *arambam*, tract) and wet land (*nirarambam* = *nir*, water, plus *arambam*), thus between dry cultivation (*punsey*) and irrigated agriculture (*nansey*) (Bandopadhyay, 1978). The superiority of irrigated agriculture in the Tamil-language context is clear in the latter terms, for *punsey* connotes meagreness (*pun-*) and *nandsey*, goodness (*nan-*) (Fabricius, 1779).

Over the millennia separating classical and modern times, *kadu* moved from the exterior into the periphery, and then into the centre of Tamil agrarian life. As it became more central, its meaning became more complex. To translate many of Subba Rao's sayings, we must rely on his knowledge of usage to decide if *kadu* means 'forest', 'waste', 'open pasture', 'closed pasture', 'dry farmland' or 'dry field' (nos.69, 85, 87, 715, 799, 861). He says *kadu* 'translated as "forests" must also be taken to include pastures', and 'in these days, as the country is filling up, these should no longer be the wild common-grazing grounds on which hitherto dependence has been placed, but also enclosed and cultivated pasture fields...' (p.86). On the saying 'To ruin a *kadu*, let loose goats' (no.861), he comments: 'the destructive results of grazing sheep and goats are alluded to, though here again the word *kadu* may either be the jungle or the field with a crop on it' (p.90), because goats eat the field stubble that could fertilise the next crop as surely as they destroy open pasture or wild scrubland.

Subba Rao's mentioning 'enclosed and cultivated pasture fields'

26 See Ludden 1985a: 91ff. Zimmerman (1987: ch.2) depicts parallel Sanskrit contrasts between *janapada*, which is both 'the country and people of the country', and *atavi*, 'primitive jungle space'. See also Stein (1980: 362).

indicates a relatively new feature of *kadu*. A number of his sayings prescribe 'fencing in' land (nos.85-93, 800), though Subba Rao claims that 'fencing is but little practised over the greater part of the country' (p.27). Where his sayings do not refer to fencing small garden plots watered by wells (normal practice today), they probably advise enclosure. No.85 would thus be rendered, 'Look at the *kadu* of a man who has closed it off [from use by others] and you will see the cattle of a man who knows how to graze livestock'. Likewise, no.87, which Subba Rao translates as, 'Sow the field and shut the gaps in the fence', is more literally 'Sow your seeds and shut the door'. People should not be allowed to use dry land as common land external to the village system of private property rights. Proverbial injunctions like this indicate that grazing rights were problematic. In the 1800s, dry land, forest and livestock became sufficiently valuable and private property rights well enough established, for forceful enclosure to become desirable for the dominant caste, rich peasant or incipient capitalist farmer: to keep the neighbours' animals off his fields, to keep his own grazing land for himself, and to make proprietary claims on commons (see Baker 1984; Ludden 1985a: 164-96).

The agricultural value of *kadu* was clearly appreciated before 1800. As pasture, its value derived from livestock. A Tamil saying (no.799) reads, 'Rich forests (*kadu*) make strong cattle; strong cattle make prosperous people; prosperous people make rich temples; rich temples make rich kings'. Heyne says that in dry upland Karnataka, cattle are the farmers' most valuable asset and everyone recognises that without cattle the farmer is ruined: '...it is the last of his property arrested by his creditors and if he owes anything to the circar they will be seized but never actually taken from him' (Heyne 1802: 78). Sayings in Tamil (no.34) and Telugu (no.485) add to this set of meanings for *kadu* without even using the word: the Telugu saying reports, 'there is no want in a house where the spinning-wheel and churn are at work'; as the Tamil saying argues, 'there is no famine (*panjam*) for a man with milk and cotton plants'. Cotton is an archetypal dry-land (*kadu*) cash crop; among dry crops, cotton is the only one that Tamil sayings claim produces 'potfuls [of money]' and cash (*panam*) (nos.730, 749-50). It is labour-intensive (no.725) and thrives with deep ploughing of rich black soil (*karisal*); it thus wants strong bulls that are well-fed.

Closely associated with rearing, buying and grazing bulls and cows throughout the peninsula, cotton-farmers provided raw material for cloth exports and then for cotton exports that doubled every few decades after 1840. Dry land used for commercial production encouraged the privatisation of dry-land property rights (Ludden 1985a: 175-88). So *kadu* became entangled in the politics of agricultural commodity production. Conflict between agriculture and trade pops up in Subba Rao's sayings: no.40 posits a natural enmity (*jenamapakai*) between parsimony (*settu*) and *velanmai*; no.81 asserts animosity in the same terms, but uses *setti* (merchant) for *settu.* The powers of *velanmai* are lost when one adopts the merchant's ways or succumbs to his control. No.78 warns that 'The man who takes loans to farm is like a tree climber who lets go with his hands'. The conflict portrayed here pits farmers against merchants, not against profit-making, for profits had long been part of *velanmai*. These sayings represent the pride and fear of dominant-caste landowners enmeshed in commodity production by the dynamics of their dominance.[27]

Company records show that by 1800 all varieties of land and capital assets were known for their commercial value and agrarian terminologies were constructed accordingly. In the Tirunelveli region, in the far south, late pre-colonial state revenue rested on a series of equations among types of land, their typical product, the market value of that product, and the most commercially sensible means for the state to realise revenue from that land.[28] Dry land (*punsey*) typically produced millets, oilseeds, pulses and cotton, and varied in value by soil type. Soil classification was basic in agricultural terminology and remained so in proverbial wisdom (Subba Rao 1896: no.123-37). Sandy soil (*manal*) was worst, black soil (*karisal*) best; and redness, clay content and saltiness determined soil quality within these extremes. Dry land area was calculated by linear measurement, with a *sangili* (chain) or

[27] Arnold (1984: 85-90) depicts enmity between the rural population and merchants at famine time, when merchants profit as people go hungry. Farming communities looked to the state to control grain prices, a function performed by the revenue system until the 1830s, but then abandoned by the colonial state. Looting in the Kallidaikurichi, Tirunelveli District, on 17 December 1876, was in fact triggered by merchants closing their shops from fear of looting. Here and elsewhere, the most violent popular hatred was directed against those who would ship grain out of localities to seek high prices elsewhere; there seems to have been popular sentiment that local provisioning should come first.

[28] For details on the discussion of Tirunelveli below, see Ludden 1985a; on 'tributary commercialism', see Ludden 1989.

rods (Bandopadhyay 1978). The length of the measuring device differed from place to place (Heyne 1802: Buchanan 1870), but everywhere people seem to have talked about dry land in terms of its linear area—a practice that may have come from estimates of land area by the number of rows that a team of bulls could plough in a day, a method reported by Heyne (1802: 89). Everywhere in the south, dry land seems to have been assessed for revenue purposes, by the eighteenth century, according to area and soil type, in cash.

Tax-related agricultural knowledge preserved by Company records indicates that Tamils systematically distinguished wet and dry land as objects of commercial evaluation and commercialised taxation. The term for 'tax on dry land' was *tirvai* in Tirunelveli.[29] From a Tamil term for 'judgment' or 'decision', *tirvai* was also used for 'toll' and 'custom duty' (Fabricius 1779) and dry land taxes were collected in cash much as custom duties were. Wet cultivation (*nansey*), by contrast, involved a terminology that coheres because rice yields vary, in the rain shadow of the Ghats, in proportion to the quality of irrigation. Wet land gained value from irrigation, not soil type; it was measured by the volume of seed sown upon it, not by linear area. Rice is the heart of wet-land terminology. Its seed-grain measure was used in the market and the kitchen and to measure a rice land area. In Tirunelveli, the measure was a *kottai,* elsewhere, a *kalam*. One *kottai* of land was the area sown by one *kottai* of grain, so that it varied in size with the volume of the measure and with seed ratios. Taxes were assessed as a share of the paddy crop, and the state would take its share (*melvaram*) in paddy to sell as prices rose after the harvest, thus securing a cash revenue that depended on rice's being a commodity. The size of the state's share was conventionalised and expressed as so many *kottai*s of grain for each *kottai* of cultivated land, so that state agents had an incentive to extend irrigation and farmers to intensify cultivation by raising the seed ratio. The term for 'land tax on wet land' was, appropriately, *varam*, 'share'. Power in the wet lands centred on shares of the paddy crop. And like the state's share, the shares (*pangu*), that signified the powers of landowners and labourers inside village society, had commodity value (Ludden 1985a, 1985b).

Similarly, garden cultivation, tree crops, houses, shops and a long

[29] See Turnbull (n.d.) and 'Census and Dehazada of the Province of Tirunelvelie'.

list of commercial as well as artisan assets were counted, taxed and described in commodity terms in eighteenth-century south India. As *kadu* travelled from the periphery towards the centre of agrarian production in the Tamil country, the landscape as a whole was slowly commodified. True, early-modern commodity production was localised. Localism determined exchange rates, weights and measures, and revenue calculations. Political relations were local. Local agricultural milieux, sustained by nature, describe a patchwork quilt designed in considerable part by physical geography. But archaic agricultural settings were also patterned by movements of people and technologies. From medieval times onwards, we can see patterns imposed by people on the physical landscape of localities everywhere. Paddy farming, for instance, everywhere had its own distinct set of technical terms and operations. For everywhere, it meant lifting, storing, channelling, draining, distributing and evaluating water; and selecting, planting, breeding and discriminating between hundreds of seed varieties (Bray, 1986). Everywhere, social relations of irrigated agriculture made sense in an institutional complex that spread throughout the south. In the centuries after 1500, agrarian commercialism seems to have pervaded localities and patterned local formations of knowledge, power and productive practice very widely indeed (Ludden, 1988).

Extinction

Old formations of agricultural knowledge became archaic as the climate of agrarian enterprise changed during the nineteenth century. This can be explained partly by unilateral state action. The colonial state did appropriate local knowledge for its own purposes and did apply to agrarian India scientific methods that made all localities look alike. It produced a new vision of agrarian space by defining regions with statistics. It made archaic what R.D. Sack (1980) calls fused perceptions of space. Formations of knowledge that fused the flora, fauna, soil, caste history, royal genealogy, astrology and rituals of specific regions became archaic as the new, scientific, imperial state codified agriculture.

But old formations became archaic in part because of changes in south Indian agrarian society. Expanding commodity production, labour mobility and education drew locals out of old, localised, agrarian milieux into the modern intellectual terrain dominated by the colonial state (see Nandy, 1983). No doubt, Subba Rao himself made this move; and dominant castes in the irrigation economies of the south were the brains behind departments of agriculture in Madras, Mysore and Hyderabad. At the same time, local dynamics of *velanmai* changed in the countryside. By 1900, *kadu* was being fenced in as private pasture and dry-crop land. Wet-land *pangus* had already become private property (Ludden, 1985b). Early-modern commodity producers used *kadu* and *pangu* in disputes that were local and political in that they involved power over land, but also productive in that they increased land productivity. Even in the early 1800s, knowing how to farm meant knowing how to engage in local conflict to establish essential rights in land, labour and capital. As profits from commercial enterprise increased under the raj, so did incentives to escape the constraints of localised formations. At the same time, government took control of the language of rights. Old land measures were standardised in the 1830s, diminishing their local variations; then, during revenue resettlement in the 1870s, they were abandoned altogether, as the state surveyed and recorded land rights for every bit of land in every village, on maps and deeds, in acres. The terminologies of power and local disputation were thus extracted from the semantic domain of agriculture, and the social praxis of *velanmai* extracted from farming. Now the imperial state and world markets defined the terms of power and productivity. Politics and economics divided up the cultural realm of agrarian power relations that connected gods and people to the land. Agriculture became economic enterprise surrounded by institutions that determined its power relations. Government could now manage agriculture in the name of science and development, in a land of farmers who increasingly managed their assets in the language of capitalism.

Table: *Agrarian regions in south India*

Regime	Rainfall (mm/yr)	Irrigation	Key food-product	Key cash-crop	Space
Pastoral dry	<500 <700	wells	millets animals	cotton	plateau forest
Dry	>500 <800	wells	millets pulses	cotton oilseed	plateau plain
Dry-wet mixed	>500 <1250	tanks wells	millets rice	cotton oilseed	plateau plain
Wet	>1000	tanks dams	rice	rice garden	plain seashore
Tropical	>1250	dams channels	rice fruit	tree garden	mountain plain

References

Arnold, David, 1984. 'Famine in peasant consciousness and peasant action: Madras, 1876-8', in Ranajit Guha (ed.), *Subaltern Studies III: writings on South Asian history and society* (Delhi), pp.62-115.

Baker, Christopher John, 1984. *An Indian Rural Economy: the Tamilnad countryside, 1880-1955* (Oxford).

Bandopadhyay, Arun, 1992. *The Agrarian Economy of Tamilnadu, 1820-1855* (Calcutta).

Benson, C., 1898. *A Collection of Telugu Sayings and Proverbs Bearing on Agriculture* (Madras).

Bernal, Martin, 1987. *Black Athena: Afro-Asiatic roots of classical civilization*, vol.I, *The fabrication of ancient Greece, 1785-1985* (New Brunswick).

Bhattacharya, Swapan Kumar, 1976. *Farmers, Rituals and Modernization* (Calcutta).

Bray, Francesca, 1986. *The Rice Economies: technology and development in Asian societies* (London).

Buchanan, Francis (Hamilton), 1870. *A Journey from Madras through the Countries of Mysore, Canara and Malabar* (Madras).

—, 1956. *Journey through the Northern parts of Kanara*, with introduction by S. Silva (Karwar).

Byres, T.J. and Harbans Mukhia (eds.), 1985. *Feudalism and Non-European Societies* (London).

'Census and Dehazada of the province of Tirunelveli', Tamil Nadu Archives, Revenue Department Sundries, no.39.

Census of India, 1961 vol.I, India, pt.IX, *Atlas* (New Delhi, 1970).

Cohn, Bernard S., 1985. 'The command of language and the language of command', in Ranajit Guha (ed.), *Subaltern Studies* IV: *Writings on South Asian history and society* (Delhi), pp.276-329.

Daniel, E. Valentine, 1984. *Fluid Signs: being a person the Tamil way* (Berkeley).

Fabricius, Johann Philip, 1779 (1972). *Tamil and English Dictionary* (revised and enlarged edition of his Malabar-English Dictionary, 1779, 1809) (Tranquebar).

Fussell, G.E., 1983. *Agricultural History in Great Britain and Western Europe before 1914: a discursive bibliography* (London).

Goldstein, Leon, 1976. *Historical Knowing* (Austin).

Greenaway, T., 1864. *Farming in India, considered as a pursuit for European settlers of a superior class, with plans for the construction of dams, weirs, tanks and sluices* (London).

Hart, George L. III, 1973. 'Women and the sacred in ancient Tamil Nadu', *Journal of Asian Studies* 32, 2 (Feb.), pp.233-50.

Heitzman, E. James, 1985. 'Gifts of Power: temples, politics and economy in medieval South India', unpublished dissertation, University of Pennsylvania.

Heyne, Benjamin, 1802. 'Correspondence to Capt. Mackenzie, Supt. of the Mysore Survey', National Archives of India, Foreign Miscellaneous Series, no.94.

—, 1814. *Tracts, Historical and Statistical on India...* (London).

Holmes, J.D.E., 1908. *Native Beliefs Regarding the Hairmarks Horses and Cattle* (Madras).

Irschick, Eugene F., 1969. *Politics and Social Conflict in Tamilnad* (Berkeley).

Jordan, Winthrop D., 1968. *White over Black: American attitudes toward the Negro, 1550-1812* (New York).

Kalakat tamil akarati, 1969. (Tamil-Tamil Dictionary) (Madras).

Karashima, Noburu, Y. Subbarayulu and Toro Matsui, 1978. *A Concordance of Names in the Cola Inscriptions*, 3 vols. (Madurai).

Lee, Albert, 1976. *Weather Wisdom* (Garden City).

Littlewood, Capt. R.W., 1936. *Livestock of South India* (Madras).

Ludden, David, 1979. 'Patronage and irrigation in Tamil Nadu: a long-term view', *Indian Economic and Social History Review* 16, 3, pp.347-65.

—, 1985a. *Peasant History in South India* (Princeton, 1985).

—, 1985b. 'The terms of ryotwari: semantics and disputes over property rights in Madras Presidency, 1800-1885'' in R.E. Frykenberg and P. Kolenda (eds.), *South Indian Studies: an anthology of critical essays on recent scholarship* (Madras), pp.151-70.

—, 1985c. 'Productive power in agriculture: a survey of work on the local history of British India' in M. Desai, S.H. Rudolph and A. Rudra (eds.), *Power and productivity in South Asian Agriculture* (Berkeley), pp.51-99.

—, 1988. 'Agrarian commercialism in eighteenth-century South India: evidence from the 1823 Tirunelveli Census', *Indian Economic and Social History Review* 25, 4, pp.493-520.

—, 1990a. 'World economy and village India, 1600-1900: exploring the agrarian history of capitalism', in Sugata Bose (ed.), *South Asia and World Capitalism* (Delhi).

—, 1990b. 'The colonial construction of agricultural development in India', in

Nicholas B. Dirks (ed.), *Colonialism and Culture* (forthcoming).
Ludden, David and M.S. Pillai, 1976. *The Kuruntokai: an anthology of classical Tamil love poetry* (Madurai).
Mackenzie, Colin, 1800-1801. 'Captain Mackenzie's Memoirs of the Northern Pargannah of Mysore', National Archives of India, Foreign Miscellaneous Series, no.92.
Manak, Elizabeth, 1979. 'Formulation of Agricultural Policy in Imperial India, 1872-1929: a case of Madras Presidency', dissertation, University of Hawaii.
Manual of Administration of the Madras Presidency, 1885. 2 vols. (Madras).
Moreland, William H., 1904. *The Agriculture of the United Provinces, an introduction for the use of landholders and officials* (Allahabad).
—, 1913. *Notes on the Agricultural Conditions and Problems of the United Provinces, revised up to 1911* (Allahabad).
Nandy, Ashis, 1983. *The Intimate Enemy: loss and recovery of self under colonialism* (Delhi).
Neale, Walter C., 1969. 'Land is to rule', in R.E. Frykenberg (ed.), *Land Control and Social Structure in Indian History* (Madison), pp.3-15.
Needham, Joseph, 1984. *Science and Civilization in China*, vol. 6, *Biology and biological technology*, pt.II, *Agriculture* by Francesca Bray (Cambridge).
Pogson, Frederick, 1883. *A Manual for Agriculture in India* (Calcutta).
Rajam, V.S., 1986. '*Anangu*: a notion semantically reduced to signify female sacred power', *Journal of the American Oriental Society* 106, 2, pp.257-72.
Ramanujan, A.K., 1966. *The Interior Landscape* (Bloomington).
Robertson, William R., 1880. *An Agricultural Class-book for the Use of Schools in South India* (Madras).
Royle, J.F., 1851. *On the Culture and Commerce of Cotton in India* (London).
Sack, Robert David, 1980. *Conceptions of Space in Social Thought* (Minneapolis).
Said, Edward, 1987. *Orientalism* (New York).
Schrottky, Eugene C., 1876. *The Principles of Rational Agriculture applied to India and its Staple Products* (Bombay).
Shah, Motilal Kashal Chand, 1988. *Principles of Agriculture for India* (Ahmedabad).
Singaravelu, S., 1966. *Social Life of the Tamils: classical period* (Kuala Lumpur).
Singh, Jasbir, 1974. *An Agricultural Atlas of India: a geographical analysis* (Varanasi).
Smith, Richard Saumarez, 1985. 'Rule-by-records and rule-by-reports: complementary aspects of the British imperial rule of law', *Contributions to Indian Sociology* 19, 1, pp.153-75.
Stein, Burton, 1980. *Peasant State and Society in Medieval South India* (Delhi).
Subrahmanyam, Sanjay, 1990. *The Political Economy of Commerce: South India, 1500-1650* (Cambridge).
Tamil Lexicon, 1935-37. (Madras).
Taylor, Philip M., 1856. *A Statement and Remarks Relating to the Expenses of Irrigation from Wells* (Bombay).
Townsend, Joseph, 1791. *A Journey through Spain in the years 1786 and 1787, with particular attention to the agriculture, manufacture, commerce, population taxes and revenue of that country*, 3 vols. (London).

'Translation of Sanskrit work on land measurement, 1797'. National Archives of India, Survey of India (Dehra Dun) Records.

Turnbull, Thomas, 'Geographical and Statistical Memoir of Tirunelvelie and its Zamindaries', Tamil Nadu Archives, Revenue Department Sundries, no.38.

Varahamihira's Brhat Samhita, 1981. M. Ramakrishna Bhat, 2 vols. (Delhi).

Wadley, Susan (ed.), 1980. *The Powers of Tamil Women* (Syracuse).

Washbrook, David A., 1976. *The Emergence of Provincial Politics: the Madras presidency, 1870-1920* (Cambridge).

Wheeler, J.T., 1862. *Handbook for the Cultivation of Cotton in the Madras Presidency* (Madras).

Wilson, H.H., 1855. *A Glossary of Judicial and Revenue Terms* (London).

Yule, H. and A.C. Burnell, 1886. *Hobson-Jobson: A glossary of colloquial Anglo-Indian words and phrases* (London).

Zimmerman, Francis, 1987. *The Jungle and the Aroma of Meats*: *an ecological theme in Hindu medicine* (Berkeley).

Chapter 3

THE PROCESS OF AGRICULTURAL COMMERCIALISATION IN EASTERN INDIA DURING BRITISH RULE: A RECONSIDERATION OF THE NOTIONS OF 'FORCED COMMERCIALISATION' AND 'DEPENDENT PEASANTRY'

B.B. Chaudhuri

This paper is intended as a critique of two crucial notions in a particular interpretation of peasant motivations behind the choice of commercial crops: (a) that agricultural commercialisation was not a 'normal' process, but a 'forced and artificial' one; and (b) that the artificiality had much to do with the 'dependence' and 'subordination' of the peasantry. The two notions form part of a wider assumption: the negative effects of agricultural commercialisation upon the small peasant economy.

The first notion means that agricultural commercialisation in colonial India was typically an 'involuntary' process distinguishable from a 'normal' one. Whatever the form of the product, a subsistence or a cash crop, goes the argument, the 'main feature of such a normal process of commercialisation lies in the fact that the increased volume of marketed agricultural produce results from increased production over self-consumption requirements by the peasants'.[1] Exchange in such circumstances, it is argued, improves the peasant's resources and also enables him to have access to a wider range of goods. In contrast, what peasants marketed as they were involuntarily involved in commercialisation was not their surplus product; they were 'forced to sell a high proportion of their output to meet immediate cash requirements'. The second notion means that the compulsion under which such transactions occurred derived primarily from the peasant's 'dependency status', that is, his subordination in various ways to dominant rural groups.

The notion of 'forced commercialisation' is not altogether a novel

[1] A. Bhaduri, 'Class relations and commercialisation in Indian agriculture: a study in the post-independence agrarian reforms of Uttar Pradesh', in K.N. Raj *et al.* (eds.), *Essays on the Commercialisation of Indian Agriculture* (New Delhi 1985), p.308.

one; it is traceable in the writings of the early nationalists. Romesh Dutt, by way of refuting the government interpretation of the noticeable expansion in the trade in food crops as evidence of peasant prosperity, called this commerce 'forced and artificial', necessitated by the obligatory remittance of colonial 'tribute' and the resultant economic drain.[2] B.M. Bhatia, later elaborating Dutt's reasoning, emphasised the influence of India's adverse balance of trade on the process of agricultural commercialisation: the inability of deindustrialised India to pay for the rapidly rising imports of British manufactured goods other than by exports of food-grains and, to a much smaller extent, of other agricultural products.[3] Dutt perceived the adverse effects of this 'forced' commerce on India's rural economy mainly in terms of the increasing instability of the food supply in the domestic market, the rising prices even of inferior grains which were all the common people could afford, and the increasing proneness of the economy to recurring scarcities.

The recent interpretation of this 'forced' commerce is strikingly different. We notice here a shift from the remote domain of India's international economy to the immediate reality of peasants' relations with the dominant groups in the village. This interpretation regards the process of agricultural commercialisation as a variant of the 'subsistence

[2] Dutt observes: 'India had to meet a heavy drain which flowed annually into Great Britain without a direct commercial equivalent; she was unable to meet this demand by manufactured articles; and she met it therefore with the food supply of the people to a larger extent than she would otherwise have exported.... Under the cheering appearance of a brisk grain trade lies concealed the fact that the homes and villages of a cultivating nation are denuded of their food to a fatal extent, in order to meet that annual tribute which England demands from India'; *Economic History of India in the Victorian Age* (5th edn.), pp.348-9. In explaining 'how this process works', Dutt emphasised the role in it of the colonial revenue demand. 'The annual Economic Drain...is met directly from the revenues of Bengal.' Later, too, the 'key link between the colonial systems of land revenue and increased agricultural production for the market' has been admitted. However, the usual line of reasoning in this regard has been questioned on the ground that it unduly exaggerates the cruciality of the role of British revenue systems in the generation of the commercialisation. Sugata Bose thus concludes: 'Although the revenue-money-commerce nexus is undeniable, the levels of monetization and commercialization reached in the immediate pre-colonial era should caution against acceptance of any notion of an exorable sequence set in train by the cash hunger of the colonial revenue establishment'; *Peasant Labour and Colonial Capital: rural Bengal since 1770* (Cambridge, 1993), p.39.

[3] B.M. Bhatia, *Famines in India, 1860-1965: a study of some aspects of the economic history of India* (Bombay 1967, 2nd edn.), chs.1-2.

agriculture' of a dependent peasantry. Amit Bhaduri, for instance, calls the trade in agricultural produce 'the familiar cycle of distress buying and selling by small peasants, usually under the compulsion of debt'.[4] In his study 'Commercialisation of agriculture in Orissa, 1900-1930', Pradipta Chaudhuri, a student of Bhaduri's, similarly concludes: 'The indebted peasants were compelled into market relations through the mechanism of debt'.[5]

The cash requirements of peasants which forced them into market relations were mostly variants, it has been argued, of their subsistence needs, such as obligatory payment of rent or revenue and payments for debt. Default in payment of rent or revenue might cost a peasant his holding, since his eviction on grounds of this default was lawful. Persistent non-payment of debt had a similar consequence, apart from the fact that peasants, once defaulting on their payments, and thus antagonising their creditors, might find their source of credit cut off for ever. Loss of land and non-availability of credit threatened a peasant's subsistence.

Bhaduri has shown how the marketing behaviour of a peasant, pressed in this manner, became inseparable from a kind of defence mechanism or survival strategy.[6] One of his explanations of the pressing cash needs of peasants applies specifically to Bengal—'commercialisation of undertenures', an altogether new development during British rule. It meant that zamindars leased out, in perpetuity or for short periods, portions of their estates at a rental much in excess of their revenue dues to government, with the lessees themselves repeating the practice. The consequent proliferation of undertenures inevitably inflated the peasant's rent burden, obliging him to borrow. The leasing practice adversely affected the peasant in another way. It now paid no one, argued Bhaduri, to invest in agriculture, particularly because of the absence of any mechanism of cost distribution. Agricultural output could not, therefore, rise fast enough to absorb the increasing rent burden. The peasant's consumption level had, as a result, 'to be forced down to generate the necessary surplus'. In fact he could not do with-

[4] A. Bhaduri, 'Class relations and commercialisation ', p.308.

[5] Pradipta Chaudhuri, 'The impact of forced commerce on the pattern of emigration from Orissa, 1901-21', in Raj , *Commercialisation*, p.184.

[6] A. Bhaduri, 'The evolution of land relations in eastern India under British rule', *Indian Economic and Social History Review* (January-March 1976).

out borrowing in order to maintain the 'bare minimum level of consumption'. Indebtedness necessitated distress sale of his crops, with the consequent depletion of his food stock obliging him to buy food from the market in the lean months of the year. This was a peculiar sort of commercialisation, since it tended to blur the usual distinction between a subsistence crop and a commercial crop.

Peasants, Bhaduri argues, could scarcely break out of this 'cycle', since 'in such a situation considerable economic stagnation would result because the sustenance of forced commercial relations necessarily requires a perpetuation of poverty of the small producers'. This is partly because of 'the peculiar nature of the peasant's exchange, exchange of paddy for paddy'. Because of his marginal access to a wider range of goods,

> the mutual dependence between agriculture and industry from the demand side could not develop. An exchange of agricultural commodities for manufactured goods would have broadened and stabilised the home market for industrial goods, and a sound industrial system would have stimulated demand from food-grains and other agricultural raw materials.[7]

Bhaduri also relates the stagnation in the rural economy to the phenomenon of interlocking credit and tenancy. The combination of the two roles made a landlord averse to innovations which would normally increase agricultural output. Bhaduri assumes that the landlord's income, from interest on loans to the tenants, might decline as a result of any increase in the output, to offset his share of the increased yield, since the tenants would borrow less as they shared the increase.

Shahid Amin, another leading proponent of the idea of forced commercialisation, concentrates on sugarcane cultivation in the Gorakhpur district of U.P. and shows how peasants found sugarcane a way out, in the context in which food crops were not dependable as a source of vital cash resources towards obligatory rent payment. The primary reason for the peasant's dependence on sugarcane, the most important cash crop of Gorakhpur, was the lack of correspondence between the peasant's rent calendar and his harvest calendar. At the time when one of the heaviest of his rent instalments became due, he had no valuable food crops to fall back upon. Hence arose his dependence on sugarcane which was then ripening: 'Caught between the imperatives of their harvest calendar and the pressing demand for prompt payment of

[7] Ibid.

rental...the small peasants fall back upon the cash-raising potential of their sugarcane.' Amin concludes:

It was this importance of sugarcane as a cash-raising and debt-servicing crop, rather than its value as a surplus accumulator, that imbued it with a special role in the small peasant economy of Gorakhpur in the late nineteenth century.... [C]ane cultivation and *gur* production were an integral part of the reproduction of a small peasant economy, and not simply a commercial activity.[8]

The badly-needed cash for the peasantry here was obtained through advances from people connected in various ways with the sugar enterprise. The peasants' willingness to cultivate jute, by far the most significant cash crop in eastern India, has been similarly explained. 'The biggest attraction...was the availability of advances at the time of the year when their food stock from the previous rice harvest went to pay off rents and interest on loans.' As a result, jute 'only helped keep the peasants at subsistence level'.[9]

The crucial elements in this interpretation of agricultural commercialisation, particularly of the peasant involvement in the cultivation of distinctive cash crops, may now be summed up. It was the expectation of cash advances towards meeting basic subsistence needs (such as payment of rent and debt) that persuaded the peasantry to grow cash crops; the sources of the advances were mostly persons interested in the peasants' produce, in their agricultural form or in various manufactured or semi-manufactured forms. Such advances differed from other types of borrowing by the peasants from professional village moneylenders or similar sources. Acceptance of advances by peasants gave the persons making the advances a decisive claim on the peasants' produce. This conditionality might not have been present in the other types of rural credit. Finally, under the conditions in which peasant involvement in cash-crop cultivation originated, it had a negative role in the small peasant economy, and peasant subordination tended to be firmly entrenched.

8 Shahid Amin, *Sugarcane and Sugar in Gorakhpur: an enquiry into peasant production for capitalist enterprise in colonial India* (Delhi 1984), pp.70-1 and 88. See also Elizabeth Whitcombe, *Agrarian Conditions in Northern India*, vol. I, *The United Provinces under British rule, 1860-900* (Berkeley 1971), pp.175-8.

9 Partha Chatterjee, 'Agrarian structure in pre-partition Bengal', in Asok Sen *et al.*, *Perspectives in Social Sciences* 2. *Three studies on the agrarian structure in Bengal, 1850-1947* (Delhi 1952), p.115.

In this paper we intend to test the validity of this interpretation, with particular reference to the most important commercial crops of eastern India—indigo, opium, sugarcane and jute.[10] The interpretation is significant in that (a) the way it looks at the phenomenon of agricultural commercialisation is distinctive, and (b) in general it serves to correct a bias in the standard explanation of the periodic variations in cash-crop cultivation solely in terms of price movements.

Here we refer only to a few representative trends in the studies on the origins of agricultural commercialisation and the nature of its implications for the regional economy. As regards the origins, the emphasis is often put exclusively on the emergence and fluctuations of the market demand for cash crops. One question recently examined is whether increasing commercialisation of agriculture can be regarded as an aspect of the alleged articulation of the regional economy with the 'world system' based on capitalism.[11] While it has seldom been denied that market demand was crucial to the progress of agricultural commercialisation, another point of view emphasises the limits to peasants' choices in responding to the market demand.[12] The particular interpre-

[10] Our exclusion of food crops from the discussion is not due to any assumption that from the point of view of their role in the peasant economy they could be clearly and invariably marked off from the usual cash crops. I do agree with a recent argument of Rajat Datta where he points out the 'futility of adhering to any rigidly antithetical notions of "subsistence" versus commercialized agriculture, of limiting oneself to the growth of so-called cash crops while studying the working of commercialized peasant production'. A major conclusion of his study 'Rural Bengal: social structure and agrarian economy in the late eighteenth century' (unpublished PhD thesis, University of London, 1990) is as follows: 'While the early eighteenth-century textile-based commerce set the tone for a long wave of commercialization, it was the late eighteenth century which provided the distinctive departure by bringing about the commercialization of the province's rice-producing economy...' (p.7). The four major 'structural changes' which, according to the study, led to this phenomenon included 'the growing demand for food in the province' and 'the apparently fortuitous cluster of dearth and famine years between 1769 and 1793' (pp.14-5). I have concentrated on a few cash crops primarily because of their direct relevance to the thrust of the argument of my paper.

[11] Sugata Bose (ed.), *South Asia and World Capitalism* (Delhi 1990).

[12] Peter Robb, 'Peasants' choices? Indian agriculture and the limits of commercialisation in nineteenth-century Bihar', *Economic History Review* xlv, 1 (1992). Particularly significant is his idea of 'interdependence within the local community'. 'Each cultivator's responsibility, even for his day-to-day activities, was circumscribed in various ways, because neither his holding nor his methods of production existed in isolation from the political structures

tation of agricultural commercialisation which we intend to examine in this paper does say that peasant response to market demand was significantly limited by the mediation of 'rural power relations'. The sources of the constraints on peasants' choices, of which Peter Robb writes, are diverse.

David Washbrook points to a constraint of an altogether different nature, in a slightly different context.[13] Here the constraint was not on peasant response to market stimulus, but on the initiative of landlords in respect of agricultural improvement. Washbrook stresses the role of the colonial state, particularly in the post-mutiny period, in checking the progress of 'agrarian capitalism'. The necessary capital for investment in agricultural development, including encouragement of cash-crop cultivation of certain types, was not lacking; what *was* wanting, argues Washbrook, were the conditions which would normally encourage the persons possessing the required capital, including landlords, to divert it to productive uses, because the existing laws relating to increase of rent, debt realisation, and so on, made them feel that investment of their capital in this manner was not really worthwhile. The colonial state, according to Washbrook, was hesitant in promoting an uninhibited market economy, free from such interventions by the state, partly out of an apprehension that exposure of the small peasant economy to such market forces might unsettle the rural society, and that the uncertainty that this would create in this society might prejudice the political stability of the British raj.

A significant perspective of many a study of the phenomenon of agricultural commercialisation is the manner in which the development affected the peasant economy, the wider economic organisation and the 'mode of production'. The model of this perspective is, notably, based on a distinctively European experience. Commercialisation of agriculture has generally been viewed from the perspective of a 'transition' in the European economy, and regarded as a 'principal element in the transition from pre-capitalist to capitalist modes of production'.[14] The point of reference for the kind of questions asked on the commercialisation issue in regard to British India is in general the Marxian notion

---

which surrounded him', p.103.

[13] D.A. Washbrook, 'Law, state and agrarian society in colonial India', *Modern Asian Studies* 3 (1981).

[14] K.N. Raj, 'Introduction', in Raj, *Commercialisation*, p.vii.

of this transition, Marx's data being mostly drawn from British experience. 'In that classic case of capitalist development the process of commercialisation had, spreading to all spheres, acted as a dissolvent of feudal relations.'[15] On the assumption that such a transition had not occurred in India,[16] the question has sometimes been examined from the associated perspective of whether 'Progressive commercialisation of agriculture by providing the springboard for division of labour in agricultural production as well as by exposing sections of the peasantry to uncertainties of the market can provide an engine for capital accumulation'.[17] For instance, in his study of the nature of the changes in the agriculture of the Bombay Presidency in the late nineteenth century, Satish Misra seeks to refute the argument of Neil Charlesworth about the rise of a 'rich peasant class' there during the period 1870-1920, and concludes that 'agriculture in late nineteenth-century Bombay remained dominated by a specialised trading moneylending class, a domination *not* seriously affected by a rising class of rich peasants producing for the market'.

This aspect of the question, related to the 'transition debate', has not always concerned historians. A major concern of some of them is the implications of commercialisation for rural agrarian structure. The question they have tried to answer is whether commercialisation necessitated a radical overhauling of the existing organisation of agricultural production. Their point of departure is disagreement that commercialisation resulted in 'agrarian transition', involving the complementary formation of a class of rising rural capitalists and a landless proletariat. Explanations of the non-occurrence of the 'transition' vary. One recent line of reasoning is that it was the efficiency and sophistication of the existing organisation of peasant production that precluded attempts at restructuring it.[18] Some historians do agree that the adverse effects of commercialisation for the peasantry, such as transfers of land

[15] Krishna Bharadwaj, 'A note on commercialisation in agriculture' in Raj, *Commercialisation*, p.332.

[16] For a discussion of a wide range of related issues see Utsa Patnaik (ed.), *Agrarian Relations and Accumulation: the 'mode of production' debate in India* (Delhi 1990).

[17] Satish Chandra Misra, 'Commercialisation, peasant differentiation and merchant capital in late nineteenth-century Bombay and Punjab', *Journal of Peasant Studies* (October 1982), p.4.

[18] C.A. Bayly, *Rulers, Townsmen and Bazaars: North Indian society in the age of British expansion* (Cambridge 1983), pp.464-5.

on a considerable scale, had much to do with the specific production relations that commercialisation had necessitated.[19] However, they argue that commercialisation had also generated a counter-mechanism which tended to neutralise such adverse effects.[20] It has thus been concluded: 'The forces of commodity production were here not destroying the peasantry, but being successfully ridden by them precisely to maintain the social status.' Commercialisation 'tempered social stratification'.[21]

This brief resumé of the broad trends in the historiography of agricultural commercialisation in British India is designed to emphasise the point that, though this development was inseparable from peasant involvement in the process, the distinctive peasant perceptions in this regard have been inadequately stressed. The particular interpretation of the phenomenon which we intend to examines seeks to redress the balance.

As noted above, this interpretation also serves to correct a bias in the standard explanation of variations in cash-crop cultivation solely in terms of price movements. This bias, largely deriving from the neo-classical assumption of a 'smoothly operating market mechanism or well-behaved production functions', consists in exaggerating peasant rationality, as reflected in the peasant's supply response. The interpretation seems to be of doubtful validity on two counts. First, agricultural commercialisation in eastern India is not invariably explicable in terms of the notion of a 'forced' change. That misrepresents the origins of the process in the case of some crops, or, at least, inadequately considers the complexities of the question. Secondly, it tends to take, particularly in its treatment of the question of 'dependent peasantry', a more-or-less determinist position. Dependence was far from an immutable attribute of the peasant relationship with dominant groups in the village. The two points are taken up separately.

The interpretation particularly misrepresents, at least in some cases, the role of the so-called 'advances' in the peasant choice of cash crops. Our argument is first that the cultivation of two crops, indigo and opium, was wholly based on advances. However, some elements of

[19] For instance Neil Charlesworth, *Peasants and Imperial Rule: agriculture and agrarian society in the Bombay Presidency* (Cambridge 1985), ch.10.
[20] Ibid.
[21] Ibid., p.294.

complexity in the situation, including the motivations of the groups providing the advances need to be considered. Secondly, in the case of jute, the most important commercial crop of eastern India, the role of advances remained marginal for a long time. It became important mostly during a period of an abrupt fall in jute prices. A marked trend was the increasing importance of the role since about World War I. Thirdly, the advances which financed the cultivation of sugarcane had normally an element of investment. This is generally ignored, and the consumption loans and production loans of peasants tend to be mixed up. Finally, in general this interpretation does not clearly distinguish between the usual advances provided by the persons interested in peasant produce and the credit obtainable from other sources. The distinction had important implications from the point of view of the control exercised over peasants by their creditors.

Poppy cultivation, usually confined to the best lands of the village and to a particular group of skilled agriculturists (Koeris), illustrates the kind of role which Amin attributes to advances. One of the major attractions of the crop for cultivators was the interest-free advances offered by the Opium Department of the government, particularly at the time when their rent instalments fell due. In fact, in distributing the advances over the entire opium season, the government normally considered the cost of cultivation as much as the timings of rent payment. The Sub-Deputy Opium Agent of Patna observed in 1860:

> there is almost a universal understanding between the zamindar and the poppy ryot, that the latter shall pay as rent to the former all the money he receives periodically from the government as advances; while the zamindar shall let him house his crops, instead of forcing him to sell them to pay up his rent, and often at ruinous rates. This mutually accommodating agreement exists...throughout the Bihar Agency, and so admirably have I found it to work...that in this district the periodical advances are purposely so timed as to fall in with the quarterly rent days.[22]

The practice continued. It was partly a deliberate design on the part of the government to secure the co-operation of powerful landlords for its enterprise.

However, it would be wrong to infer that the Opium Department sacrificed 'sound commercial principles' in order just to please land-

[22] *Report of the Indigo Commission* (1860), Appendix, Part 2, Opium Agent of Patna to Opium Agent of Bihar, 12 June 1860, para.15. (The extract is taken from an earlier letter of his, dated 10 September 1857.)

lords. The amount of the advances distributed was strictly determined by basically commercial norms, so that the amount of advances was reduced, or the advances altogether suspended, if the state of the opium production by cultivators justified the measure.[23] Moreover, the particular motivations of the government in providing the advances also need to be considered. The advances were an effective instrument towards sustaining the monopoly control of the government over the poppy cultivation and the sale of opium. No cultivator could grow the poppy without the formal permission of the government. Only those cultivators registered with the Opium Department, and accepting its advances were allowed to grow it.

As for indigo, cultivators themselves admitted how the prospect of interest-free advances motivated them, at least initially, in growing the crop.[24] However, the *continuity* of the cultivation is not explicable by the inducements of advances, since, as planters themselves admitted, most cultivators could not produce enough to cover the advances and were therefore deprived of further advances. Two elements of complexity in the situation need to be considered. First, since indigo had no domestic market, cultivators had no interest whatsoever in growing it unless they were sure of the sale of their produce, and the advances constituted part of the price they expected for their produce. Secondly, as in the case of opium, advances gave planters a lever of control over cultivators. Indigo planters could not be sure of securing the required supply of indigo plant, unless the cultivators were tied to the European indigo system through advances. Planters wanted indigo grown by small peasants (*ryoti*) and not through a system of wage labour (*zerat*), largely because, in the way the system of advances was made to work, the ryoti cultivation of indigo came to be based on a nearly unpaid labour process. Moreover, the advances also ensured planters' control over the land and labour of peasants, particularly in the context of recurring inter-planter clashes, at least till 1835 (when the dominant 'indigo concerns' formally agreed to define their respective spheres of

[23] Ibid., Annexures, p.370. The Sub-Deputy Opium Agent explains how the amount of advances varied with the nature of the peasants' crop.

[24] The Indigo Commission thus observed: 'advances...are in some instances willingly and even greedily accepted; some are in debt, others want to spend money, and all like money without interest', para.57 of the *Report*.

influence).

Jute cultivation only partially depended on cash advances. The following observation of the Bengal Jute Commission (1873), the first official agency to investigate the question of the extent of acceptance of advances by cultivators, was not based on any careful statistical survey, but on the evidence of the jute-growers and jute-traders: 'There is no demand for advances, and a good number of ryots in easy circumstances carry on the cultivation without taking any advances from *mahajans* or dealers'.[25] It seems the observation is true more or less of the period before the First World War. Writing of the system of rural credit in Dacca, F.D. Ascoli, the Settlement Officer, thus concluded in 1914: 'I doubt if one cultivator in 1,000 is dependent in any way on *aratdar*, *bepari* or jute merchant for loans and advances. The experiment has been tried in the district but has been abandoned.'[26] World War I was a setback for jute growers, who then faced a slump in raw jute prices. It was then that dependence on advances noticeably increased:

> *Dadani* forms of credit started cropping up in areas where they were wholly unknown or extremely rare in the ante-bellum period. By the end of 1916, *dadan* had become an important form of credit in the Narainganj subdivision of Dacca; it was gaining a foothold in Pabna and Rangpur.... In Mymensingh, too, loans were now contracted to be repaid in jute at prices less than the market rates.[27]

Despite the increased dependence of jute-growers on advances, the size of the jute cultivation directly financed by advances was still not very impressive even by 1926, as an official enquiry in that year revealed.[28] The role of advances was then limited even in the most important jute-growing districts, such as Mymensingh, Rangpur and Dacca.[29] While in some parts of Rangpur the system of advances did prevail, it is striking how acceptance of advances did not necessarily

[25] Hem Chandra Kar, *Report on the Cultivation of and Trade in Jute* (Calcutta, 1877), Appendix. For the evidence of jute-growers themselves, see statements nos. 10, 11, 12, 14, 19, 22, 25-8, 31-2, 38-9, 48-9, etc.

[26] Quoted in Omkar Goswami, 'Agriculture in slump: the peasant economy of east and north Bengal in the 1930s', *Indian Economic and Social History Review* (July-September 1984), p.338.

[27] Ibid., pp.339-40.

[28] *Marketing of Agricultural Produce in Bengal, 1926* (Calcutta, 1926).

[29] Ibid.; for Dacca, see p.38; for the Jamalpur subdivision of Mymensingh, see p.41.

oblige the growers to sell their jute at prices lower than those prevailing ones at the time of sale.[30]

Two Bengal Jute Enquiry Committees (1934 and 1939) investigated, among other things, the role of advances in jute cultivation. The findings show how widely the operation of the system varied from place to place. However, some general trends are identifiable. Two features stand out in jute growers' credit relations in the 1930s, the period of pervasive trade depression and understandably untypical. First, while jute-cultivators, like the growers of other crops, desperately needed credit, the credit supply tended to dry up, with the professional moneylenders sharply reducing their loan operations. As a result, jute cultivators increasingly relied on the agents of jute mills and jute-traders. Secondly, though the demand for raw jute continued, albeit at a considerably smaller level, the erratic market behaviour and the uncertainty about the prospects of a trade revival made these agents generally reluctant to continue making advances to cultivators. They had also to consider the uncertainty produced by the fast-changing political temper of the time—the growth of peasant resistance to payment of rent and debt, and the readiness of the local government to intervene in favour of the cultivators in distress. The background to the peasant unrest was the rising real value of cultivators' rent and debt obligations in the context of the abrupt fall in agricultural prices. The sense of alarm caused among the jute-mill agents and traders by the growth of 'no-rent', 'no-debt' mentality among cultivators was naturally heightened by the measures of the local government, such as the Bengal Agricultural Debtors' Act (1935) and the activities of the Debt Settlement Boards. However, in some regions at least, cultivators, still keen on jute, could not, in the circumstances, do without advances from jute mills or jute-traders.

Contemporary reports on raw jute marketing do give an impression that in normal circumstances the role of advances in jute cultivation was far from crucial. Most peasants, we are told, sold their produce in local markets (*hats*). Occasionally, particularly at the time of a rising demand for jute, small traders (*farias*) collected it from the cultivators' houses. Previous advances on a large scale would have precluded such open-market operations. The description of raw jute marketing by the Bengal Jute Enquiry Committee (1934) does not suggest much of a

[30] Ibid., p.59.

change in this regard.[31]

Official descriptions of the way in which jute slumps affected jute-growers provide further evidence of the marginality of the role of advances in jute cultivation.[32] The slumps had some common features: a booming jute trade reflected in an upswing of raw jute prices persuading cultivators to plan a big increase in their jute cultivation, and slump in the jute prices abruptly hitting them hard. An effective system of advances would have saved the growers from such a predicament.

During the slump of 1872/3 the unsold quantity of jute in the districts of the 24-Parganas and Hugli was estimated at one-quarter to three-quarters of the total production. In Dacca and Mymensingh, two major jute districts, 'everywhere...quantities of the jute grown last year had been left abandoned on the fields. The falling market in Calcutta seems to have created a panic amongst the cultivators.'[33] A comparable slump occurring in 1882/3 hit the jute growers as hard. Prices fell by 29 per cent in Dacca. Faridpur experienced a steeper fall of two thirds. In Tippera 'many ryots declared that it would not pay them to cut and prepare jute owing to the low prices'.[34] During a similar crisis of 1890/91 the raw jute prices in Purnea fell by 42.1 per cent. In Noakhali jute 'fetched little more than half the expected price'. In Tippera the 'fall has marked the history for the year'. In Dinajpur 'indeed the fibre was hardly saleable. Large areas were not even cut.'

It may be argued that the crop was not harvested because of unwillingness of the creditors to collect it because of the dull market. This was highly improbable, since they would have incurred a bigger loss as a result. Moreover, creditors suffering a loss in a slump year would normally have been eager to make up for it in the following year, provided the slump did not persist, by continuing the advances. The considerable reduction in jute cultivation which usually followed a jute slump contradicts this assumption.

[31] *Report of the Bengal Jute Enquiry Committee*, 1934, vol.1, pp.12-13.

[32] For a detailed study of the question see Binay Bhushan Chaudhuri, 'Growth of commercial agriculture in Bengal and its impact on the peasant economy', *Indian Economic and Social History Review* VII, 2 (1970), pp.238-44.

[33] H.C. Kar to Government of Bengal, 10 April 1873; Bengal Agriculture Proceedings, November 1873, File 1-85/86.

[34] Annual General Report, Chittagong Division, 1882/3, 30 June 1883, para. 38.

We need to explain the limited role of advances in jute cultivation. Jute-growers did require credit, at least partly because jute cultivation was far more labour-intensive and therefore more expensive than rice cultivation, since at least part of the labour process had to be paid for directly. Yet cultivators could do without advances from the usual jute interest groups because of the availability of alternative sources of credit. This was because jute seldom constituted a distinct and exclusive sector in the economy. As growers of rice and other crops cultivators could borrow, in times of need, from the village moneylender and other sources. They preferred to do so, because they could thereby avoid the consequences of the acceptance of advances, particularly the hypothecation of the jute crop to the creditor at a price much below the market rate. In fact it made no sense to them to pledge the crop at a loss, when they felt reasonably assured that the market would take whatever they produced.

On the other hand, jute traders and manufacturers did not necessarily have to rely on making advances in order to get the jute they wanted. The prevailing jute prices, they often found, were enough inducements for cultivators to bring their produce to the market. There were exceptions, for instance in 1905/6, the year of an exceptional boom in the jute trade. The consequent arrival in the jute market of a large number of purchasers and the possibility of a further rise in the jute prices made many a trader eager to make advances to jute-growers in an attempt to bind them to pledge part of the jute crop. In normal market conditions, however, they preferred to do without making advances, partly because of the difficulty in effectively exercising the kind of control necessary to the administration of a system of advances over the widely scattered jute-growers.

Peasants thus did produce jute even without advances. How much of it they would choose to grow, and the area variations from time to time, were largely determined by the level of jute prices and actually by the jute-rice price ratio. We may take two periods of a sharp fall in the jute prices, where the effects on the jute area were dissimilar because of the differences in the jute-rice price ratio: the First World War and the depression of the 1930s. Jute-growers reduced the jute area soon after raw jute prices started to plummet during the war, largely because the magnitude of the fall in the rice prices was much smaller. This did not happen during the depression, at least to the extent that the jute price slump would normally have justified. This was partly due to the limited opportunities for peasants to change over to other crops. The decisive reason for this was the continuing depression in the prices of rice, the major option for cultivators in times when jute lost its charm for them.[35]

Some explanations of the choice of jute by cultivators, particularly in the early twentieth century, insufficiently stress this role of the complex market situation we have indicated above, and emphasise more structural factors, such as demography.[36] 'The logic of the switch from the cultivation of a subsistence crop to a commercial crop for the world market on peasant smallholdings', it has been argued, 'must be seen in the context of agrarian overpopulation.' What role did jute play in this specific context? To the cultivator 'faced with a rapidly deteriorating land-man situation jute held the attraction of not only a higher gross income, but also higher per-acre expenditure of labour energy'. To the extent that jute made possible the remunerative employment of part of the labour and other resources of the peasant family, which otherwise would have remained under-utilised, the family's preference for jute is understandable. This did have a role *in sustaining a certain level of cultivation.* However, the demographic factor alone does not adequately explain the frequency in the variations in the jute area.

The available evidence about the exact role of advances in sugarcane cultivation is rather frail. Official reports, the most valuable sources of information on the question, are not always precise on the point. The Government of Bengal once (1884) observed: 'The cultivation of sugarcane, though expensive, is lucrative. It is confined to the well-to-do ryots, and the aid of a moneylender is seldom sought.'[37] Local reports, however, contradict the assumption that only the affluent cultivators grew sugarcane. The growing demand for sugar in the domestic market increasingly attracted peasants of even restricted

[35] This I have explained in a paper 'Commercialization of Agriculture', in Sirajul Islam (ed.), *History of Bangladesh 1714-1971*, vol.II, *Economic history* (Dhaka 1992), pp.370-427. See also Saugata Mukherji, 'Some aspects of commercialization in Eastern India, 1891-1938', in Sen *et al.*, *Three Studies*.

[36] Sugata Bose, *Agrarian Bengal: Economy, social structure and politics, 1917-1947* (Cambridge 1986), pp.52-4.

[37] India Agriculture and Horticulture Proceedings, May 1884; Government of Bengal to Government of India.

means, who did borrow to carry on sugarcane cultivation.

The pertinent question is why normally most cane-growers could not do without credit. Was the reason the same as Amin found in Gorakhpur: pressing rent obligation in the context of a particular harvest calendar? The sequence of the harvests in Bengal, as in U.P., was ecologically determined and therefore immutable under the given agricultural technology. The nature of the distribution of the rent burden was, however, different in the Bengal Presidency, the heaviest rent instalment being payable after the major winter harvest (*aman*, mid-November to mid-December). The moneylender's willingness to lend to cultivators normally depended on the prospect of this crop. The role of sugarcane as a 'cash-raising and debt-servicing' crop was thus more limited in Bengal than in U.P.

Cane-growers still depended on credit. The emphasis in Amin's study on the purely subsistence requirements of the growers (obligatory rent payment in this case) as their crucial motivation in this regard, tends altogether to ignore an element of 'investment' in the utilisation of the credit. The cost of cane cultivation and of making *gur* or *rab* was to a considerable extent met out of this credit, and cane-growers borrowed obviously with an expectation of gains from this enterprise.

Borrowing towards the financing of cane cultivation was mostly unavoidable because of its expensiveness. 'Of all his various crops', admits Amin, 'sugarcane demanded the most time, labour and attention of the U.P. peasant. The crop occupied the land for ten months or so, and the manufacture of *gur* engaged the peasants for another ten to twelve weeks.' Extraction of juice and making of *gur* were similarly expensive processes. The Behia mill, a new cane-crushing iron mill, initially cost the growers far more than the traditional implement for the purpose—the wooden *kolhu*. The huge boiling pan was an expensive input too. Loans for all these purposes were distinguishable from borrowing towards rent payment.

The particular attitude of cultivators towards a productive use of part of their credit is partly evident from the extraordinary rapidity with which they opted for new technology in the form of Behia mill (named after a Shahabad estate whose proprietors Thomson and Mylne, invented it). The new mill, which replaced the old *kolhu* (stigmatised by the inventors as 'the barbarous and wasteful method of cane culture'), was superior in many ways. It could crush nearly three times as much sugarcane as the old mill, and that more efficiently. The juice extracted by it was far cleaner, and the yield of sugar higher by nearly 25 per cent. The inventors found that, from the beginning, the new mill recommended itself to the cane-growers as a labour-saving device. The Rajshahi Commissioner estimated that it reduced the cost of sugar manufacture by about one-third.[38]

That the invention did not take long to be popular with cultivators was largely due to its coincidence with the period of a flourishing sugar market. As early as 1879-80, the number of such mills in use in Shahabad was 9,000.[39] In 1883, the 'old-fashioned village *kolhu*' was found to be in use largely in the Siwan subdivision of the district. There, in the course of only four years, as the Patna Commissioner reported, 'Behea mills are superseding the country-made mills all over the Siwan subdivision.'[40] In the south of Bhagalpur, it was found in 1886, 'the old country mill had become a thing of the past'. While compiling his *Bihar Peasant Life* (1885), Grierson 'found it impossible to find a single one [*kolhu*] in the Sadar subdivision of the Patna district which was in a fit state for a photograph to be taken of it. They had all been broken up for firewood, only a few dismantled ones being available.'[41] The new mill became so popular, that in 1887 its inventors had to ask government for better legal protection from fraudulent imitations. Sometimes resourceful peasants jointly purchased the new mill. More often they hired it from a wealthy moneylender, and co-operated in using it.

An important methodological question which the particular interpretation of agricultural commercialisation under review seems to ignore, is the possibility of credit availability from multiple sources. Peasants cultivating cash crops need not necessarily have borrowed solely from the agencies directly interested in peasant produce. Many interpretations also tend to ignore another vital issue—the possibility of a change in the nature of peasant dependence. They seem to rest on an assumption of its immutability. The element of time is here nearly frozen, and the interpretation inadequately stresses the specificity of the context in

[38] Annual General Report, Rajshahi Division, 27 June 1883, para.10.
[39] Report on the Administration of the Irrigation Department, 1879-80, para.76.
[40] General Annual Report, Patna Division, 1886-87, 30 June 1887, para.31.
[41] G.A. Grierson, *Behar Peasant Life* (Calcutta 1885), p.50.

which domination over peasants could be effective, or could lose much of its effectiveness.

In fact, peasant subordination did not preclude manoeuvres to improve or to stabilise their economic position, or even to get rid of their subordination altogether. There is abundant evidence that the dominant groups could not always sustain their effective domination over the growers of cash crops. Peasant initiatives or manoeuvres to contest it were inseparable from the particular organisation of peasant society and economy. There could be either large-scale organised resistance to the exercise of domination or 'everyday forms of resistance', that is, peasant reactions constituting what has been called the 'middle ground of peasant politics between passivity and open, collective defiance'.[42] Reactions of the second kind 'typically avoid any direct symbolic affront to the authority...[and] though less dramatic...than uprisings are more or less constant and constitute the normal means by which the peasantry resist claims on its surplus'.[43]

Peasant reactions to the exercise of monopoly control by the Opium Department of the government illustrate the second kind of initiative. The control could not always be ensured everywhere, even by the exercise of the enormous legal powers of the Opium Department, or by the collaboration of powerful local groups, such as zamindars, *khatadars*,[44] and village headmen. It is striking how under certain conjunctions of market forces the exercise of the legal powers or the dependence on the local power groups largely ceased to be effective. The government had then to come to terms with the reality of the altered market scene, and, if it was keen on sustaining the cultivation at a certain level or on increasing it, it had to adjust the cultivators' prices for opium to a suitable level.[45] This was done, for instance, after the Mutiny which induced an abrupt rise in commodity prices and wages, while the

[42] James C. Scott and B.J.T. Kerkvliet (eds.), *Everyday Forms of Peasant Resistance in South-East Asia* (London 1986), 'Introduction'.

[43] Ibid.

[44] *Khatadars*, literally meaning holders of books (*khatas*), were the persons through whom the Opium Department, which did not deal with individual cultivators at all, distributed opium advances. An important source of their power over the cultivators was their status as moneylenders. Kemble, the Bihar Opium Agent, called the *khatadar* the 'village banker'.

[45] For a detailed discussion of the question of fixation of prices to be paid to opium-growers, see Benoy Chowdhury, *Growth of Commercial Agriculture in Bengal, 1757-1900*, vol.1 (Calcutta 1964), pp.25-41.

opium-growers' prices remained fixed. Cultivators steadily changed over to other crops such as oilseeds, cereals and potatoes, the demand for which was then greatly stimulated by the 'demand for foreign markets and the presence of a large European force'.

Even where cultivators chose to continue opium cultivation at the earlier prices, the monopoly tended to be occasionally vulnerable to various strategems of cultivators, such as the clandestine sale of opium to itinerant traders at prices much higher than the monopoly price, a strategy amounting to a silent war with the monopoly system. Another possibility was withdrawal from poppy cultivation of the best land of the village normally devoted to it, and substitution of relatively inferior land, or a kind of mixed cropping—that is, the growing of minor crops along with the poppy, a practice the Opium Department discouraged because it tended to reduce the productivity of opium lands.

It is amazing how the apparently formidable apparatus of control over the opium-growers virtually broke down where they were determined to change over to other crops because of new opportunities in this regard. The collaboration of the usual rural groups with the Opium Department seldom enabled it to arrest the drift of reluctant cultivators away from opium. In fact, the *khatadars*, one of the main props of the opium monopoly, had little to gain from keeping the cultivators tied to an unremunerative cultivation. The viability of their primary occupation, moneylending, was based on the stability of the peasant economy. As for zamindars, opium advances ensured the security of their rental income as long as cultivators were keen on opium cultivation. Moreover, there was a limit to the exercise of their authority as zamindars to make cultivators grow a crop against their will.

Indigo cultivators' reactions to the European indigo system combined both the types of resistance mentioned above. A typical 'everyday form of resistance' was to neglect to cultivate with as much intensity as planters insisted on. It was not at all a case of malicious intent on the part of cultivators to harm the enemy: it was part of their survival strategy, since they knew cultivation of their other crops, including food crops, would be irreparably damaged without this alleged 'misdemeanour'. A reaction of a different kind was evident in stray cases of violation of indigo contracts, most often done in collusion with rival planters. The violations, though actually meaning transfer of allegiance from some planters to others, so that the domination of planters as a class was not undermined at all, are still significant in that they show how under certain circumstances cultivators could assert their will by severing ties with their old tyrants. A qualitative transformation of the resistance came about when peasants of the major indigo districts rallied round the radical programme of overthrowing the indigo system altogether. Such resistance in Bengal (1859-62) nearly crippled the indigo system there.

The collapse of planters' domination, institutionalised in the particular coercive system they had devised, was inseparable from new market forces: on the one hand, new opportunities were available for peasants to grow remunerative crops which made the ties with the indigo system far more unbearable than before, and on the other hand the indigo market was shrinking, so that planters were obliged to impose harder terms on the cultivators.

The distinct trend in the organisation of indigo production in Bihar—the increasing importance of *zerat* cultivation, a cultivation carried out on the basis of grossly underpaid wage-labour on the planters' demesne lands—exposed the limits of planters' control over the local peasant community, and pointed to their realisation of the unworkability of the old peasant mode of indigo cultivation. The viability of the *zerat* system largely depended on the combination of special circumstances, particularly the powers that the planters exercised in their capacity as rent farmers (*thikadars*) of the estates of local zamindars, who were mostly indebted to them.

We have emphasised the role of certain conjunctions of market forces in creating conditions where dependent peasants could exercise their collective will more effectively than on other occasions because the usual characterisation of agricultural commercialisation as a 'forced and artificial' process, tends to underestimate this role. The market forces were then to a considerable extent autonomous, in the sense that their operation was now possible without the mediation of the established power relations. In our analysis of that question we have concentrated on two crops, indigo and opium, because peasant subordination here was admittedly more pronounced than in other cases.

Chapter 4

PEASANT PRODUCTION AND AGRARIAN COMMERCIALISM IN A RICE-GROWING ECONOMY: SOME NOTES ON A COMPARATIVE PERSPECTIVE AND THE CASE OF BENGAL IN THE EIGHTEENTH CENTURY

Rajat Datta

Introduction

This paper examines the dynamics of peasant production and agrarian commercialism in eighteenth-century Bengal, a region deeply involved in inter-regional and international commercial exchanges of growing complexity between the seventeenth and eighteenth centuries.[1] These circuits made the peasantry of Bengal operate in a complex matrix of production designed to meet a whole range of requirements, from those generated in the tiny village market (the *haath*) to the burgeoning demand for some of their produce in the international market. All this happened within the overarching primacy of a rice-producing economy. It is therefore crucial that one specifies the end-purpose of peasant production to make any sense of the nature of the interaction, if any, between commercial processes and commercialisation of agricultural production.

Historically, peasant production has been much more than the culti-

[1] This growing complexity in commercial involvement can be seen from the case of silk, the most prized of Bengal's exports during this period. In the seventeenth century, the market value of this commodity, and the returns to Bengal by its sale, were determined by interregional demand. Witness, for instance, the following assessment of Bengal's silk market in 1661: 'Silk sells in Agra as the price of it in Cossimbuzar riseth or falleth. The exchange of money from Cossimbuzar to Pattana and Agra riseth or falleth as the silk findeth a vent for it in Pattana or Agra, Register of Papers Relating to the English and Dutch East Indies, British Museum; Additional Manuscripts 34123, f.42 (hereafter BM). However, in the eighteenth century we see international demand establishing its primacy in place of inter-regional demand in Bengal's trade in silk. This is clearly demonstrated by the fact there was a sharp fall in the exports of silk from Bengal to Western India throughout the second half of the century. Simultaneously, there was an equally spectacular increase in the East India Company's investment for silk in the same period; compare India Office Library (hereafter IOL), Mss.Eur.D.281, f.14; and *Parliamentary Papers, Reports of the Select Committee of the House of Commons*, Ninth Report, 1783, p.112.

vation of a variety of crops. It has been the central point of an agrarian economy around which various other social relations are articulated. Thus the state of the entire agrarian economy, defined as the intermeshing of natural conditions, social relations and institutional networks in the countryside, was determined by the fluctuations in the domain of agricultural production and peasant enterprise.[2] The primacy of agriculture was all the more vital in such economies because even non-agricultural, or craft production, was undertaken largely as a part of agriculture. The artisans, in most cases, were primarily peasants (that is, they cultivated land mainly with their own or their domestic labour) who entered non-agricultural production as an extension of their principal enterprise as cultivators. Resnick's typology of peasant activity in South East Asia, as the 'effort devoted to the production and cultivation of crops and the time spent on a multitude of home or artisan activities',[3] is applicable in general outline in the context of pre-industrial/pre-colonial societies. Thus in Bengal, salt was manufactured by the 'tillers of the land' who used the slack agricultural season to acquire supplementary incomes by working in the salt-pans (called *khalaries*),[4] while the weavers were said to have the 'double resource of tilling the lands'.[5]

It is being increasingly realised that there was a clear commercial context of agricultural production in the eighteenth century, and Bengal was not alone in this respect. C.A. Bayly highlights the intricate connections between peasant production and an integrated produce market created by the formation of intermediate and large marketing centres (*bazaars* and *ganjs*) in much of northern India in the late eighteenth and early nineteenth centuries.[6] David Ludden discusses the transformation

[2] For an excellent overview of this centrality in the Chinese and Japanese context, see Francesca Bray, *The Rice Economies: technology and development in Asian societies* (Oxford 1989); also 'Patterns of evolution in rice-growing societies', *Journal of Peasant Studies* (hereafter *JPS*) 11, 1 (October 1983).

[3] S.A. Resnick, 'The decline of rural industry under export expansion: a comparison among Burma, Philippines and Thailand', *Journal of Economic History* 30, 1 (March 1990), p.51.

[4] See Bengal Revenue Consultations (hereafter BRC), India Office Records (hereafter IOR) P/49/52, 29 May 1775: and ibid., P/50/3, 12 August 1977; ibid., P/51/45, 5 August 1789, for statements regarding the combination of agriculture with artisanal production in the salt sector.

[5] West Bengal State Archives (hereafter WBSA), Controlling Council of Revenue at Murshidabad (hereafter CCRM), vol.8, 5 December 1771.

[6] C.A. Bayly, *Rulers, Townsmen and Bazaars: North Indian society in the*

of traditional peasant production in South India into 'commercialized peasant agriculture'—that is, production for a market on an increasing scale—during the course of the eighteenth and nineteenth centuries, by a close nexus between peasants with assets in land and merchants in towns.[7] There is, therefore, substantial and growing evidence to show that much of the South Asian economy in the late pre-colonial epoch was organised in similar 'hierarchic interdependency',[8] between peasant production and a tiered network of markets of ascending importance.[9]

Yet there are firm indications that despite this integration of production with the market and a remarkable spurt in overseas demand for Bengal's textiles (symptomised by the fact that the province supplied more than 80 per cent of the English East India Company's Asian exports of cotton piecegoods through the century),[10] there were remark-

---

age of British expansion, 1770-1830 (Cambridge 1993), pp.74ff.

[7] David Ludden, *Peasant History in South India* (Delhi 1989), pp.9, 162 and passim.

[8] The term is from Frank Perlin, 'Proto-industrialization and pre-colonial South Asia', *Past and Present* 98 (February 1983).

[9] See the various discussions in S. Subrahmanyam (ed.), *Merchants, Markets and State in Early Modern India* (Delhi 1989); and S. Bose (ed.), *South Asia and World Capitalism* (Delhi 1990). Such indeed also happened beyond the frontiers of South Asia. For instance, the formation of a 'national market' in bulk commodities in eighteenth-century China has been seen as the outcome of similar linkages between various local and regional production and marketing centres; Susan Naquin and Eleanor Rawski, *Chinese Society in the Eighteenth Century* (New Haven), pp.98-100. And its impact is easily discernible in the spread of cotton cultivation in the provinces of Fukien, Kwangtung and Kiangsu in the Ming and Ch'ing periods. See Kang Chao, *The Development of Cotton Textile production in China* (Cambridge, Mass. 1977), pp.13-24. Also, the eighteenth-century upturn in European grain prices apparently led to the rapid spread of maize in the Ottoman peasant agriculture, which also underwent a mini 'cotton boom' under the influence of the French and American revolutions; R. Kasaba, 'Incorporation of the Ottoman Empire, 1750-1820', *Review* 10, 5 and 6 (supplement) (Summer/Fall 1987), pp.807-28; also M. Cizacka, 'The incorporation of the Middle East in the world economy', *Review* 7, 3 (Winter 1985), pp.353-77.

[10] See K.N. Chaudhuri, *The Trading World of Asia and the English East India Company, 1660-1760* (Cambridge 1978), p.509 ff., for figures for the first half of the eighteenth century, and 'Report of the Select Committee of the Court of Directors to the East India Company upon the Subject of the Cotton Manufactures of the Country, 1793', appendix 2, IOR Home Miscellaneous (hereafter HM), vol.401, pp.138-41, for figures pertaining to the second half of the eighteenth century.

ably few concerted attempts by peasants to increase the area under cotton cultivation on a matching scale. On the contrary, evidence indicates an actual decline in the cultivation of the eighteenth-century 'cash' crops par excellence, cotton and mulberry for silk, and a greater concentration on the production of cereals—paddy and rice and other foodstuffs.[11] This broad trend is paradoxical; and it apparently goes against the notion of a 'rational' peasant as developed by Samuel Popkin in his study of the Vietnamese peasantry,[12] though it may also indicate an obvious danger in limiting any analysis of agrarian commercialism to 'cash crops' alone.

The rationality of peasant production

Figure 1 (p.128) represents the sectoral contribution of different crops in the total agricultural output of the province.[13] Food grains (rice, paddy and millet) comprised 45.5 per cent of the output, whereas the cash crops (cotton, mulberry and tobacco) contributed 20.5 per cent. Pulses and oil seeds, comprising about 20 per cent of total production, were certainly designed for sale, but they were not on a par, in terms of exportable and commercial importance, with crops like cotton and mulberry, and therefore may be seen as part of the overall production of food. The features of agricultural production as revealed by Colebrooke's estimates are also confirmed by some other contemporary observers,[14] and from scattered evidence available from specific districts.[15]

Such data certainly suggest that the agricultural landscape in eighteenth-century Bengal was one where oceans of paddy fields were

[11] For a detailed discussion see Rajat Datta, 'Rural Bengal: social structure and agrarian Economy in the late eighteenth century', PhD thesis, University of London, 1990.

[12] For Popkin, a rational peasant is *homo economicus*, an individual who is continuously calculating how to improve his well-being, whatever the situation. See *The Rational Peasant: the political economy of rural society in Vietnam* (Berkeley 1979).

[13] H.T. Colebrooke, *Remarks on the Husbandry and Internal Commerce of Bengal* (first published 1793; Calcutta 1884), pp.15-16.

[14] William Tennant, c.1797, gave a much smaller proportion of the cultivated lands occupied by these cash crops: 'most valuable crops' like 'tobacco, sugar, indigo, cotton, mulberry and poppy' did not occupy more than a 'twentieth part of the land' under cultivation. *Indian Recreations* (3 vols., Calcutta 1804-1808), vol.2, p.11.

[15] Datta, 'Rural Bengal', pp.260-2.

dotted with a few islands of commercial crops. The situation was not substantially different even in those areas which constituted the core suppliers for Bengal's textile industry. Thus in the major silk-producing district of Rajshahi 'there is no Ryott that does not hold Grain lands and in greater proportion than Mulberry ones'.[16] Peasant production near Qasimbazar was similarly structured. A survey of lands held by eight peasants living near the Company's factory at Qasimbazar conducted in 1788, showed that of a total of 22.4 acres of land held by them, mulberry lands accounted for four acres and paddy for ten.[17]

Yet Bengal supplied the overwhelming portion (more than 80 per cent) of the Company's annual trade in cotton piecegoods and silk textiles, and this certainly is a paradox. It appeared so to contemporary observers. The almost monotonous complaints regarding the 'indolence' of the 'native' peasant as a barrier to raising the productive capacity of the land and introducing 'new' (that is, European) methods of cultivation, were essentially designed to show the intrinsic agricultural backwardness and technological deficiency of the average Indian peasant and to counterpose the supposedly superior European (specifically British) methods of cultivation and land-use. Central to this exercise was the assumption that the British cultivator was a rational economic agent who responded to the stimulus of the market, whereas the Indian (or Bengali) peasant was not.[18]

Robert Kyd, the superintendent of the Company's Botanical Gardens at Sibpur on the outskirts of Calcutta, had the following to say about the possibilities of introducing new horticultural techniques in Bengal:

> I can give the most positive assurances of this climate [of Bengal] admitting the resource derivable from the cultivation of Peach, Plum, Apricot, Cherry; with the Chinese fruits...which succeed in perfection [in Bengal] and may in the process of a few years be *disseminated throughout the Company's possessions* if the Factory at Canton execute the Commission, lately sent, in furnishing a proper supply.... I may further add that the culture of the English Melon and grapes of every kind, may be introduced with great success, altho' the latter will succeed better within the precincts of the Town [Calcutta], than in the

[16] BRP, IOR, P/71/10, 25 June 1789.

[17] BRC, IOR, P/51/27, 28 November 1788.

[18] See, for instance, Colebrooke, *Remarks*, pp.27-30; Tennant, *Indian Recreations*, vol.2, pp.8ff.; Francis Gladwin to Council, BRC, IOR, P/53/55, 31 January 1798; W. Roxburgh, 'An account of the Hindoo method of cultivating the sugar cane and manufacturing the sugar and jagary in the Rajahmundry Sircar', in IOR, HM, vol.210, p.150.

neighbourhood of Calcutta, which from the lowness of the soil, intense moisture and heat of the atmosphere in gendering innumerable insects which destroy the foliage in such a degree as to prevent their bearing. Add to this, the defect of the soil, which bears testimony of the element from which it has apparently emerged, apparently at no distant period of time, being strongly impregnated with salt.[19]

Kyd's remarks reveal that there were no deeply structured 'cultural' barriers to the introduction of new products among the cultivators; after all the principal vehicle of Kyd's 'positive assurances' of disseminating the new horticultural products 'throughout the Company's possessions' would have been the peasant. Kyd additionally demonstrated that the receptiveness, or otherwise, to new products depended mainly on climate and soil, two vital elements in the rationale of peasant production all over the world.

To the European eye, the average peasant in Bengal appeared extremely cautious in choosing the crops to be sown and the extent of the land which was to be under a particular crop at any point in time. Cultivators 'seldom engage in extraordinary speculations' was how their initial response to indigo cultivation was described in 1790.[20] From the cultivator's point of view, this cautious approach to indigo was entirely justified for, apart from the coercion which its cultivation entailed, the production and marketing of this commodity was far from efficient, at least in our period: capital invested in indigo was often precarious, the interest rates were too high to be attractive, there was an absence of a ready market for indigo *within* Bengal, and the sales of this commodity in European markets often suffered from excessive delays.[21] Hence caution originated from the peasants' perception of the risks involved in production, and their response was either an extremely slow acceptance in official circles as 'indolence',[22] or a conscious anti-

[19] BRC, IOR, P/51/7, 8 June 1787 (emphasis added).

[20] C. Blume, 'Short sketch of the measures adopted for the introduction of indigo and promotion of agriculture in Bengal between 1779 and 1790', IOR, HM, vol.443, p.600.

[21] Ibid. The cultivation of poppy for opium was also largely unattractive to the average peasant in our period because poppy was an extremely precarious crop: 'the poppy is of so delicate a texture before it arrives at maturity that an accident of the weather such as too dry a season, or a tempest of wind and hail coming on before the plant has gathered sufficient strength to resist its force may utterly destroy the produce of a whole pergunnah', BRC, IOR, P/49/55, 11 August 1775.

[22] Thus, according to William Roxburgh, 'amongst the natives of India, the

pathy, and the latter was misconstrued.

An antipathy of this type was demonstrated by the peasants to the Company's efforts to find an Irish solution to problems of famine and dearth in Bengal, namely the cultivation of potatoes on an extensive scale. Despite the apparently strenuous efforts made by officials to encourage the peasants to take to its culture by 'offering seeds to any who would engage to cultivate them', they clearly failed 'to prevail upon the Ryots to assist Government in its kind intentions [sic] in their favour'. The fact that the peasants did not take to potatoes with the speed the Company desired, immediately brought forth the charge of 'native ignorance', and their 'natural' unwillingness to adopt a new crop. The real reasons for peasant antipathy were entirely different.

First, the potato seeds were extremely expensive: a maund of seed potatoes was priced at Rs.4 in October 1798. Second, the cultivators were not convinced of the suitability of potatoes to local soils. Those of Birbhum clearly stated that the soil was 'unfavourable of the plant thriving, and being brought to maturity in this district; the soil in general being formed of a hard substance, and mixed with small stones or conker [*kankar* or gravel] renders it uncongenial for the culture of potatoes'. The peasants of Dhaka objected on grounds that 'the soil is in general unfit for it and that in Bowal the only part of the district where good potatoes might be raised, the number of wild hogs is so great that their potato fields would be continually liable to the devastation of those destructive animals.' Third, Bengal abounded in a number of:

> Esculent roots of a farinaceous and nutritive quantity which are a common food among the lower class of people [such] as yams, the Spanish potatoes and the numerous varieties of Caychoo [local tubers]...all extremely productive and ...capable of being stored and preserved for a greater length of time by due attention to prevent the access of humidity.

These local tubers were naturally considered 'more congenial and better adapted to the soil, and climate, and [therefore] less precarious in point of produce than potatoes'. Yet some peasants did express their willingness to cultivate potatoes. In Coch Behar the raiyat appeared 'extremely willing to extend the cultivation of potatoes, provided they

---

transitions from one stage of improvement to another are so exceedingly slow, as scarce to deserve the name.... [H]owever when they see a certain prospect of gain, with little additional trouble, they have been frequently known to adopt our practices', IOR, HM, vol.210, p.150.

are supplied with a sufficient quantity of seed in the first year'.[23]

Thus one of the key components of peasant production was their perception of risks. This attitude was entirely reasonable in the context of the late eighteenth century because of the almost continuous threats from famine and dearth to their subsistence and production. Risk avoidance was certainly one important reason behind the preponderance of rice and paddy in cultivated lands. Corn (that is, rice and paddy), noted Colebrooke:

> Though not equally profitable with dearer articles [cash crops] serves to alleviate the risk attending the cultivation of them; for they [the 'dearer' articles] seem precarious in exact proportion to the greatness of the profit which they are expected to afford. On the failure of his mulberry or his sugar cane, the peasant, had he no corn, must suffer the extremities of want; but raising in that and other grain a sufficiency for mere subsistence, he can wait the supply of his other wants from the success of other cultures; or, he can reserve a hoard from the crop of a successful year to meet the difficulties of one that is calamitous.[24]

Nevertheless, it would be wrong to assume that Bengal's agriculture was entirely a risk-prevention enterprise. Under conditions of relative normality, average peasants could be seen attempting to try their hands at cultivating the 'dearer' and 'precarious articles' of production. In Jangipur, for instance, cultivators often strove to strike a workable balance on their lands between cash crops and the cultivation of essential staples. Here jute and paddy were sown on the same lands 'adjoining to each other' because '*paat* [jute] will bear a length of Drought which would destroy the Grain, nor does it suffer from excessive rains', so that 'in case of the failure of one, from casualties of weather, the whole together may yield him a profit'. Interestingly, there was very little mulberry grown in Jangipur despite the soil's being 'well adapted to the Mulberry plant'; and the reasons for this were that mulberry required relatively larger amounts of capital, and that its cultivation was susceptible to frequent losses caused by weather or by 'disasters to the silk worms'.[25] Peasants realised the commercial potential of different crops

[23] The discussion of potatoes in the eighteenth century is based on the evidence in BRC, IOR, P/53/55, 29 January 1798, 12 February 1798, and 23 March 1798; ibid., P/53/58, 28 September 1798. In Midnapur the cultivation of *shakkarkand* (sweet potatoes) was an important part of agriculture. The seeds for this tuber were sown in the month of *Kartic* (October-November), and reaped in *Pous* (December-January), ibid., P/51/15, 28 January 1788.

[24] Colebrooke, *Remarks*, p.67.

[25] M. Atkinson, Resident, Jangipur Factory, 21 November 1792, IOR, HM,

and, circumstances permitting, attempted to cultivate for a market. An inquiry into the agricultural conditions of the district of Baldakhal in eastern Bengal had the following to say about the peasant's choice of crops: 'Cotton in this district is an article of universal cultivation, for the Ryot whose land is not favourable to the produce, provides himself with cotton land in some other place more proper for its growth.'[26]

In 1788-9, the Company made inquiries about the cultivation of plants yielding a yellow dye 'possessing the Quality of Madder'. The description of this dye from Birbhum is revealing:

> From the small fibre of the root of the tree cultivated in this district, called Auch, a yellow dye is extracted; and of which should it be found to answer the purpose of Madder, large quantities may be obtained, as on its becoming an Article of Demand, the *Ryotts will not fail to increase the cultivation of it.*[27]

Similarly, in September 1792, the Board of Revenue ordered the district collectors to make the peasants and merchants more aware of the commercial advantages of sugarcane since 'in future the demand for this commodity will be such as to yield to them an ample profit on any quantity that may be brought to the market'. A positive response to the stimulus of a market was anticipated:

> *If the profit accruing from this increased price is in the first instance secured to the Cultivators*, there can be no doubt of their extending their plantations to such an extent as in the course of a few years will enable this country to export large quantities of sugar not only to England, but to different parts of India where the commodity is in demand.[28]

Though the exact details of governmental efforts in the cultivation of sugarcane are unknown, we can be reasonably certain about the expansion in the cultivation of another cash crop, jute (*sona paat*), in Rangpur in our period. The first inquiries regarding the state of jute cultivation in Rangpur were made in 1794, and were rather imprecise both about output and lands under cultivation.[29] For the greater part of our period at least, *paat* was cultivated almost entirely by the fishermen on small spots of land for making nets,[30] but by the first decade of the nineteenth century the cultivation of jute had doubled in Rangpur

vol.375, p.318.

[26] 15 June 1789, IOR, HM, vol.385, p.325.

[27] BRP, IOR, P/71/10, 15 June 1789 (emphasis added).

[28] Ibid., P/72/7, 19 September 1792 (emphasis added).

[29] IOR, HM, vol.375, 9 June 1794, p.118.

[30] Ibid., p.288.

because of the 'demand from Europe and the advances given by the Company'.[31] The case of betel-nut cultivation in Chittagong is also revealing in this regard. There, despite its being a highly-taxed crop, its cultivation was spreading rapidly in the last decades of the century: 'Owing probably to the great number of Burma boats that now come here and supply themselves...in place of importing the betel from Sumatra, every man is planting the [betel-nut] tree around his house'.[32]

Thus the average peasant of our period blended an attitude of extreme caution with a healthy regard for commercial profit. But these peasants, like all producers, functioned in a material milieu which was usually beyond their power to control, and the influence of this environment was critical in determining the specific features of agricultural production and its social content.

The so-called 'inefficiency' of Bengal's agriculture

For Western observers, agricultural production in our period was extremely inefficient. Bengal's peasants did not use manure in adequate quantities to increase productivity, they did not practise fallowing in order to restore the natural vitality of the soil, and they had no knowledge of crop rotation of the type (fodder, cereal, fodder) which was practised by the peasants in eighteenth-century Europe. These were listed by observers as some of the major 'imperfections of husbandry',[33] in order to explain what they considered to be the essential backwardness of Bengal (and Asian) agriculture.

The fact that the greatest portion of cultivated land, under rice, was not extensively manured,[34] except naturally, was very much in keeping with the intrinsic nature of rice. Rice cultivation depends almost exclusively on adequate water supplies for most of its nutrients, and the area under wet-rice therefore is not dependent on large supplies of manure. Additionally, the fertilising power of water enables the fields to be cultivated continuously without fallowing.[35] Deposits of silt and rotting

[31] Buchanan, 'Ronggoppur', IOL, Mss.Eur. D.75, vol.2, book 2, f.42.

[32] BM, Add.Mss., 19286, f.4.

[33] See, for instance, Colebrooke, *Remarks*, pp.22-5; Tennant, *Indian Recreations*, 2, pp.11ff.

[34] F. Buchanan, 'An Account of a Journey Through the Provinces of Chittagong and Tipperah, 1798', BM, Add.Mss. 19286, f.2.

[35] D.H. Grist, *Rice* (5th ed., London 1975), p.38; F. Bray, 'Patterns of Evolution', p.9.

vegetation left by the seasonal inundations were enough to give successive rice harvests without the need for any additional, or artificial, manures.

Nevertheless, manuring was practised for those crops which needed to be manured. Unfortunately, very little is known about manuring practices in different districts, but a comparison of Robert Kyd's survey of the western side of the Hughli river with Buchanan's survey of Rangpur does show certain common features. Leaving the roots of a previous harvest to rot in the ground was a commonly-used means of natural fertilisation for the cultivation of *aman* rice. Burning the rooted stubble of one crop and then ploughing in its ashes for the next was another widespread practice for the cultivation of rice and lentils. More elaborate and costly forms of manure were used for cultivating sugarcane, cotton, tobacco and betel-leaf. Here expressed oil-seed cakes (especially of mustard) were mixed with the top soil, and then evenly distributed over the bed and rows in order to provide additional nutrients to the plant and to replenish the soil subsequently.[36] Marling, or dressing lighter soils with clay or loamy earth in order to enable them to retain water and manure near the surface, was not in vogue ('the use of the earth from marshes and ditches is very little, if at all, practised'), probably because of the costs involved, which could be prohibitive;[37] and this can perhaps be ascribed as one likely reason for the limited spread of cotton cultivation in our period.

'The Indian allows [the land] a lea, but never a fallow' was Tennant's way of describing the absence of fallowing by the peasants of Bengal.[38] Apart from showing that lands in Bengal were intensively cultivated, Tennant's statement also demonstrates the essential divergence of Bengal's (or South Asia's) agriculture from that of contemporary Europe. In Europe, fallowing was essential to restore the fertility of

[36] See IOL, Mss.Eur. F.95. f.30-35 and Mss.Eur. D.75, vol.2., book 4, f.82-83.

[37] IOL, Mss.Eur. D.75, F.82; see J.D. Chambers and G.E. Mingay, *The Agricultural Revolution* (London 1966), pp.62-3 for marling in late eighteenth-century English agriculture; it seems to have expanded, and along with manuring, drainage and convertible husbandry, constituted one of the four major sources of agricultural improvement Before the mid-eighteenth century it had not been extensively used to improve the quality of lighter soils because of the almost prohibitive costs; see C.G.A. Clay, *Economic Expansion and Social Change, England, 1500-1700*, vol.1, *People, Land and Towns* (Cambridge 1984), p.112.

[38] Tennant, *Indian Recreations*, vol.2, p.16.

the heavy and wet clay soil after two or three years of successive cropping. There fallowing was more than giving some respite to the land; it also entailed successive ploughings to break the surface thoroughly and get rid of weeds.[39] Obviously this process tended to raise costs of production and, by the end of the seventeenth century, the rigid rotation of two crops and a fallow had come to be recognised as one of the major impediments to the raising of the productivity of land in Europe.[40] By contrast, fallowing was not an important feature of Bengal's agriculture because the soil here was more amenable: it could be broken up more easily by 'scratching the superfices of the earth to make it yield'.[41] In such a situation, fallowing would mean an unnecessary loss of output and unduly high average costs of production.[42]

Nevertheless, the awareness that land tends to be exhausted by repeated cultivation, and that it needs some rest to recuperate, was well developed in Bengal. Crops like sugar-cane, cotton and mulberry sapped the land after three successive harvests, and lands bearing these crops were rejuvenated in a four-yearly cycle. In areas bordering upon the Hughli river, such lands were fallowed for about a year, during which 'the ground is restored with manure and fresh soil...and exposed once a month by the plough to the influence of the Atmosphere'.[43] In parts of Rangpur:

> Four years *fallow* are usually allowed after 2 crops. The first year the ploughing is rather difficult, and the grass roots require to be collected and burnt; but on the whole the trouble of cultivation on such lands is exceedingly small, and the crops are good. The land is fallow for two or three years, and produces abundance of grass, but it is of a wretched quality.[44]

However, in comparison with Europe, the practice of fallowing was undertaken on a much smaller scale in Bengal. There only a tiny fraction of lands producing selected crops was actually fallowed, whereas

[39] Chambers and Mingay, *Agricultural Revolution*, p.49.

[40] Ibid., p.49; also W. Abel, *Agricultural Fluctuations in Europe: from thirteenth to twentieth centuries* (trans. Olivia Ordish; London 1980), pp.109, 208; Clay, *Economic Expansion and Social Change*, vol.1, p.113.

[41] IOR, HM, vol.47, 4 September 1762, p.43.

[42] Significantly these were precisely the considerations which were increasingly reducing the importance of the fallow in English agriculture from the late eighteenth century (see Chambers and Mingay, *Agricultural Revolution*, passim).

[43] IOL, Mss. Eur. F.95, ff.36-37.

[44] IOL, Mss. Eur. D.75, vol.2, book 4, f.84 (emphasis added).

in Europe fallowing meant that from a quarter to a third of the arable was left unused for up to three or even four years.[45]

The reasons why some European observers (for instance, Colebrooke and Tennant) noted an absence of crop-rotation in Bengal are unclear. Contemporary European agricultural practices appear to have been conducted basically on a three-course rotation of crops (cereal, fodder, cereal), though in some parts, for instance in Flanders, twelve-course rotations had been introduced by 1800; the so-called Norfolk system of rotation (turnips, barley, clover, and wheat) was being gradually extended to the rest of England along with an increasing tendency to cultivate part of fallow with fodder crops.[46] Evidence from Bengal does suggest a fairly widespread existence of crop rotation. The simplest form was the rotation of rice and lentils on relatively lower lands; on higher lands rice and mustard were sown in regular succession. *Sih-fasli* (three-harvest) lands, usually of the best intrinsic quality, produced rice with either two crops of oil-seeds or lentils, or they were used for producing successive crops of mustard, cotton and lentils in a regular annual rotation.[47] In fact, Buchanan listed 'alternate cropping' as one important agricultural practice employed to prevent soil-exhaustion in parts of Rangpur.[48] Interestingly, alternate cropping, or the sequential use of fodder crops and corn crops to obviate fallowing (also called convertible husbandry), had only just started spreading to the enclosed and fairly large farms of England in the late eighteenth century.[49]

Facets of agrarian commercialism in late pre-colonial Asia

The exact fit between agriculture and commercialism in the late pre-colonial period is certainly problematical. Obviously, much depended on the overall milieu in which producers functioned. P.J. Marshall suggests that the balance of advantages and disadvantages in the cultivation of various, particularly cash, crops in Bengal's agriculture was more complex than the cultivator's response to an underlying profit motive. More crucial were the considerations of the relative pitch of revenue and the availability of ready cash. C.A. Bayly argues that the cultivation of cotton in early nineteenth-century Bundelkhand was

[45] Chambers and Mingay, *Agricultural Revolution*, p.49.
[46] Abel, *Agricultural Fluctuations*, pp.207-8.
[47] BRC, IOR, P/51/15, 28 January 1787; also IOL, Mss. Eur. F.95, ff.30-35.
[48] IOL, Mss. Eur. D.75, vol.2, book 4, f.83.
[49] Chambers and Mingay, *Agricultural Revolution*, p.61.

determined by a complex combination of the available labour surplus in a peasant family, quality of land, the revenue regime, the terms of the available cash advance, and the prevailing prices on the world market; the mere existence of high prices was not a sufficient inducement for the 'middle or poor peasants' to take to the cultivation of cotton. Ludden views peasant production in pre-colonial south India as a function primarily of a given ecology—namely, soil, water and drainage—which imparted its particular blend of crops (agricultural specialisation) to communities in their respective agricultural zones.[50] Additionally, producers often combined two occupations, agriculture with artisanal production, in their respective households.

Yet it would be difficult to argue, as Marx did, that the combination of agriculture with artisanal production in a peasant symbolises the working of a pre-capitalist 'natural' economy; nor would it be acceptable to see this phenomenon in a Chayanovian perspective of a labour-consumer balance in a peasant household.[51] Also, the fact that peasants consciously resisted expanding the domain of 'commercial' farming (for example, cotton in Bengal) and sought to intensify the cultivation of rice, should not be treated as an irrational or perverse economic choice. Pre-colonial Asian agriculture was far from homeostatic. In fact, the typical feature of the late pre-colonial peasant economy of Asia was the nexus between atomistic peasant production (in personal holdings) and a hierarchy of demand-requirements (including those of revenue/rent) which stimulated peasant production and was serviced by it, producing an ascending chain of internal and global markets.

Caglar Keydar finds that peasant-household production in late Ottoman Turkey was 'immersed' in a series of 'interlocked' factor and product markets.[52] Studies in the economy of Ming and Ch'ing China have

[50] P.J. Marshall, *Bengal: the British bridgehead, Eastern India, 1740-1828* (Cambridge 1987), p.170; Bayly, *Rulers, Townsmen and Bazaars*, pp.245-6; Ludden, *Peasant History*, pp.52ff.

[51] Karl Marx, *Pre-capitalist Economic Formations* (ed. E.J. Hobsbawm; London 1978); also Karl Marx, *Capital: a critique of political economy*, vol.3 (Harmondsworth 1982), p.931. For the Chayanovian perspective, see Basil Kerblay, 'Chayanov and the theory of peasant economies', in T. Shanin (ed.), *Peasant and Peasant Societies, selected readings* (2nd edn.; Harmondsworth 1988), p.178.

[52] Caglar Keydar, 'Small peasant ownership in Turkey: historical formation and present structure', *Review* vii, 1 (Summer 1983), p.5; Halil Inalcik, 'Capital formation in the Ottoman Empire', *Journal of Economic History* 29, 1 (March

conclusively demonstrated the articulation of similar markets in China with overseas trade, flows of bullion and taxation in cash. Such studies have also documented the effects these flows had on the creation of regional specialisation: the different provinces divided into grain and cash-crop producing areas in the seventeenth and eighteenth centuries.[53] The peasant economy of Tokugawa Japan, which was relatively 'closed', nevertheless in this period (1600-1868) underwent a phase of intensive commercialisation of agriculture which was achieved largely by the proliferation of rural periodic 'peasant' markets linked into a network of markets of ascending importance leading to the large urban centres.[54] Towns meant a substantial urban presence which demanded bulk rural-urban trade in rice.[55]

Japan's economic experience of the eighteenth century is of great importance in any analysis of peasant production in Asia because the entire commercial edifice during the Tokugawa period was constructed mainly on the basis of rice and by the *commercialisation of its rice-producing economy*.[56] Peasant initiative in expanding the acreage devoted to rice for subsistence and for the market was the prime mover and hallmark for this commercialisation, which was accompanied by the transition from rent-in-kind to rent-in-cash in the Japanese countryside.[57] There is no reason to assume the uniqueness of Japan in the eighteenth century, for, as Clifford Geertz shows, it was not before late in the nineteenth century that one can effectively speak of the diver-

---

1969).

[53] D.W. Perkins, *Agricultural Development in China, 1368-1968* (Chicago 1967), pp.144-8; W.S. Atwell, 'Notes on silver, foreign trade and the late Ming economy', *Ch'ing-shih wen-t'i* 3, 8 (December 1977); R.M. Myers, 'The "sprouts of capitalism" in agricultural development during the mid-Ch'ing period', *Ch'ing-shih wen-t'i* 3, 6 (December 1976); Naquin and Rawsky, *Chinese Society in the Eighteenth Century*, ch.4.

[54] F.W. Moulder, *Japan, China and the World Economy: toward a reinterpretation of East Asian development* (Cambridge 1977), pp.36-7; Tsuneo Sato, 'Tokugawa villages and agriculture', in Chie Nakane and Shinzaburo Oishi, *Tokugawa Japan: the social and economic antecedents of modern Japan* (tr. Conrad Totman; Tokyo 1991), pp.70-1.

[55] Gilbert Rozman, 'Castle towns in transition', in M.B. Jansen and G. Rozman (eds.), *Japan in Transition: from Tokugawa to Meiji* (Princeton 1988), pp.274-5, 318.

[56] See Yasukazu Takenaka, 'Endogenous formation and development of capitalism in Japan', *Journal of Economic History* 29, 1 (March 1969), pp.141-62; also Bray, *The Rice Economies*, pp.210-7.

[57] Francesca Bray, 'Patterns of evolution', pp.18, 21.

gence of the Japanese 'path' from that trodden by other countries of Asia.[58]

The example of rice production in China also demonstrates links between commercial impulses and the so-called subsistence sector in agriculture. Here, state initiative in the early medieval phase encouraged the cultivation of better strains of rice which fitted well with the enterprise of the primary producers. The result was a steady expansion in agricultural production up to the end of the eighteenth century,[59] characterised by a high degree of commercialisation in which most farmers 'produced in part for the market and some, probably along the coast, produced primarily for the market'.[60] 'Market expansion' and 'market conditions' have been seen as the important variables in stimulating agricultural production in southern China in the eighteenth century.[61]

The cases of Tokugawa Japan and Ch'ing China amply demonstrate the intricate connections between cereal production, peasant economies and commercialised agriculture in late pre-colonial Asia at the typological level; they also show the futility of adhering to any rigidly antithetical notions of 'subsistence' versus 'commercialised' agriculture, of limiting oneself to the growth of so-called 'cash' crops while studying the working of commercialised peasant production. Unfortunately,

[58] Geertz shows on the basis of Javanese agriculture that the main reason for this divergence was the colonisation of other Asian countries and Japan's escape from this experience; Clifford Geertz, *Agricultural Involution: the process of ecological change in Indonesia* (Berkeley 1963), pp.130-43. Another argument advanced is the concurrence of the growths in agricultural and non-farm productivity in the nineteenth century which did not occur elsewhere in Asia. See T.C. Smith, 'Farm family by-employments in pre-industrial Japan', *Journal of Economic History* 29, 4 (October 1969); B.F. Johnston, 'The Japanese "model" of agricultural development: its relevance to developing nations', in C.R. Wharton (ed.), *Subsistence Agriculture and Economic Development* (London 1970). It must, however, be emphasised that 'by-employment' was an adjunct—a 'side occupation'—of the peasantry which did not reduce the overarching primacy of rice cultivation: 'They were auxiliary and supplementary to the farm household'; O. Saito, 'The rural economy: commercial agriculture, by-employment and wage work', in Jansen and Rozman (eds.), *Japan in Transition*, pp.405-6.

[59] Bray, 'Patterns of evolution', p.16; and*The Rice Economies*, p.207.

[60] D.W. Perkins, 'Persistence of the past', in D.W. Perkins (ed.), *China's Modern Economy in Historical Perspective* (Stanford 1975), p.4.

[61] Evelyn Sakakida Rawski, *Agricultural Change and Peasant Economy in South China* (Cambridge, Mass. 1972).

there have been few attempts to break these rigid, and largely incorrect boundaries in the context of agricultural production in eighteenth-century South Asia: the reason is the total absence of a comparative *Asian* perspective and an excessive preoccupation with the impact of the early colonial regime on the South Asian economy. The latter preoccupation is most pervasive in the historiography of the province of Bengal. There, the second half of the century is interpreted as a period of agricultural recession brought about by the revenue burden of the English East India Company and by the severe famine of 1770, and commercial production is located exclusively in those sectors (like opium and indigo) which were structurally in the service of the Company's commerce.[62] This kind of argument apparently assumes the existence of a dual economy in the province, a view which is difficult to sustain as there is now fairly substantial evidence to show the existence and strengthening of an integrated provincial economy in this period, operating in the main through a hierarchy of markets and trading networks.[63]

Significantly, the central indicator used to measure the health of this economy was the price of rice and not of any other commodity. It was also generally accepted that the prevailing price of rice influenced other agricultural and non-agricultural prices;[64] and that it was access to rice, and not to any other 'cash' crop, which was crucial in ensuring physical survival during a crisis of subsistence.[65] This was perhaps natural. The

[62] See B.B. Chaudhury, 'Agricultural growth in Bengal and Bihar, 1770-1860: growth of cultivation since the famine of 1770', *Bengal Past and Present* January-June 1976; 'Eastern India', in D. Kumar and M. Desai (eds.), *Cambridge Economic History of India* (Cambridge 1982; Indian edition, Delhi 1984; hereafter *CEHI* 2); and Chaudhury, *Growth of Commercial Agriculture in Bengal, 1757-1860* (Calcutta 1964).

[63] Marshall, *Bridgehead*, pp.13-14; Datta, 'Rural Bengal', chs.1 and 6, and also 'Merchants and peasants: the structure of local trade in late eighteenth-century Bengal', *Indian Economic and Social History Review* (hereafter *IESHR*) 23, 4 (October-December 1986); D.H. Curly, 'Rulers and merchants in late eighteenth-century Bengal', unpublished DPhil. thesis, University of Chicago 1980.

[64] Colebrooke, *Remarks*, p.67f.

[65] Rajat Datta, 'Some notes on the causation of dearth and famine in late eighteenth-century Bengal' (mimeo.), paper presented at a seminar on 'Famine in India: eighteenth and nineteenth centuries', School of Oriental and African Studies, September 1989; also 'Subsistence crises, markets and merchants in late eighteenth-century Bengal', *Studies in History* 10, 1 (1994). That Bengal was not alone in attaching such a degree of importance to rice in determining

product value of the different varieties and qualities of rice and paddy accounted for more than 45 per cent of the gross agricultural output. It must, however, be borne in mind that the dominance of the foodgrain sector in total agricultural production, and the influence of cereal prices on the prices of other farm and non-farm products, was not limited to Asian agriculture.[66] Rice in Bengal was far from a subsistence crop.[67] Large portions of the land under rice (sometimes as high as 70 per cent in some districts) were cultivated by peasants under a complex system of cash advances made by the grain merchants. These loans were aimed at denying the peasantry any direct access to local markets, and ensuring the advance hypothecation of their product *prior to the completion of the production cycle.*[68] This feature, the extent of mercantile penetration in primary agriculture, perhaps imparted a distinctive thrust to Bengal's rice production, thereby making it substantially more commercialised than most contemporary economies in eighteenth-century

---

the prices of agricultural products is borne out from other similar developments in China and Japan. Cf. W.S. Atwell, 'Some aspects of the "seventeenth-century crisis" in China and Japan', *Journal of Asian Studies* 45, 2 (February 1986), pp.230-3. For a good discussion of the economic significance of rice prices in China see Yeh-Chien Wang, 'The secular trend of prices during the Ch'ing period', *Journal of the Institute of Chinese Studies* (Hong Kong) 5, 2 (1972).

66 Fluctuations in the price of wheat exerted a most critical pressure on the prices of other farm produce in England throughout the eighteenth century; in fact, the Corn Laws were a device adopted by the state to reduce the massive oscillations in the price of wheat in that period; see Chambers and Mingay, *Agricultural Revolution*, pp.107-9. The main trend of French agriculture during the seventeenth and eighteenth centuries was to increase the cultivation of land under the cultivation of cereals; and the causal connections between cereal and non-cereal prices were most sharply brought into focus during the severe harvest failures in the 1770s and 1780s which exacerbated the crisis of the *ancien regime*; compare Jean Jacquart, 'French agriculture in the seventeenth century', in P. Earle (ed.), *Essays in European Economic History* (Oxford), pp.166-82; and Peter Jones, 'The French peasantry of France on the eve of the French revolution', *History of European Ideas* (special issue, ed. Jones, *The European Peasantry on the Eve of the French Revolution*) 12, 3 (1990), pp.335-61.

67 Subsistence production is defined here, following Wharton, 'as a situation where the fruits of an individual or group productive efforts are dictated more toward meeting immediate consumption needs...without any or few intermediaries or exchange...'; Wharton, 'Subsistence agriculture: concepts and scope', in C. Wharton, *Subsistence Agriculture*, p.133.

68 For a fuller discussion see R. Datta, 'Agricultural production, social participation and domination in late eighteenth-century Bengal: towards an alternative explanation', *JPS* 17, 2 (October 1989), pp.92ff.

Asia.

In Tokugawa Japan, cash-cropping was often the monopoly of the big feudal *hans*, while rice was produced in peasant farms. The system of cash loans for rice was practically undeveloped, but an extremely elaborate system of mercantile credit and investment of capital did exist for other products and commodities.[69] The merchants' function was to purchase grain from the lords' or the farmers' markets, or to commute the peasants' rice revenue into cash, and to move their stock for sale to the larger cities and to the capital.[70] In Ch'ing China the marketing of rice was apparently the main item of local trade (so much so that rice was often used as the standard to reckon the price of other goods along with existing metallic currencies), but the process of moving grains and other commodities from villages of origin was accomplished 'largely through inequalities in the form of rents and surpluses for marketing', a system which favoured the landlords.[71] Merchants did invest in the cultivation of tea, cane sugar and tobacco,[72] but remained largely dissociated from the production of rice.

The extent of mercantile involvement in the production of rice in Bengal appears to have been unique also within the Indian subcontinent. Two studies of regional grain-trading in eastern Rajasthan and Bihar demonstrate the existence of an elaborate hierarchy of markets and specialist grain merchants,[73] but provide no evidence to show whether these networks had any influence on peasant-merchant relationships. In the case of eastern Rajasthan such developments were probably stymied by the extremely intrusive role of the state in the marketing of grain which, as Bajekal argues, forced these traders to operate more in the nature of state agents than as autonomous merchants. In

[69] Takenaka, 'Endogenous formation', p.144ff.; Satoru Nakamura, 'The development of rural industry' in Nakane and Oishi (eds.), *Tokugawa Japan*, p.91. It is possible that the Japanese peasants had access to these sources of credit which may have reduced their dependence on loans which (as in Bengal) were often usurious.

[70] G. Rozan, *Urban Networks in Ch'ing China and Tokugawa Japan* (Princeton 1973), p.134.

[71] Ibid., pp.85-7.

[72] Naquin and Rawski, *Chinese Society*, p.99.

[73] Madhavi Bajekal, 'The state and the rural grain market in eighteenth-century Rajasthan', and Kumkum Banerjee, 'Grain-traders and the East India Company: Patna and its hinterland in the late eighteenth and early nineteenth centuries', in Sanjay Subrahmanyam, ed., *Merchants, Markets and the State in Early Modern India* (Delhi 1990), pp.90-121, 163-89.

Bihar, grain-trading was virtually free of state control, but the grain merchant had very little involvement in the processes of agricultural production. Production loans were apparently provided, not by specialist traders, but by a coterie of peasant small-traders, the *grihastha-byaparis* (literally, the householder trader), who are described by Banerjee as the 'dominant peasants [who] used their position in the village society to branch out into trade'.[74] This was perhaps (and here one can only speculate) due to the overarching influence of caste in the social hierarchy of rural Bihar.[75] The existence of relatively comfortable peasants, connected to their counterparts by ties of caste, may have prevented the grain merchants from intruding into agricultural production in Bihar. By contrast, in Bengal, where the caste factor in production was much weaker, and where relative affluence was based on control over resources and labour and not on the caste-based exploitation of a destitute peasantry,[76] it was perhaps easier to exploit production in the context of an expanding trade in grain.

Yet situations like Bihar and Rajasthan may not disprove the fact that there was, in the eighteenth century, an increasing incidence of agrarian commercialism, and that this process was caused by the sizeable integration of production with the market. A contrary view has recently been advanced by Peter Robb who argues, from the historical location of Bihar's agriculture in the nineteenth century, that: (1) (empirically) Bihar's commercial growth in the nineteenth century (and most of the eighteenth century too) was based on a 'wide range of controls related mainly to the extraction of rent' in the midst of an extremely weak market, and that 'the agriculture of commerce was far from commercial'; and (2) (conceptually) that many market-centred theories of commercialisation in India, especially pre-colonial India (like the one being advanced here), are flawed since they do not take into account the non-market (extra-economic) dimensions in production, which for Robb are of determining economic importance, and because they ignore the existence of a number of imperfections in the market.[77]

Robb is perhaps overstating his case a little by assuming that those

[74] Banerjee, 'Grain-traders', p.166.
[75] Marshall, *Bridgehead*, p.37.
[76] Datta, 'Agricultural production, social participation', pp.77-83.
[77] Peter Robb, 'Peasants' choices? Indian agriculture, and the limits of commercialisation in nineteenth-century Bihar', *Economic History Review* xlv, 1 (1992).

historians who assign a certain primacy to the market in production in late pre-colonial India do not recognise its many imperfections. The argument on the contrary is of an economy (or economies if one takes regional perspectives) producing for a hierarchy of markets;[78] and nowhere is it even suggested that these markets were perfectly competitive. In fact, merchants often succeeded in imposing oligopsonic controls over local resources.[79] The burdens imposed on the cultivators by the pressure of revenue naturally aided these merchants. Since revenue was overwhelmingly collected in cash (at least this was so in Bengal), the produce had to be sold, and the quantity of cash which would accrue to the producer was directly related to the price line. Witness, for instance, the following description of this connection given by Colebrooke (c.1794):

> When the crops of corn are very abundant, it is not only cheap, but wants a ready market. As the payments of rents are regulated by the season of the harvest, the revenue is due and must be paid, whether there be or not be a vend for the produce.... Thus the eagerness of the vendors [further] reduces the price.[80]

Merchants would buy stocks almost at distress prices, store them in their *golahs* (granaries), and then sell them when prices were again favourable from their point of view.[81] The price situation was therefore of critical importance in all market transactions; but it would be unreasonable to argue that both prices and transactions were determined by an untrammelled operation of the laws of supply and demand, or that everyone benefited equally. The village was at a relative disadvantage in such a system of trade. Since each of these markets was connected with the next rung in an increasing scale of commodity transactions, the primary focus of trade naturally tended to shift away from the village. Given this situation, no single peasant-household could have

[78] Datta, 'Merchants and peasants' and 'Agricultural production, social participation'; and C.A. Bayly, 'Indigenous social formation and the "world system": north India since circa 1700', and D. Ludden, 'World history and village India, 1600-1900: exploring the agrarian history of capitalism', in Bose, *South Asia and World Capitalism*. Obviously, production for market is quite different from the concept of a fully-developed market economy in a capitalist setup where supply and demand operate with a great degree of autonomy, and where prices are by and large determined by the two without any or much external regulation.

[79] Datta, 'Rural Bengal', pp.29-33.

[80] Colebrooke, *Remarks*, p.67.

[81] Datta, 'Merchants and peasants', pp.29-33.

had an appreciable impact on the formation of prices even at the level of the village *haath*.[82] Moreover, the oligopsonic power of the local merchants over trade in agricultural produce, and the monopsonic control of the East India Company in sectors like cotton piecegoods and silk (later to be supplemented by opium and indigo), made for a situation where the formation of prices depended on command and coercion to an extent equal to the influence exerted on them by the interplay of demand and supply.

It is precisely because of imperfections that these markets were more than areas of contact between potential buyers and sellers. They reflected prevailing social inequalities in the sphere of exchange which were crucial in the processes of accumulation in pre-colonial India. In fact, as I have argued elsewhere, the various devices adopted by merchants to restrict the peasants' independent access to such markets were part of an elaborate system which (1) ensured that the surplus in kind was extracted by the revenue, tax and rent, and was converted into surplus and into cash; and which (2) facilitated the intrusion of merchant-capital into peasant production through the advance-contracts or debt-servicing.[83] Thus, instead of the peasants going to the market, the market seems to have come in a fairly big way to the peasants of eighteenth-century Bengal.[84] A symptom of this was the rapid mushrooming of intermediaries between the merchant and the cultivator in rural Bengal in this century.[85]

A proliferation of mercantile intermediaries was necessary in order to minimise the degree of risk involved in operating in an economy where production and procurement centres (the villages) were often

82 This is a modified version of Keyder's argument regarding late Ottoman Turkey ('Small peasant ownership in Turkey', pp.53-5).

83 Datta, 'Rural Bengal', pp.329-38.

84 Somewhat analogous developments have occurred in Bihar in the same period for three reasons. First, Bihar apparently had a more developed town life than Bengal in the eighteenth century. Second, there seems to have been a substantial degree of integration of peasant and artisanal production with the local markets along with an interregional and overseas flow of goods. Third, there was a robust trade in grain and peasant involvement in the market and a network of trading intermediaries (Marshall, *Bridgehead*, pp.38-9; Banerjee, 'Grain-traders'). The effects these had on the structure of peasant production is unfortunately not clear; but this lack of clarity may not necessarily symbolise economic 'backwardness'; it may equally reflect the pressing need to do more detailed research on late pre-colonial Bihar.

85 Datta, 'Merchants, markets and subsistence crises', pp.85-8.

widely dispersed. Crucial also was the role these intermediaries performed in keeping a close check on prices in order to forestall, or at least minimise, the effects of variations in price, especially those massive fluctuations which accompanied both bad and plentiful harvests.[86]

Situating eighteenth-century Bengal

The above accounts have tried to indicate that perhaps there were unique features in the processes of agrarian commercialism in eighteenth-century Bengal. The complexities involved in the process must have involved a number of variables, of which three seem to have exerted fundamental influence. First, there were the linkages established in the seventeenth century. These were (1) a shift from spices to textiles by the European trading companies; (2) the transfer of imperial tribute from Bengal to Delhi in the form, among others, of high-value textiles; (3) the export of foodgrains to other provinces on a regular basis both by sea and by land.[87] The significant development of the eighteenth century was the emergence of an integrated provincial market in foodgrains, which extended and broadened the range of commercial production and commercialised transactions initially set into motion by items like silk and cotton.[88]

Second, there was the integration of Bengal in a global economic network by the expanding world economy of the eighteenth century. The political economy of this integration had the following dimensions:

> To keep in motion the great machine of the Company commerce, it is absolutely necessary that the provision of Cargoes for the return of ships be ensured. Hence the origin of previously investing Funds at the Aurangs [manufactories];[89] hence also the necessity of being secured in due realisation of goods for which money is advanced.... The accession of the Company to the Government of the Country [to the Diwani in 1765] did not change these [earlier] principles and usages, tho' [there was] the immense increase of their Funds and of the wealth of their Servants, all to be centr'd finally in Europe and to be conveyed thither only in goods...

The impact of such an integration was quite fundamental, as it 'exceed-

[86] Datta, 'Rural Bengal', pp.306-8.

[87] For a good overview of these changing patterns of European trade, see Om Prakash, 'European trade and Asian economies: some regional contrasts, 1600-1800', in L. Blusse and F. Gaastra, *Companies and Trade* (The Hague 1981), pp.189ff.; for the other aspects of the seventeenth-century developments, see BM, Add.Mss.,34123; IOR, H.M., vol.456F.

[88] Datta, 'Rural Bengal', chs.1 and 6; Marshall, *Bridgehead*, passim.

[89] The *dadni* or the advance contract system.

ingly enlarged the demands for the Market of Bengal'.[90]

Third, there was the political dimension. Bengal was the first province in India to undergo colonial conquest whereby its revenue resources were placed at the disposal of the Company to invest in trade and empire-building. Profits accruing from monopsonic controls over silk and piecegoods were, in 1772, buttressed with a declared official monopoly over salt. Monopsonic controls were subsequently extended over opium and indigo. On the other hand, the Company's declared intention was to ensure 'fair trade' in agricultural produce, in those commodities which were not directly connected with the international flow of commerce. Thus, after 1757, state intervention assumed apparently contradictory forms of rigorous control in the marketing of certain commodities, and a relatively striking non-interference in the movement of others; the intention in the latter case was to ensure that the local merchant was at 'liberty to carry his merchandise wherever he thinks proper for sale'.[91]

The other aspect of the Company's rule was revenue, mainly its assessment and timely collection in cash. In fact, this was the culmination of a process already set into motion by the previous regime, the Nizamat. But there was a major difference: the Company's state persistently refused all suggestions to revert to revenue in kind during periods of price slumps in order to reduce the pressures on the peasantry, which was a major departure from one of the abiding principles of Mughal revenue policy.[92]

The British conquest of Bengal had some fundamental effects on the province's economy. Between 40 and 45 per cent of the agricultural output taken as revenuewas annually converted into cash.[93] Obviously, a greater degree of appropriation in money would depend on the facilities available for the circulation of cash and commodities. Secondly, there were henceforth significant changes in the structures of local

[90] Report of Commercial Occurrences, 6 August 1789, IOR, HM, vol.393, pp.279-80.

[91] The term 'fair trade', and the policies to be followed in connection, were established by Warren Hastings in a Minute dated 9 March 1773, IOR, HM, vol.217, pp.444-9; also see BRC, IOR, P/49/38, 23 March 1773.

[92] James Steuart, 'Memoirs of the Coinage of Bengal', IOR, HM, vol.62; John Shore's Minute in BRC, IOR, P/51/50, 28 October 1789; also Datta, 'Rural Bengal', pp.55-9.

[93] Minute of John Shore, 18 June 1789, in W.K. Firminger (ed.), *The Fifth Report*, vol.II (Calcutta 1917) (hereafter *FR* 2).

markets, and in the relationship between the state and these markets. Crucial in this respect was the policy followed by the Company between 1773 and 1790 of dissociating landed proprietors (especially the zamindars) from collecting local taxes on trade (*sa'ir jihat*) in their respective territories.[94] This removed, to a large extent, one of the principal problems hampering the easy movement of goods under the Nizamat. The fact that the Company tried to break the existing bottlenecks in local marketing systems is an example of a major goal of early colonial policy: 'to create markets where they [did] not exist before, or to regulate existing markets'.[95] One immediate effect of this (essentially political) decision was the remarkable geographical proliferation of markets in Bengal during the second half of the eighteenth century. Sources are replete with descriptions of the 'extension of hauts and bazars' all over Bengal in this period.[96]

The scale with which markets in Bengal appear to have expanded in this period was almost spectacular. A comparison with China is revealing. The Ch'ing period, which was characterised by an expansion of trade and the growth of 'periodic markets',[97] was undergoing a general 'accretion of markets'. For instance, in Sechuan the number of standard markets increased from four in 1622 to thirteen in 1875.[98] Compare with Dhaka where in one district alone there were 536 markets (*haaths, bazars* and *ganjs*) in about 1765; these had increased to 650 in 1791.[99] Obviously, areas previously deficient in markets were now brought under their purview. This facilitated peasant market integration which, as is being argued here, was one of the distinctive developments in the eighteenth century. Skinner sees the proliferation of marketing centres in Ch'ing China as the 'intensification of the rural landscape'. In China this meant an increase in the volume of trade in an average market day

[94] See IOR, HM vol.217, pp.44-9; BRC, IOR P/52/13, 11 June 1790; Board of Revenue, Miscellaneous Proceedings, IOR P/89/37, 27 June 1792; Datta, 'Rural Bengal', pp.36-40.

[95] Cyril S. Belshaw, *Traditional Exchange and Modern Markets* (Delhi 1965), p.74.

[96] Datta, 'Rural Bengal', pp.39-40.

[97] Rawski, *Agricultural Change*, p.69.

[98] W.G. Skinner, 'Marketing and social structure in rural China', *Journal of Asian Studies*, part 2, 25, 2 (February 1965), p.195.

[99] Compare John Taylor, *A Sketch of the Topography and Statistics of Dacca* (Calcutta 1840), p.203; Board of Revenue, Miscellaneous, IOR, P/89/37, 27 July 1791.

at already existing markets and/or an absolute increase in the number of fixed market days.[100] Considering the phenomenal expansion in Bengal, it seems likely that these features would be present there; and indeed they were. Apart from the *ganjs*, which were fixed daily markets, the *haaths*—the periodic (usually weekly) village market—in rural Bengal were beginning to be assembled every alternate day in certain localities.[101] Even in low-lying areas prone to inundation by the seasonal overflowing of rivers, markets were held 'for four and sometimes in some places for six months of the year *on board of boats*'.[102]

In the light of such evidence, it is certainly possible to say that there was a tricuspid integration—of peasant-household, agriculture and trade—with production for the market in Bengal. Of this the integration of the peasant-household, both as seller and as producer, was of vital importance because it was this which determined (despite the many burdens and pressures on it) the amount of agricultural produce which could actually be marketed. It also shaped the various tenurial and subtenurial arrangements, particularly share-cropping, which apparently mushroomed in Bengal during this period, and made the growth of a share-cropping peasantry in the eighteenth century an instrument of commercialised agriculture rather than of a certain (and largely undefined) subsistence security-system of a 'pre-capitalist' variety, or of exploitation under a nebulous sort of 'feudalism'. Therefore, it will be worthwhile to explore some of the central features of share-cropping and its connections with commercialised agriculture.

Share-cropping as an index of agrarian commercialism

There are essentially two views of share-cropping. The first suggests (on the basis of the *kedokan* in Java) that share-cropping in the past, and its stubborn persistence in some contemporary societies, are features of a 'pre-capitalist' system of risk-sharing in a largely subsistence-oriented economy.[103] The second views share-cropping as a '[semi]-feudal' social relation; and this has been applied in the case of the

[100] Skinner, 'Marketing and social structure', p.196.
[101] Taylor, *Dacca*, p.294.
[102] Board of Revenue, Miscellaneous, IOR, P/89/36, 15 September 1776 (emphasis added).
[103] B. Hart, W. Turton and B. White (eds.), *Agrarian Transformation: local processes and the state in South East Asia* (Berkeley 1989), p.256.

Philippine *kasama* and the *bargadari* relationship in eastern India.[104] Discussions of share-cropping arrangements in pre-colonial (specifically eighteenth-century) or early colonial India are few, but they nevertheless incorporate the two dominant views of share-cropping. For instance, C.A. Bayly suggests that, in origin at least, the institution of share-cropping may have been a system of mutual protection and interdependence in the cash-strapped economies of northern India in the eighteenth century.[105] Another analysis, specific to colonial Bengal, has described share-cropping as a semi-feudal social relationship of exploitation of a low-caste (and landless) peasantry under conditions of abject indebtedness.[106] A recent study sees share-cropping as a sub-regional phenomenon (limited to north Bengal) in the context of an agrarian structure sharply polarised between the jotedars who controlled the credit and product markets and the bargadars who had 'no recognised long-term rights to the land they tilled'.[107]

These approaches do not consider whether or not share-cropping can be a social typology of a different kind. On the other hand, the Marxian position that it is a method of surplus appropriation, transitional between feudalism and capitalism, is also problematical because of its unequivocal insistence that share-cropping will wither away with the growth of capitalist relations in the countryside.[108] There is a specifically Eurocentric concept of capitalism ingrained here, and, therefore, very little effort to come to grips with the specific features of share-cropping in different social formations. At best, as Husken points out, the nexus between share-cropping and commercialism is seen as a process of incorporation, that of a largely pre-capitalist relation (share-cropping) into a newly-developing commercialised village economy.[109]

[104] For a discussion see F.V. Aquilar Jr., 'The Philippine peasant as capitalist: beyond the categories of ideal-typical capitalism', *JPS* 17, 1 (October 1989). For the Indian context see the influential essay of Amit Bhaduri, 'A study of agricultural backwardness under semi-feudalism', *Economic Journal*, March 1977.

[105] C.A. Bayly, 'Indigenous social formation', p.121.

[106] Rajat and Ratnalekha Ray, 'Zamindars and jotedars: a study of rural politics in Bengal', *Modern Asian Studies* 9, 1 (1975).

[107] Sugata Bose, *The New Cambridge History of India*, III:2, *Peasant Labour and Colonial Capital: rural Bengal since 1770* (Cambridge 1993), pp.93-5.

[108] For a good but reiterative discussion in this perspective, see R. Pearce, 'Share-cropping: towards a Marxist view', *JPS* 10, 2 and 3 (January/April 1983).

[109] F. Husken, 'Landlords, share-croppers and agricultural labourers: chang-

Also the implicit assumption in most Marxist writings is that wage-labour is economically (and politically) superior to most supposedly non-wage forms of labour; and this has hampered an analysis of share-cropping as a spearhead (or even as a symptom) of agrarian commercialism *sui generis*.[110] It is therefore important to consider the possibility that share-cropping—its proliferation and spread—was the tenurial and social relation arising out of the commercialisation of agricultural production in a small-peasant economy.

The evidence for share-cropping in Bengal is quite clear about its increase—in terms of both its labour use and numbers—in the economy during the century, particularly during its second half.[111] One explanation, that the proliferation of share-croppers was the result of a landless low-caste peasantry looking for jobs,[112] is unsustainable, for share-croppers were not limited to any one particular caste (or castes) and abject landlessness was certainly not one of the major agrarian problems of the eighteenth century. The problem, on the contrary, arose from the fact that peasant holdings were small: there was a *relative insufficiency* of land held by individual peasants, which created numerous obstacles in the way of small peasants in meeting the requirements of household subsistence and in generating tiny domestic surpluses. As I have suggested,[113] share-cropping in our period can largely be explained in terms of a conscious strategy adopted by the small peasant either to avoid protracted impoverishment or to acquire a supplemen-

---

ing labour relations in rural Java', *Journal of Contemporary Asia* 9, 2 (1979), p.147.

[110] Caballero has convincingly argued that share-cropping may in fact be a more effective way of utilising labour than pure wage-labour because the latter has a greater propensity for lowering profits through acts of shirking and wastage by the workers. Under share-cropping, shirking lowers the output which defrauds the share-cropper to the same extent as it does the landowner. Under conditions of wage-labour, where the worker has no share in the output, shirking does not necessarily lead to a contraction in wages, but raises the costs of supervision, thereby raising labour and product costs This may lead to a reduction in labour hiring. It is therefore entirely feasible that 'the wage system may finally result in less effective labour actually being put in than under share-cropping', 'Share-cropping as an effective system: further answers to an old puzzle', *JPS* 10, 2 (January/April 1983), p.108.

[111] Unless otherwise stated, the following discussion is based on Datta, 'Agricultural production, social participation', pp.89-92, and Datta, 'Rural Bengal', ch.4.

[112] Ray and Ray, 'Zamindars and jotedars'.

[113] Datta, 'Agricultural production, social participation'.

tary income for consumption or production-oriented purposes.

Three dimensions reinforced the connection between share-cropping and agrarian commercialism. First, it entailed a specific relation between investment and labour, and between investment and redistribution of the agricultural product. Investment was the job of one social group (zamindar or merchant), while the labour-power utilised belonged to another (the *bargadar* or *adhiar*). Economically, this separation was significant. Since the capital invested was often a loan with heavy rates of interest, it introduced the potential of expropriation (realised in the nineteenth century) within the *bargadari* relationship. This separation between the two productive functions was crucial in the distribution of the product, for the investor invariably managed to get the greatest share of the produce through the debt-servicing mechanism.

Second, in as much as share-cropping involved remuneration for labour provided, it entailed the utilisation of a notional wage, located in the distribution of the agricultural product, the terms of which were naturally regressive and biased against the share-cropper. This emerges clearly when the share-croppers were remunerated in cash (itself a significant commercial fact) in place of the product in kind. In Burdwan and Bishnupur not all share-croppers' contracts were produce-based. Some were paid 'a certain amount of money calculated upon the price which the same article [the crop sown] may bear in the adjacent markets'.[114] Francis Buchanan's description of similar arrangements in Rangpur is revealing. Here, under a system called *khorakee*,

> the usual money advanced is 6 rupees; and this is called a year's wages...; but in fact the term for which he labours always extended beyond a year, and from 1 to 10 months are added for interest, according to the proportion between the number of employers and those who consent to take service.[115]

Third, sharing the produce between the two agents did not necessarily mean a physical division of the rice heap. In most cases the product had to be sold, and the responsibility for doing so rested with the share-croppers. But the value of the share (established at the beginning of the share contract) had to be paid in cash immediately after the harvest. The resultant close connection between the market (where the produce was sold), the share-cropper and the conversion of the product into cash, is clearly an instance of the tricuspid integration (peasant-household,

[114] BRC, IOR, P/51/21/, 20 May 1788.
[115] F. Buchanan, 'Account of Rongoppur', IOL, Mss.Eur. D.75, f.112.

agriculture and trade) with the market which was highlighted above as one of the central features of agrarian commercialism in our period.

Therefore share-cropping cannot be seen solely as the legacy of a land tenure carved out by the British in Bengal through the Permanent Settlement, as has been suggested by some historians.[116] By strengthening the zamindari, the regulations of 1793 certainly intensified the exploitative relationships embedded in share-cropping; but they were in no way responsible for its genesis. Its spread during our period needs a separate explanation. Cooper notices a rise in the incidence of share contracts during the years of high food prices and recurring food shortage in the 1930s and 1940s in Bengal.[117] The case of eighteenth-century China is also significant in this context. There share-cropping became the dominant tenurial relation in the midst of a secular upward trend in the price of rice and a steady aggregate expansion in the output of food.[118] These instances are important for they indicate a connection between share-cropping and the state of agricultural prices, though this is not to imply that share-croppers emerge because small peasants are able to take advantage of an upward swing in prices by switching to share-cropping.

In fact, Keyder has suggested a contrary connection in his study of Turkey. In his view, agricultural prices (especially those prevailing in the world market) exert a crucial influence in the 'pendular motion' between independent peasant production and share-cropping. Rising food prices favour the consolidation of independent peasant cultivation (by making the terms of trade move in favour of agriculture), whereas falling prices tend to shift the pendulum in favour of share-cropping (by shifting the terms of trade away from agriculture); and this represents a step backwards into 'a lower level of subsistence...[and] a lower degree of commercialization'. Thus, 'greater market integration, commerciali-

[116] For a recent reiteration of this old view, see W. van Schendel, *Three Deltas: accumulation and poverty in rural Burma, Bengal and South India* (New Delhi 1991).

[117] A. Cooper, 'Share-croppers and landlords in Bengal, 1930-1950': the dependency web and its implications', *JPS* 10, 2 and 3 (January/April 1983), p.231.

[118] Compare T.J. Byres, 'Historical perspectives on share-cropping', in ibid.; Yeh-Chien Wang, 'The secular trend of prices during the Ch'ing period (1644-1911)', *Journal of the Institute of Chinese Studies* (Hong Kong) 5, 2 (1972), p.231; James Lee, 'Food supply and population growth in Southwest China, 1250-1850', *Journal of Asian Studies* 41, 4 (August 1982), p.739ff.

zation and possibly expansion in output correspond to the entrenchment of an independent peasantry, while declining commercial opportunity and subsistence orientation implies [sic] that sharecropping gains ground against independent peasantry'.[119]

The chief problem with Keyder's 'pendular' shift is that it presumes the absolute autonomy of agricultural production and tenurial relations. It is impossible to conceive of commercialised peasant production (independent or otherwise) without markets or trading intermediaries or landed proprietors placing numerous barriers between the raw output from the peasant's field and its monetary value from the movement of prices. It is also quite feasible that the intervention of the world market will not alter this relationship in any fundamental way as the huge gap between local prices (the price of procurement) and the world market-prices is usually mopped up by the landed and mercantile intermediaries. Bengal's jute in the late nineteenth and early twentieth centuries is a classic example of peasants losing out on prices for a commodity for which there was a large global demand.[120] The central thrust of this paper is in fact opposite to the one offered by Keyder, for it is argued here that share-cropping in eighteenth-century Bengal was part of agrarian commercialism and expansion in output, and that this occurred in the midst of rising food prices.

Agricultural prices and the dynamics of agrarian commercialism

The price history of eighteenth-century Bengal (and most of India) is largely non-existent. A common orthodoxy is to argue that Bengal underwent a sharp rise in agricultural prices through much of the seventeenth century (largely owing to the import of bullion by the European traders), which was followed by a protracted deflation in the late eighteenth century because of the British conquest and the drain of silver from the province.[121] One trouble with this explanation is that it com-

[119] Caglar Keyder, 'The cycle of sharecropping and consolidation of small peasant ownership in Turkey', *JPS* 10, 2 and 3 (January/April 1983), pp.130ff.

[120] For a discussion see Saugata Mukherjee, 'Some aspects of commercialization of agriculture in Eastern India, 1891-1938', in A. Sen *et al.*, *Perspectives in the Social Sciences* 2: *Three studies on the agrarian system in Bengal, 1850-1947* (Delhi 1982); also for a recent overview of the vulnerabilities faced by a jute producers from the vicissitudes of the market, see Bose, *Peasant Labour and Colonial Capital*, pp.52-63.

[121] Irfan Habib, *Caste and Money in Indian History* (Bombay 1987), pp.35-8.

pletely ignores the price situation in the first half of the eighteenth century. The other problem is that it has been developed on very inadequate data.

Figure 2 (p.129) provides the unweighted index of the average prices of common rice and sugar in Bengal between 1700 and 1800, and is based on the data compiled and interpolated by Akhtar Hussain from the ledger and account books of the Company.[122] Despite the numerous crests and troughs there is a fairly clear long-term upward trend in price. The average increase in the price of rice was about 250 per cent during these 100 years, while sugar prices were inflated by a relatively modest 190 per cent. The increase in price appears to have accelerated from 1740 and reached its peak between the 1760s and 1790s, precisely at a time when large amounts of specie were being drained out of the province. Thus, instead of having a deflationary effect, the drain of wealth from Bengal seems to have been accompanied by a sharp inflationary spiral. The other significant feature of Figure 2 is the very pronounced increase in the price of rice in comparison with that of sugar. This development exerted a strong influence on the commercialisation of rice production.

The significant increase in the price of rice, particularly during the second half of the century, can also be seen from Figure 3 (p.130) showing the unweighted index of common rice prices in Calcutta between 1754 and 1800 computed from the series of agricultural prices in Calcutta provided by W.B. Bailey in the *Asiatick Researches* (1816). Common rice in the city seems to have undergone an average increase of about 219 per cent in the 46 years between 1754 and 1800; and once again the sharp inflationary swing between the 1760s and 1790s is clearly visible. Moreover, the provincial tendency for rice prices to escalate at a faster rate than those for sugar is also visible from Figure 4 (p.131; based on data provided by W.B. Bailey), which shows the comparative price indexes of common rice and jaggery in the city of Calcutta between 1754 and 1800.

It would thus appear that the rise in prices in the eighteenth century may have been for reasons other than the ingress and egress of bullion in the economy. In any case, and despite some vague statistical evi-

[122] A.S.M. Akhtar Hussain, 'A quantitative study of price movements in Bengal during the eighteenth and nineteenth centuries', unpublished PhD thesis, University of London, 1977, pp.277-8.

dence, the role of bullion in the seventeenth-century price increases is far from clear, as the exact transmission mechanism (of bullion to inflation) remains unexplained. In fact, a detailed study of Bengal's economy in the seventeenth century has indeed shown the negligible connection between the price line and the imports of bullion by the Dutch V.O.C., suggesting instead increases in output and employment as the two other likely outcomes of these imports.[123]

The fact that rice prices grew at a much faster pace than the price of sugar, over most of the eighteenth century, can be taken as an indication of the state of relative prices in the cash-crop (sugar) and the food-crop (the so-called subsistence) sectors; and this would (at least logically) have two related outcomes: (1) that the inter-sectoral terms of trade in agriculture would tend to move in favour of rice; and (2) that a greater effort would now go into the production of rice and its marketing.[124] How this was achieved would differ from region to region as well as from household to household, but one can broadly identify two trends. The first is the switching-over trend, and the second we may call the spreading-out trend.

Dharm Narain's analysis[125] of the impact of price movements on agriculture shows that a perfectly rational productive choice for a peasant (operating in the context of a differential increase in the prices of agricultural produce) would be to endeavour to make the most of the available resources by switching to crops which were relatively the most profitable, even if it meant abandoning (for varying degrees of time) the higher income from a relatively high-value but high-cost crop. In the context of the two crops being discussed, the switch from sugar-cane to rice could profitably be made as rice was easily cultivable, and at a much lower cost, on land meant for the cultivation of sugar-cane, while the reverse was not always feasible because of higher initial and operating costs. Therefore it would not be unreasonable to see peasants

[123] This is striking indeed, because in Bengal up to the early years of the eighteenth century, the Dutch were the main traders who paid for their purchases overwhelmingly in bullion; Om Prakash, *The Dutch East India Company and the Economy of Bengal, 1630-1720* (Delhi 1988), ch.2.

[124] It should also be stressed that the prices of commodities like opium and, especially, indigo are not being considered in this discussion: despite being cash crops, they were cultivated by the peasants under compulsion, and the prices offered for them by the Company were mostly dictated and arbitrary.

[125] K.N. Raj *et al.* (eds.), *Studies in Indian Agriculture* (Delhi 1988), ch.1.

preferring to switch from sugar-cane to paddy cultivation on their lands, at least in some localities of the province. This indeed happened in Purnea and Mymensing.[126] The fact of switching over, even on a limited scale, cannot be taken to be a reversion to subsistence agriculture (de-commercialisation) or a perversity of economic choice. However, switching over on an extensive scale could have emerged as the general trend only in the context of an actual decline in the price of sugar. Given the fact that prices of the two actually rose, but at different speeds, the efforts to extend the cultivation of rice could be sustained only under the spreading-out trend.

It is not often realised that in a favourable land-labour environment, that is, a situation of excess land to labour (like that of Bengal in the eighteenth century), rising agricultural prices do not necessarily encourage efforts to raise the intrinsic productivity of an existing peasant holding. Equally plausible responses are 'the greater use of land through bringing more land under cultivation and the use of more capital in the form of land improvement'.[127] There is overwhelming evidence to show that the eighteenth century was characterised, among other things, by a wave of agricultural reclamation in Bengal. Though the pace of reclamation was faster in some areas (for instance in the eastern and south-eastern parts of the province), it was of general significance. It was predominantly financed and activated by local landed proprietors, by the holders of charitable grants, and by the grain merchants, through an elaborate system which comprised the direct provision of agricultural loans and the grant of leases of cultivation (*pattas*) on time-bound concessional terms of revenue. The latter mode of encouraging the extension of land was limited to landed proprietors, and the advancing of loans to needy cultivators was a device adopted principally by the grain traders, who were also the people who seem to have profited most from the bargain.[128]

[126] BRP, IOR P/72, 8, 29 October 1792, and P/71/10, 26 November 1792.

[127] J.W. Mellor, 'The subsistence farmer in traditional economies', in Wharton, *Subsistence Agriculture and Economic Development*, p.219.

[128] See Datta, 'Rural Bengal', ch.1 for a general discussion of agricultural reclamation in this period; Datta, 'Agricultural production, social participation', notes.60-92, for references to Company documents on reclamation of land, and pp.87-9, 92-102, for the participation of landed proprietors and merchants in reclamation. For a detailed discussion of the role of landed proprietors in the agrarian economy, see Datta 'Rural society and the structure of landed property: zamindars, ta'alluqdars and *la-kharaj* holders in eighteenth-century Bengal', in

Conclusions: agrarian commercialism in eighteenth-century Bengal

A recent study suggests 'dependent commercialisation' as a typological categorisation of Bengal's economy during the late eighteenth and early nineteenth centuries. Basing itself on the model of dependent commercialisation developed by Philip C. Huang for northern China,[129] this study characterises dependent commercialisation as the control exercised by the intrusive 'foreign' merchant capital which brings the 'agricultural production process firmly under its sway but stops short of capitalist accumulation and consolidation of land', which in the case of Bengal was brought about by the Company-engineered integration of Bengal's textiles with the world market.[130]

There is undoubtedly some justification in seeing a degree of dependency in the case of Bengal's textiles in the eighteenth century because of connections with the world market, but one must also remember that the Company's investment in cotton seldom cornered more than ten per cent of the output from Bengal. The remaining 90 per cent was traded and consumed within the country, and the Company had no control over the structure of local trade.[131] The evidence on cotton also shows that there was a decline in its cultivation within the province through the eighteenth century, and that districts producing cotton cloth depended on a very 'large importation from the banks of the Jumna and the Dakhin'.[132] For instance, the cotton industry of Nadia required 16,800 maunds of de-seeded cotton per year. Of this, local procurement amounted to 4,000 maunds, while the rest (12,800 maunds) was imported via Patna, Mirzapur and Surat.[133]

Therefore, as suggested earlier in this paper, the dynamics of agrarian commercialism, at least in the eighteenth century, become explicable if we look at the entire process from the standpoint of a regionally-

---

Peter Robb (ed.), *Agrarian Structure and Economic Development: landed property in Bengal and theories of capitalism in Japan* (Occasional Papers in Third World Economic History no.4, London 1992).

129 Philip C. Huang, *The Peasant Economy and Social Change in Northern China* (Stanford 1985).

130 Sugata Bose, *Peasant Labour and Colonial Capital*, pp.42-4; quotation from p.44.

131 Datta, 'Rural Bengal', pp.254-5; also Datta, 'Merchants, markets and subsistence crises', pp.88-9, for the Company's failure to control internal trade.

132 H.T. Colebrooke, *Remarks*, p.84.

133 BRP, IOR, P/72/10, 15 June 1789.

integrated market. The evidence of cotton also suggests that the integration of the market went much further than the provincial level; the existence of a comprehensive interregional nexus in agricultural produce was a symptom of this integration.

The developments in rice production in Bengal were extremely significant for four reasons. First, rice alone amounted for more than 40 per cent of the total agricultural output, and developments in that sector would naturally have a major bearing on the agrarian economy. Secondly, the prices of rice escalated at a faster rate than the prices of sugar (a cash crop) through the eighteenth century. This had a significant effect on the inter-sectoral distribution of crops in favour of rice. Third, the response of the cultivators was in tune with the state of relative prices. Greater effort seems to have been expended by cultivators upon the cultivation of rice both in absolute terms (that is, by reclaiming new land), and in relation to cotton and sugar-cane.[134] Fourth, new tenurial relations emerged in the eighteenth century as a response to an increase in output and the prices of agricultural produce. These included share-cropping and the jotedars at the other end of the tenurial scale.[135]

A reinterpretation of the existing notions of the divide between commercial and subsistence agriculture is perhaps necessary in the light of such developments. Dependence on the cultivation of rice does not interfere with the growth of petty commodity production and spread of 'commercial' farming in a small peasant economy. This is amply demonstrated from the case study of Bengal in the eighteenth century. Evidence from China and Japan lend support to this generalisation.[136] To a large degree the face of rural Bengal in the eighteenth century was shaped by the critical intermeshing of agricultural production with internal demand, the market and agricultural prices. The intervention of the state, especially in the second half of the century, intensified these interlinkages. While it is difficult to assign a descriptive label to this process, the fact that there was a silent transformation of Bengal's rice-producing economy in the eighteenth century appears indisputable.

[134] See Datta, 'Rural Bengal', pp.229-58, for a discussion of patterns of agricultural production in the province.

[135] For a discussion of the share-cropping and jotedari in this period, see Datta, 'Agricultural production, social participation', pp.83-7, 89-92.

[136] Francesca Bray, 'Patterns of evolution', passim.

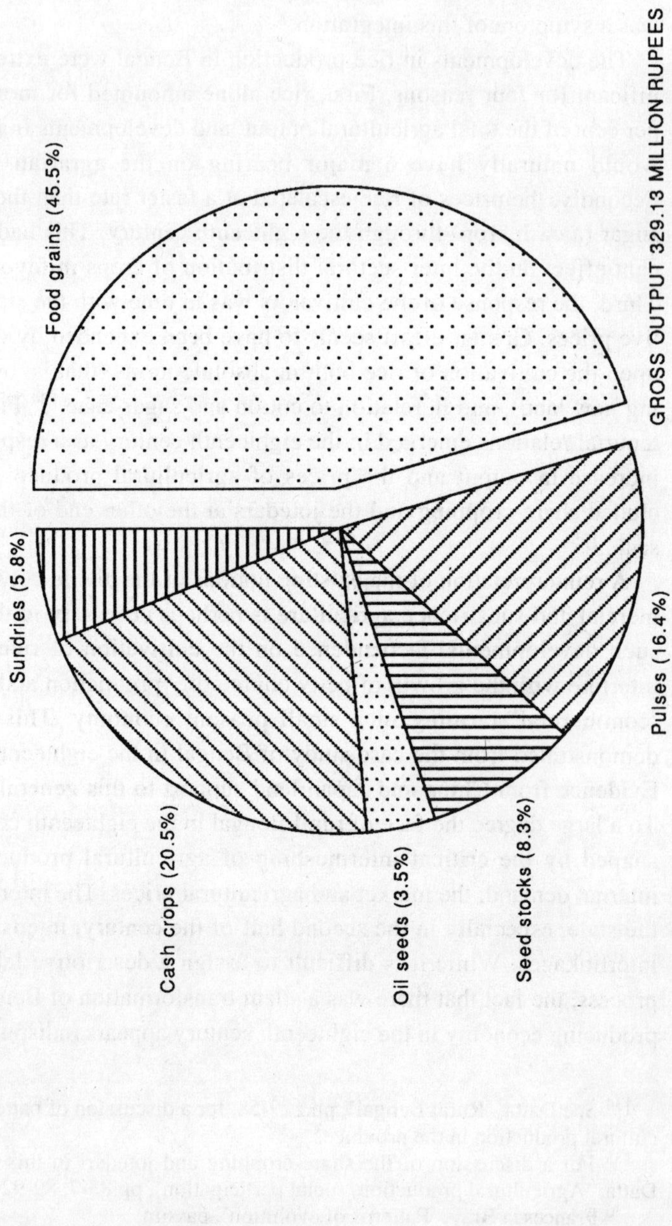
Food grains (45.5%)
Pulses (16.4%)
Seed stocks (8.3%)
Oil seeds (3.5%)
Cash crops (20.5%)
Sundries (5.8%)
GROSS OUTPUT 329.13 MILLION RUPEES

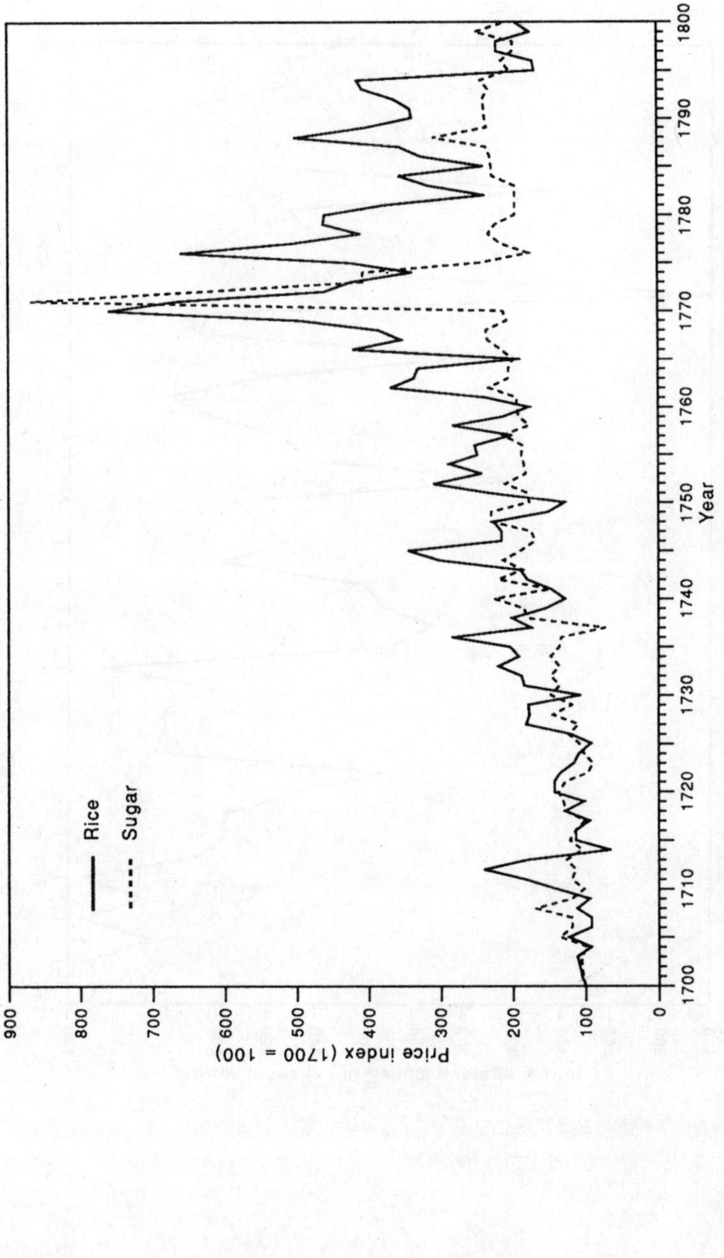

Figure 2. Average prices of common rice and sugar in Bengal, 1700 to 1800
(unweighted index; 1700 = 100)

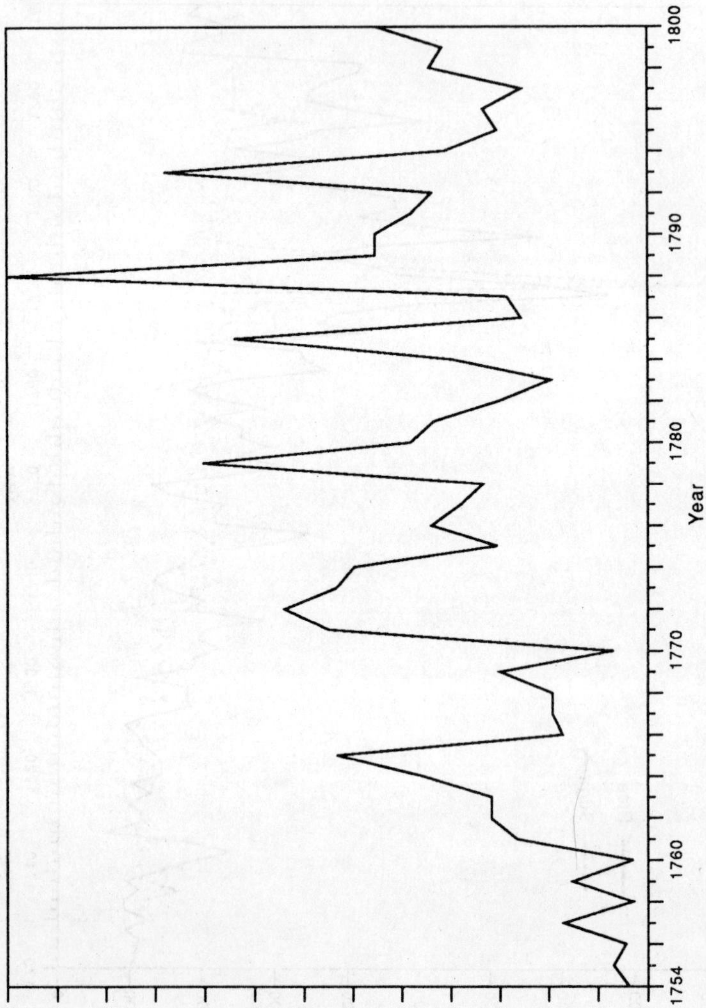

Figure 3. *Common rice prices in Calcutta, 1754 to 1800*

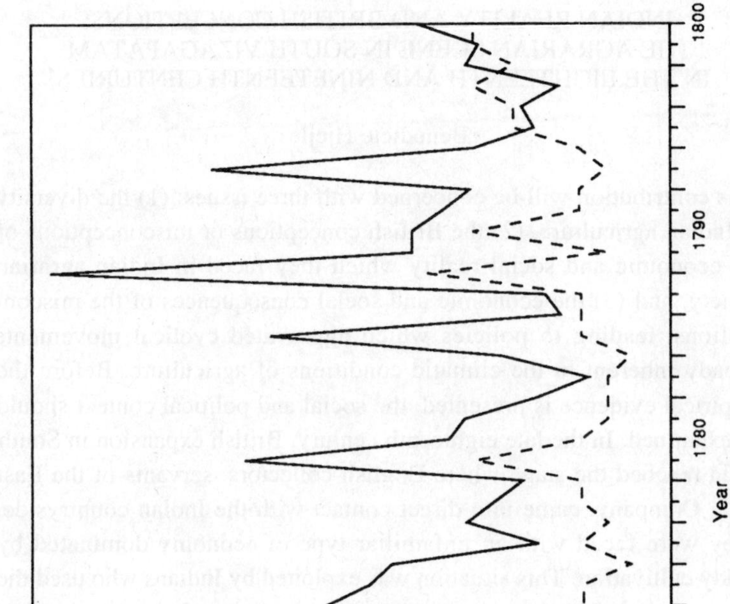

Figure 4. Comparative price indices of common rice and jaggery, Calcutta city, 1754 to 1800

Chapter 5

INDIAN REALITY AND BRITISH CONCEPTIONS: THE AGRARIAN SCENE IN SOUTH VIZAGAPATAM IN THE.EIGHTEENTH AND NINETEENTH CENTURIES[1]

Benedicte Hjejle

This contribution will be concerned with three issues: (1) the diversity of Indian agriculture, (2) the British conceptions or misconceptions of the economic and social reality which they faced in Indian agrarian society, and (3) the economic and social consequences of the misconceptions, leading to policies which aggravated cyclical movements already inherent in the climatic conditions of agriculture. Before the empirical evidence is presented, the social and political context should be examined. In the late eighteenth century, British expansion in South India reached the stage where English collectors, servants of the East India Company, came into direct contact with the Indian countryside. They were faced with an unfamiliar type of economy dominated by paddy cultivation. This situation was exploited by Indians who used the figures in the old revenue accounts to put in claims for ownership and control of land on which they had formerly only possessed a right to a share in the produce. By manipulating the accounts they added to their influence and income. They were defending the interests of the local community against the revenue claims of the British collectors, but they incurred a risk of undermining old practices and customary law. The British conception of an agrarian economy was one of reasonable stability, limited fluctuations in harvest produce, measurement of area, and legally provable ownership of the land. They had difficulties comprehending the complexities of Indian agriculture. The old Indian revenue system was geared to heavy fluctuations in harvest produce, and a sophisticated system of crop distribution existed between the partici-

[1] The general account of the political and administrative history of South India has been drawn from a larger work in preparation on the history of the Vizagapatam district in the eighteenth and nineteenth centuries. Detailed references have been avoided as taking up inappropriate space. The location of the statistics are recorded in Tables 5-8. I should like to acknowledge the courteous cooperation of the India Office Library and Records in London.

pants in the work process and various levels of the social order. Indian advisers came to understand British conceptions long before the British were able to comprehend Indian reality. As a result, a flexible model of the agrarian economy was created which could be utilised in various ways to minimise the pressure of taxation.

It should be remembered, however, that a false picture of reality carries a danger. If and when it becomes enshrined in official thinking it may act as an obstacle to revision and reform when unforeseen forces of development occur. The long-term effects of fluctuations in population or changing monsoons are not easily dealt with in a world of make-believe. Evidence will be presented showing how the model was created. By the time of Lord Wellesley's permanent settlement (of the revenues) in 1802, it reached a stage of perfection which secured the Brahman community a safe and increased income as an economic base from which to expand their influence. Their gains were offset by the losses suffered by the warrior elite, who played an important role as agrarian entrepreneurs under the old regime.

The model was used and re-enacted at different times and for different purposes according to local circumstances. It should therefore be analysed in a regional setting. An important point about the disclosure of the model is the more realistic picture of the paddy economy which emerges when the agrarian accounts are re-examined. The cultivated area in the villages can be divided into safe, uncertain, and marginal lands which correspond with different demands in capital and labour. The havoc of floods and droughts re-emerges, and the fluctuations in harvest produce can be studied. This bears directly on the subject of the 'meanings of agriculture'. Such meanings are easily obscured in a colonial economy where intermediaries hide the reality.

The evidence chosen is taken from the villages and towns along the Sarada river in the Southern Division of the Vizagapatam district. In the second half of the eighteenth century most of the district came under the immediate supervision of the Pusapati family, who held sway over the Vizianagram kingdom. The relationship between the Pusapatis and the British began as an alliance which secured the independence of the kingdom until the British army arrived in 1794 to take control of the district. This meant a sudden change from Indian to British administration. Comparisons can therefore be made which are more difficult to attain where a political unit has been subject to a prolonged period of dual administration during a transfer of power.

The primary data on which the paper relies were presented in two surveys made about the time when Lord Wellesley's permanent settlement was introduced in Madras. They were both detailed, but cannot be taken at their face value. The purposes for which they were presented have to be analysed and understood, and the figures compared with contemporary information. In the instructions which were forwarded to the collectors by the Madras government prior to the investigation (in October 1799), two points were stressed which did not escape the intelligent observation of the Brahman informants and interpreters. One was that uncultivated land (which had formerly been cultivated) would be freely donated to the proprietors who bought the estates which were to be created out of the former *hawili* and zamindari land which had escheated to the British government in Madras. Secondly no measurements of the village lands should be undertaken. To these instructions Wellesley personally added the order that no inquiry should take place into the *inam* tenures, this complicated question being referred to the new law courts which were shortly to be established. They and only they would be permitted to decide the legality or illegality of the tenures. This order came out only after some information had been collected. As a result of the instructions nearly all cultivable land in the villages was included in the surveys which were conducted prior to the settlement whether in fact it was cultivated or not. If not, it came at the lower end of the income scale, providing us with some useful information about the amount of marginal land.

The division between *sarkar, sanad,* and *mirasi* land was a headache to all British collectors, who never spent enough time in the villages to grasp the principles according to which the division was made. *Sarkar* land was the fully assessed land divided into plots and cultivated by resident cultivators with full rights to their holdings, provided they paid their dues to the government. *Sanad* land was land held mostly by resident Brahmans attached to the local pagodas. It was held on *sanad* (legal documents which showed the nature and extent of the donation). It was cultivated by the resident cultivators on favourable terms, while the *sanadars* paid a quit rent to the government. The *mirasi* land was land which was allotted to village servants such as craftsmen (smiths and carpenters), service castes (barbers, tailors, and washermen), message-bearers, overseers and controllers. This land was not particularly valuable, but provided the holders with a chance to participate in the blessings of a superior harvest. In a somewhat different category was the land held by *karnams* (accountants), and *saraffs* (money-changers). The plots were small, but apparently fairly well situated. A quit rent was also attached to the *mirasi* land.

The early collectors came to understand only one essential feature of the paddy economy. Where the produce fluctuated, the harvest was divided between peasants and tax collectors in shares, different amounts according to the quality of the land. On good land cultivated by resident cultivators the peasants received one-third of the produce, while the government received two-thirds. On land at some distance from the irrigation works where more labour was required, the peasants received two-fifths and the government three-fifths. If non-resident cultivators were called in they received one-half of the harvest produce. The fact that land was granted on tenures confused the collectors and made them think of the *sanad* and *mirasi* lands in terms of area measures with set and unalterable quit rents. In general they failed to realise that the quit rents were also measured out as shares of the produce. The failure to understand how the *sanad* and the *mirasi* land was incorporated within the village economy led to some costly mistakes.

The British collectors in the Northern Sarkars were made responsible in about 1794 for the collection of revenues, which immediately began to dwindle. Clearly, the warrior elite which participated in revenue collections, irrigation maintenance, and the grain trade withdrew. The British collectors had to rely on the inferior administrative service, and attempts were immediately made to safeguard the interests of the local community against the demands of the alien rulers. A favoured measure was a misrepresentation of the village accounts, where *sanad* land with its easy revenue burden was expanded at the expense of the fully assessed *sarkar* land. As the revenues dwindled and the financial position of Madras grew more desperate, a measure of relief became urgent. If a permanent settlement was carried out, the collector would be able to withdraw from the local scene, and others with more accurate information could be placed between local society and the British government.

But a sum of money had to be settled for, distributed and agreed. As a final service the collectors were to collect the information on which the distribution was to take place. After six years of experience in revenue collections, they remained ignorant of basic information concerning *inam* tenures. There were, however, Indian advisers at their disposal who had learnt a great deal about British conceptions during the six years which had passed.

The instructions issued to the collectors prior to the introduction of the permanent settlement stressed that no measurements of land were to be taken. As a rough measure paddy lands can be divided into safe, uncertain, and marginal lands, according to their position within the irrigation system. The yields differ even in years of plenty. During a bumper harvest an acre (a measure which corresponds almost exactly with the local *visam* used in Vizagapatam) produced from 1.5 to 1 *garce* (hereafter 'gc') on the most favourably situated land, while more distantly situated fields produced from 0.9 to 0.75gc, and marginal lands produced from 0.666 to 0.5gc. These differences could be exploited, more certainly so because any long drought would lead to no crop at all on the last two categories. Thus poor *sanad* land could be exchanged for good *sarkar* land. The only problem about this construction was that poor *sanad* land did not exist, as only the best land within the village would be donated to the Brahmans. A solution to the problem was not beyond the intelligence of Indian administrators; it will be described shortly. It was expressed in a survey of 'a good year' (after the repair of the irrigation works) which they handed over to collector Brown, who willingly and proudly forwarded it to the revenue authorities in Madras.

The strange and interesting point about this survey is that it offers a perfect sketch of the distribution of village lands and the division between safe, uncertain, and marginal lands. As we pass on to the situation of *sarkar*, *sanad*, and *mirasi* land something seems to have gone wrong. We are dealing with a sophisticated game of musical chairs. It is clear that the situation of different categories of land had changed. Under these circumstances can the evidence provided in the surveys be used? The answer is yes. They provide a quite reliable picture of the economic opportunities of the villages, and, if the game of musical chairs can be unravelled, the proportionate division of the land between *sarkar*, *sanad*, and *mirasi* can be reconstructed with reasonable accuracy.

A model of paddy cultivation

Before we look into the actual accounts, the basic features of the paddy economy should be described. This is done in Table 1 below (the tables begin on page 162). We are looking at a small *kasba* town where 881gc of paddy is produced on 1,103.33 *visams* of land. The area is divided into safe, uncertain, and marginal lands according to its proximity to the irrigation channels, which is reflected in the area yield. This means that 540 *visams* are situated so close to the channels that the fields are certain of receiving a regular and timely supply of water; 263.33 *visams* are situated at some distance from the channels, where the fields are likely to be irrigated if the water supply is abundant or can be supplemented from local tanks, while 300 *visams* are situated at the tail end of cultivation. In years of abundance they will carry a paddy crop, but if the local rain fails to supplement the water in the channels they will have to do with a dry crop. As the paddy frontier is continually moving, the division between safe, uncertain, and marginal is not as rigid as shown below; but a rough distinction would be made and known by inhabitants, accountants and revenue collectors. Adjustments could be made, but the distinction would be useful.

The second division is that which was made between *sarkar*, *sanad*, and *mirasi* land. *Sanad* land was incorporated with *sarkar* land in the same fields and they were situated at the top level, where the harvest was safe and high yielding. One-quarter of the safe land was alienated and given on *sanad* to the local Brahmans. When the harvest took place and the crops were measured, a quarter of the crop went to the *sanad* holders. Out of this they paid a quit rent (*kattubadi*) of one-sixth of their share to government, and a half of their share to the peasants who cultivated on their behalf, keeping one-third of their share for their own consumption. When the peasants cultivated their own holdings situated in the *sarkar* lands they received only one-third of the produce, handing over two-thirds of the paddy to the revenue collector. They gained a larger share on the cultivation of *sanad* land than they were allowed on *sarkar* land.

Out of the 520gc produced on the safe lands shown in Table 1, the peasants received 195gc, the *sanad* holders 43.33gc, and the *sarkar* 281.67gc, corresponding to: *sarkar* $(3/4 \times 2/3) + (1/4 \times 1/6) = 13/24$; *sanad* holders $1/4 \times 1/3 = 1/12$; peasants $(3/4 \times 1/3) + (1/4 \times 1/2) = 3/8$. Out of 24 heaps of paddy, 13 would be allotted to the government, 2 to the *sanad* holders, and 9 to the peasants.

The *mirasi* land was generally situated close to the village under conditions of uncertainty and more limited yield. The *mirasi* holders paid a *kattubadi* of a quarter of the produce to government; but they seem to have managed cultivation by family labour so that they could keep three-quarters of the produce for their own consumption. As we pass on to the marginal lands the efforts of cultivation increased, and the reward diminished, and the burden of taxation was reduced. If cultivation was carried on by peasants with full cultivation rights they were permitted to keep two-fifths of the produce, and if peasants were called in from neighbouring villages to take on cultivation they were allowed one-half of the produce. It was important to keep all the village land under regular cultivation, and, in order to secure the necesssary investment, the tail-end lands were offered to those who could arrange for its cultivation. This could be undertaken by cultivators from neighbouring villages at half shares as already mentioned, but the land could also be cultivated by the village *karnam*, or by the renter employed to collect the revenues.

If we attribute the revenues of the *kasba* town according to custom, the collections would amount to the sums which are shown in Table 1.2, the commutation rate being set at Rs.30 per *garce*. The accounts seldom look quite so simple. In most towns and villages paddy production was supplemented by dry crops and garden cultivation, and in the border area dry and wet cultivation were intertwined. The simplicity of the share system was obscured by the manner in which the revenues were imposed. On the paddy lands *anchanas* (measurements) were taken, but depending on the security or insecurity of the water supply they were carried out in two different ways. On the secure land the *anchanadars* (generally consisting of some experienced peasants and one or two representatives of government) agreed by a simple review of the crops in the field and no actual measurements were taken. Then a commutation price was agreed with due regard to the market price. If the harvest eventually exceeded the amount on which the agreement was made an extra cess could be imposed, but this was not wantonly done. When fields were removed from the immediate neighbourhood of the main channels, or where fluctuations in harvest produce were known to be the rule rather than the exception, the measurement took place only after the paddy was harvested. But the real headache to people who were not familiar with the local scene was the position of the *sanad* and *mirasi* land.

How a misrepresentation could be made is given in Table 1.3 below. It required the accounts of a bumper harvest to be incorporated in a survey modelled on the crops produced during that season, succeeded by the accounts of a good but less abundant season incorporated in a second survey modelled on the crops then returned. By moving the position of the *sarkar*, *sanad*, and *mirasi* land, the *sanad* land could be expanded in area, while the *mirasi* land could be moved to a safer position, and the area of *sarkar* land contracted. The area distribution would be altered while the produce remained the same. When the area figures had been established, a second survey could be undertaken, backed up by the produce figures of a good but less abundant season. The second survey figures could be arranged so that a new and larger proportion of the safe land went to the *sanad* holders, and the *karnam* land at the tail end could be moved up to a higher yield and safer position. The result is summarised in Table 1.3, but will be more easily understood with reference to figures A to D in Tables 2.1-3.

The produce figures of the bumper harvest correspond to those of Brown's survey (described in the next section) while the area figures have undergone some significant changes. There is a considerable gain to the *sanad* holders while *sarkar* land has shrunk. The small loss in *mirasi* area was well compensated by its position. As we move on to the less abundant season, the loss in produce (amounting to 92.5gc) on *sarkar* land is seen as closely connected, with roughly half of the produce at the 0.6gc level being out of cultivation, and with 30gc of the *sarkar* share at the tail end at the 0.5gc level. In the second survey, however, the produce of the good but less abundant season is distributed across the full area so as to cover the uncultivated land; only 37.92 *visams* have been chopped off. Incidentally they come close to the original *karnam* cultivation. There is one essential feature which should be noticed. The harvest which differed from season to season was measured out and the principle of proportionate distribution could not be dispensed with. It was ingrained within the paddy economy, and a redistribution, even one which was formally expressed in area measures, had to be worked out in proportions of produce to make sense from a practical point of view. During the bumper harvest we find the crop which was produced on safe land divided between *sarkar* and *sanad* in the proportion of 3:1. In the second survey, by Alexander, when the new dispensation was achieved, the *sarkar* and *sanad* ratio equalled 1:1. It is easier to understand the details with reference to figures A to D in Table 2.1-3. As long as no cadastral survey was undertaken, no collector would have been able to check the accuracy of this type of information. What is expressed in the figures could hardly have been challenged without access to an area survey. Ninety years were to pass before anything approximating a cadastral survey was carried out. The figures would have looked convincing enough to a British collector with reference to the accounts presented.

Four kasba *towns on the Sarada river*

We should now examine some actual data. When Wellesley's orders concerning the introduction of a permanent settlement arrived in the provinces, the collector, William Brown, was responsible for the revenues of the Southern Division of Vizagapatam district. He had spent about four years in the district, first as assistant to collector Snow, who proved a failure, and then, after a break of one year as secretary to the Madras Board of Revenue, he returned to the district to initiate the new policy of village settlements favoured by the Board. Brown worked hard, but the revenues declined even in years of abundance. In December 1799 Brown decided to carry out a major enquiry of the resources of his division before submitting his recommendations to the Madras authorities. All *karnams* were called to his *kachari* on a particular day and asked to bring the accounts for the last fourteen years. Among the accounts which were handed in, those for *fasli* 1197 turned out to be particularly full and detailed. In his final report Brown therefore showed two sets of figures, one which gave the average produce of fourteen years, and another which gave the produce of 'one good year' after a thorough repair of the irrigation works. The last set of figures will be called collector Brown's survey. The produce figures were clearly modelled on those of *fasli* 1197. In his recommendations to the Board of Revenue, Brown suggested a revenue about half-way between the average of 14 years' revenue and the revenue worked out in the 'one good year' survey. In September 1800, Brown left Vizagapatam, for an appointment as collector of Ganjam, in which district he proceeded to carry out another survey on similar lines. Brown was succeeded by Robert Alexander, who had served with the Chief and Council at Vizagapatam for more than ten years. Shortly after his arrival, Alexander told the Board of Revenue that he considered collector

Brown's survey too high; and he was invited to make his own recommendations. In May 1801 he set out on a tour of the Southern Division and in 1802 he forwarded an extensive report on his researches. In the appendices to his report details are shown of the *sarkar* area under cultivation and the revenues which dry and wet lands were supposed to yield. To the total sum of *sarkar* revenue Alexander added the *sanad* and *mirasi kattubadis*. The *mirasi kattubadis* equalled those of the best year he had observed, most likely corresponding with *fasli* 1210, and the *sanad kattubadis* were said to have been worked out as an average over ten years. No information was provided about *sanad* and *mirasi* area, but Brown's area figures remained unchanged with minor modifications on *sarkar* land. If we compare the revenue figures for *sarkar* land in Alexander's survey with the figures for *fasli* 1210 in the estate records, they come close enough to suggest that his survey figures were modelled on the figures in the renters' accounts for *fasli* 1210 which he searched during his tour. It is the connection between the surveys made by Brown and Alexander which we are going to examine.

Produce and area figures for four *kasba* towns on the the Sarada river are given in Table 1.4 below. Anakapalle was a provincial town which served as the residential seat of the Pyakarao family, who managed the zamindari of Anakapalle and Sathiavaram until the death of Appala Pyakarao in October 1776, when the zamindari came under the influence and administration of Sittaram Raju of Vizianagram. It was situated about 20 miles west of Vizagapatam. In 1799 the number of inhabitants was reckoned at 9,000. In terms of irrigation Anakapalle was extremely well-placed. Most of its irrigation came from the Sarada river by means of sluices which made irrigation safe and abundant. In addition to this, Sittaram Raju constructed a reservoir connected to the Anakapalle lake. There was some danger, however, that this could overflow. In years of shortage it had the advantage of producing a *pyra* (spring) crop. The town also possessed a large area of marginal land capable of producing black paddy.

Munagapaka was situated four miles south of Anakapalle. It contained about two thousand inhabitants with many castes and professions represented. There was a score of gentry households, 78 Brahman households and nine *racavar* households, representatives of the principal warrior caste in the Northern Sarkars. Munagapaka received its water supply from the Krishnamaraju channel, just south of Anakapalle, and from the Anakapalle lake. There was also a major reservoir within the boundaries of the town capable of containing a large amount of water, but apparently empty in the dry season.

Kondakarla was situated three miles south of Munagapaka at the Kondakarla lake, which acted as a reservoir for the Krishnamaraju channel. It contained about 1,200 inhabitants with several Brahman and *racavar* households. Some of its paddy land suffered from flooding, which was caused either by an overflow from the lake or from an overflow in some of the Munagapaka channels. Some of the dry land was poor and suffered from regular drought. Cultivation in Kondakarla was more vulnerable than in Anakapalle and Munagapaka.

Melupaka was situated on the Sarada river about nine miles southwest of Anakapalle. The inhabitants totalled 1,236, with a number of Brahman and *racavar* households. It was mostly irrigated by the Rallagedda, which was a continuation of the Krishnamaraju channel. But it was vulnerable to flooding when the river banks caved in, a quite frequent occurrence. About two miles north of the town the Mamidivakagedda ran into the Sarada river. This acted as an advantage during some seasons, while it caused damage in others. Melupaka was equally subject to flooding and drought.

The produce and area figures for the two seasons and the two surveys which should be compared are marked in Tables 1.4 and 2.2-3. The produce figures for *sarkar* land in *fasli* 1197 and in collector Brown's survey correspond in Anakapalle and Kondakarla, with some discrepancies in Munagapaka and Melupaka. The figures for Anakapalle come very close to the theoretical model presented and exemplify it. It is more difficult to interpret the figures for Munagapaka. Here much of the area of adjacent dry and wet land remained uncultivated even in good seasons. When a bumper harvest occurred, the land could be cultivated by those who had the means—that is, the *sarkar* (generally meaning *paikaris* invited in by the headmen), renters, or village *karnams* employing their own labour force. In collector Brown's survey the *sanadars* and the village *karnams* probably wanted to make certain that their rights were recorded and accordingly put down their claim for cultivation done in that year by the *sarkar*. They explained the difference between the *sarkar* produce in *fasli* 1197 and the survey figures by a deterioration in a stone sluice at the opening of the Krishnamaraju channel. But what was lost to the *sarkar* was gained to the *sanadars* and *karnams*. In Melupaka there was also some difference this time in favour of the *sarkar* and to the loss of the *sanad* holders. The loss was compensated on the dry lands, while the *sarkar* gain in area turned out an illusion owing to flooding. Looking at the area figures we find that the area cultivated by the *sarkar* had shrunk in Munagapaka, Kondakarla and Melupaka.

As we move on to a comparison of the produce figures in *fasli* 1210 and in collector Alexander's survey we find that they are closely related, except in Kondakarla where 80gc were flooded but nonetheless included in the survey accounts under *sarkar* produce. The difference of 5gc in Anakapalle can be explained by the transfer of sugar cultivation from dry lands to wet lands in Alexander's survey. In the figures here quoted they have been returned, but paddy is likely to have been grown on some of the sugar land, explaining the addition of the five *garce* to the wet produce. Something similar had happened at Melupaka. The actual distribution of the revenues in Alexander's survey between various yield levels do not correspond with the yield levels in Brown's survey, but the final figures compare only too well. A correspondence of the yields was clearly out of the question if the final area figures were to correspond with the area figures in Brown's survey.

The connection between the two surveys is revealed in the area figures for *sarkar* land. They are very close in Anakapalle, while some *sarkar* land was written off as land out of cultivation in Munagapaka, Kondakarla and Melupaka, where they were added to *sanad* and *mirasi* land to correspond with the new distribution. An important point about the new distribution was that it was proportionate. The stacks of grain could still be divided in terms of shares.

The figures shown for the four *kasba* towns on the Sarada conform with the theoretical model, but the true significance is revealed by a study of the figures in Tables 5.1, 6.1, 7.1 and 8.1. The figures bring out the details and show how the land was moved about, creating a new proportional division between *sarkar*, *sanad* and *mirasi* land.

If we look at the land distribution in Anakapalle in *fasli* 1197, and compare it with the land distribution in Brown's survey, we can see how the *sanad* land is moved down while the *mirasi* land is moved up. When the new area distribution was expressed in Alexander's survey the *sanad* land had doubled in the high-yielding area, while the *mirasi* holders had taken over half of the land at the level of medium yield.

The *sarkar* was left with a large majority of the low-yielding and marginal land. The *sanad* gain was substantial. There is one point, however, which cannot be explained merely by looking at figure D. If we are thinking in terms of shares it is unlikely that the *mirasi* holders should have taken over land formerly held by *sanadars*, except perhaps on the edges of safe cultivation. According to figure D, 74 *visams* of land, at the yield level of 0.833gc per *visam*, producing 61.67gc, are put down as *mirasi*. An explanation can be offered if we look at the dry lands. Here the *sanad* holders in collector Brown's survey claimed 102.5 *visams* of dry land. When this area was transferred to collector Alexander's survey of dry land they produced 70.08gc. We should remember that the *sanad* holders were Brahmans who did not participate in cultivation, and had no interest in dry land, which required a different type of tenant. We can therefore infer that some swapping was intended so that the *mirasidars* could take hold of the dry land while the *sanad* holders made arrangements with their tenants for the cultivation of the wet land. The *sanad* holders almost doubled their income and moreover paid no *kattubadi* on their gains. In Alexander's survey the *kattubadi* was worked out at Rs.1,439, which was a moderate increase on the Rs.1,352 paid during the bumper harvest. The *mirasi* land was moved to a safe position, and a substantial part of it held ready for a later transfer to high-yielding dry land. All vulnerable land was left to the *sarkar*. The revenues settled in Alexander's survey would be realised only during a bumper season, while 217gc would be lost if the marginal lands went out of cultivation.

The same trends can be observed in Munagapaka. The *sanad* holders had doubled their area and their produce, while the *sanad kattubadi* decreased from Rs.575 in Brown's survey to Rs.551 in Alexander's. The *mirasi* land was moved up and the produce almost doubled. This would not have been possible if Brown had not been persuaded by his *kachari* staff that a permanent decrease had occurred in the *sarkar* produce owing to a failure in the water supply. In other words, this was embodied in the survey and paved the way for the more favourable distribution of the shares. As a result the *sarkar* was left with 125gc on uncertain lands and 144gc on marginal lands.

As we pass on to Kondakarla we have to deal with a new complication. The town was subject to flooding, and flooding had clearly occurred in *fasli* 1210. The flooded land was registered as *sarkar* land in the survey, and the 80gc which it should have produced was retained in the accounts. They were unlikely to be realised unless extensive repairs were undertaken. The *sanad* holders were compensated for their losses on flooded land by the larger proportion of land and produce which came their way. The replacement of *sarkar* land by *mirasi* land at the yield level of 1gc per *visam* is not fully in tune with the model and should be examined. It is likely to have been a temporary measure. The *sarkar* was to bear the brunt of uncertainty in terms of inundation and drought, while the *sanad* holders and *mirasi* holders remained in a safe position. If and when the sluices were repaired, the *sanad* holders were likely to claim 136 *visams* producing 156gc, and the *mirasi* holders 20 *visams* producing 17gc, while 24 *visams* were restored to the *sarkar*. The *mirasi* holders would be generously compensated for their losses on wet land by their gains on the dry land.

The same pattern existed in Melupaka too. This time 61.5 *visams* of *sarkar* land were merely written off by Alexander, corresponding to 20 *visams* at the top level, where flooding had occurred, and 41.5 *visams* at the level of 0.667gc per *visam*. The 60 *visams* of *sarkar* land at the top level, which had been flooded, were included in Alexander's survey although they carried no crops, leading to a general deterioration in yield level. The *sanad* and *mirasi* holders were transferred to safe lands and all the marginal lands, amounting to 146gc, were left to the *sarkar*.

The figures for flooding in Kondakarla and Melupaka are reconstructions made on the assumption that the figures in Alexander's survey conformed to the model. But how reliable are they? Flooding is continually mentioned in the local reports, the danger points along the Sarada can be studied in the maps, and in later years they are confirmed in the papers of the irrigation department. Moreover flooding had happened in *fasli* 1205 where the renters' accounts are available so that confirmation can be sought. When the flooded land was shown to collector Alexander during his visit of the district, he must have accepted the information that the flooded land was *sarkar* land. The device of adding the area without the produce, thereby reducing the yields to a much lower level, shows the confusion created by the concentration on area size rather than produce.

Dry cultivation

Although paddy cultivation was the major agricultural activity along the Sarada, dry crops and garden cultivation provided a useful addition to the resources. They were included in the surveys and should be examined. Collector Brown's survey of dry lands and garden cultivation is almost as accurate as his survey of wet lands in terms of yields and crop composition, with one minor reservation. If we look at the survey and compare it with the records of the bumper harvest in *fasli* 1197, there is a contradiction which can, however, be explained. On garden land, sugar was the most valuable produce. It was a labour-intensive crop and one which was regularly shifted between the fields as it exhausted the soil. During a bumper harvest there was a shortage of hands and sugar cultivation did not reach its maximum level. The *karnams* registered all the sugar land in the survey, which meant that the increase in produce had to be compensated by a reduction of yield on the ordinary dry land. At the same time, a claim was made to all cultivable land in the village, which meant that marginal land was included and recorded as bearing low-yielding dry grains such as *allu* (*paspalum scrobiculatum*) and *chamalu* (*oplismenus trumencatus*). Despite these increases, the comparison with *fasli* 1197 works. It is not too difficult to explain. The difference between a good season and a less favourable season is reflected in the different proportion of the two ordinary dry crops like *chollu* (*eleusine corocana*) and *gantelu* (*penicellaria spicata*). In a good season more *chollu* will be grown as it requires more water; in an ordinary season *gantelu* will prevail. High-yielding *chollu* is generally grown on wet *ragada*, while *gantelu* is grown on ordinary *ragada.* If the higher yield produced during 1197 (*chollu* rather than *gantelu*) is transferred to the sugar lands and the low-yielding dry grains, the total income on dry land would correspond to the survey figures in Brown's survey. *Allu* and *chamalu* were generally sown on what was called *konda garuvalu*. The area was clearly exaggerated in the survey accounts. If we keep this in mind there is little reason to challenge the area figures and the yields mentioned. We should note, however, that the *mirasi* and *sanad* lands were with a few exceptions placed at the marginal level, with yields which corresponded to that of *allu* and *chamalu*. They were added as a fictive claim on all cultivable land in the village. It is inconceivable that Brahman *sanad* holders should hold land producing low-yielding dry grain. The manner in which the dry land was taxed made manipulations in the accounts easy. A *kaul* (cultivation agreement) was issued to the peasants at the opening of the season, stating the area they were going to cultivate and the revenues they were supposed to pay. This sum could, however, be increased if the harvest became abundant. In the accounts this was expressed in terms of *sistu* (first assessment) and *malavati* (additional assessment).

The final produce figures of *fasli* 1197 were used to back up the claims expressed in collector Brown's survey of dry land and garden land, while a careful reading of the figures suggests the reservations already expressed. The dry produce of *fasli* 1210 was short owing to an unfavourable early season with late showers, but it was well suited for the type of arithmetic used in Alexander's survey. The diminution of the revenues was considerable. The connection between the two surveys reminds us of the connection already shown on the wet lands. The *mirasi* and *sanad* lands were moved from the margin to a safer position, and all marginal land left to the *sarkar*. The figures in the two surveys are shown in Table 4.1-2. The lands are classified as safe, uncertain, and marginal.

In Anakapalle and Munagapaka it is fairly easy to see what has happened. *Sanad* and *mirasi* land was moved up from the marginal levels to a better position and *sarkar* land was moved down. In Kondakarla the dry crop in *fasli* 1210 was too small to cover the land which the *kachari* servants wanted registered, and some uncultivated land was included in *sarkar* land bringing the yield down further. In Melupaka dry produce was down at approximately half level and the model was made consistent by writing off 24.75 *visams* of land.

Before we leave the question of the model on which the surveys were arranged, we can point to a simple equation. What was lost to the *sarkar* was gained to the holders of *sanad* and *mirasi* land. If we deal with the towns and villages on the upper Sarada, we can use this equation to show exactly how much revenue was lost to the *sarkar* during a bumper harvest. It is hardly necessary to look at the detailed distribution of land. We can simply subtract the final revenue figure in Alexander's survey from the figure in Brown's survey. As we pass on to the towns and villages in the middle reaches of the river, which were more easily exposed to flooding and drought, we must be more careful as some lands which had gone permanently out of cultivation or which required extensive repairs were incorporated in Alexander's survey. Where this is the case we have to draw up the figures of land distribution to show how much produce was lost to the *sarkar* and how much was gained by the holders of *sanad* and *mirasi* land.

There is one curious fact which should be mentioned. When the Commission for the permanent settlement looked at the surveys conducted by Brown and Alexander, they ended up by choosing an average of 14 years' revenue as their guideline for the final sum to be levied. This does not detract from the importance of Alexander's survey. The figures came close and Alexander's survey presented a breakdown of the land revenue distribution at village level. It spelled out the intentions of Brown's area distribution.

Except for a more general statement on the sources, detailed references have been left out. The documents from which the figures have been extracted are fully quoted in Tables 5 to 8. These contain the *kham* accounts for *fasli* 1197, the figures extracted from the estate records in Brown's survey, and the revenue distribution shown in Alexander's survey. Alexander showed the revenues not the produce, but the produce figures can be reconstructed on the assumption that the principle of customary shares was fully respected. The extraction of the produce figures for the *kasba* towns and villages in Brown's survey is time-consuming, and minor mistakes may have been made in the figures, particularly in the size and situation of marginal land, but they are unlikely to challenge the overall picture.

We have chosen the *kasba* towns rather than the large Brahman-dominated villages in the neighbourhood of the Krishnamaraju channel, as the renters' accounts for *fasli* 1197 and 1205 were included in Brown's research material for the *kasba* towns only. It is nonetheless possible to reconstruct the division of lands in the large paddy-growing villages. The irrigation was safe enough to secure a large surplus produce and one-third rather than a quarter of the safe land was given on *sanad* during the pre-British regime. The new distribution created in Alexander's survey altered the proportion of *sanad* land from one-third of the safe lands to half. This happened in the villages of Nagulapalle, Vompolu, Totada, Jaggannadhapuram, and Mulajampa.

Land and labour

We should now try to relate the agrarian scene to the social order in the villages. Who were responsible for the revenue collections on *sarkar*, *sanad*, and *mirasi* lands? In general, renters were appointed to make the collections on behalf of the Vizianagram administration. They were men of substance who undertook the revenue collections of a small region for a series of years. They were granted a commission of ten per cent on the revenues, and, if the collections increased under their management, they would be offered a larger area next time or a more favourable rent. If they took up neglected lands and made the necessary investments the land remained tax free for a number of years. They were *racavaru* or Brahmans attached to the Vizianagram court. They used to put in their sons or other family members as subrenters at vulnerable points in the irrigation system or where close supervision was required. In *fasli* 1197, Anakapalle was held by a Brahman renter while Munagapaka and Dimila were held by members of the Vatchavoy and Dantla families. These belonged to the top *racavaru* attending the Vizianagram *darbar*. That they were men of substance and credit is clear as they paid the first instalment of revenues in the early season long before any grain had been harvested. Their position as grain dealers was equally important. They were always prepared to receive the *sarkar* share of the revenues in kind. They had the grain delivered at the ports where they kept godowns. They owned ships and traded with Madras. By their command over the grain-pits they are likely to have prevented heavy speculation during harvest fluctuations. What we are dealing with is what is commonly characterised as a customary-command economy. The revenues were obviously high but they were levied by people who were themselves agriculturalists. The *racavaru*, even those who attended the *darbar* with great regularity, resided in villages where they had a home farm. They cultivated large tracts of land and held large herds of cattle.

In the records the peasants are often called 'cultivating inhabitants' ignoring the distinction between *kadims*, *kadim paikaris*, and *paikari* strangers. In Brown's research material the division is shown for *fasli* 1209, when the inquiry was made. For most of the towns and villages Brown also mentioned the names of the *pedda kapus*, who were the headmen in charge of the *kandams* (subdivisions) in the village. The *pedda kapus*, who supervised the cultivation of the paddy lands, were influential and generally managed to retain the best lands, including the *sanad* holdings. How and by whom the *kadim paikaris* were employed is not explicitly stated in the documents, but their presence generally coincides with the presence of *racavaru* in the village rolls. They were resident in the villages. The *paikari* strangers were cultivators who came in from neighbouring villages to cultivate for the season.

An attempt should be made to compare cultivation figures in Brown's survey with the number of ploughs kept by *kadims*, *kadim paikaris*, and *paikari* strangers. If we work on the assumption that each plough produced four *garce*, correlation seems possible. Four *garce* is obviously a rough measure, only reasonably accurate where there was a positive land-labour ratio. After three years of famine (1790-2) and war (1794) the labour force was much reduced. Brown's survey was made in *fasli* 1209, and the labour force was recorded for that year. The season was favourable on wet land as well as dry. The comparison is made between safe, uncertain, and marginal land in the *kasba* towns, and the number of ploughs quoted in collector Brown's survey. The result is shown in Table 4.2.

In Anakapalle most of the marginal land was out of cultivation. The figures show a reasonable correlation. Uncertain land could be managed by *kadim paikaris* while *paikari* strangers could deal with what was left on the safe lands and the small amount of marginal cultivation. In Munagapaka a minor proportion of marginal land was apparently out of cultivation. The *kadims* required some help in order to deal with the produce on safe land. Some of the land which had been registered as safe in accordance with Brown's survey was probably liable to flooding. One of the hamlets under Munagapaka was named Timmarajupeta. It was situated in a favourable position south of the town at a spot where we know from later reports in the irrigation department that flooding occurred, but the yields were high. The figures for Kondakarla suggest that all marginal land was out of cultivation, as was that which remained flooded when Alexander worked out his survey. A large share of the labour force was drawn in from the neighbouring villages. This implies that uncertainty depended largely on seasonal fluctuations. The number of *kadim paikaris* was limited, but large enough to deal with the flood-ridden area. In Melupaka only a minor percentage of marginal land remained in cultivation. The correlation between different categories of land and different categories of the labour force is reasonably close. Complete harmony between the figures cannot be expected as the labour force could be influenced by factors other than the availability of different categories of land, but a connection between the two can hardly be questioned.

Finally we should consider the character of the agrarian economy described in the survey figures. The two towns situated at the upper reaches of the Sarada river (Anakapalle and Munagapaka) had a flourishing economy with a large proportion of safe and high-yielding produce. Uncertain and marginal land existed but constituted a minor proportion. During a fortunate season it would be cultivated but mainly by *kadim paikaris* or *paikaris* brought in from neighbouring villages. Further down the river or the channels derived from it, uncertainty increased and flooding and drought were regular visitors. More than half of the produce in Kondakarla was obtained from uncertain and marginal land, and the produce in Melupaka was equally precarious. The income lost by neglect of the irrigation works was considerable. Famine and civil war had preceded the surveys, and it is remarkable that the *kadims* and the number of ploughs at their disposal correspond fairly closely with the land characterised as safe. In other words we can distinguish between a safe core of land with a stable produce and labour force and an outer circle of uncertain and marginal land with a mobile labour force which nonetheless made a substantial addition to the overall income. While the investments in the repair of feeder channels, in cattle and in agricultural implements could be managed by the village *kadims* on the safe lands, the risks to be borne on uncertain and marginal lands are likely to have been out of their reach. The uncertain land can be seen as an additional and regular source of income to be safeguarded by careful attention to the irrigation works and timely provision of *takavi* and *maramat*. The *racavaru* seem to have been the men who were ready to perform this duty. With a mobile labour force at their disposal, seedgrains, cattle and small sums of money, they could keep up cultivation, and benefit by the commission they gained on the revenues. As uncertainty increased so did the importance of *racavar* assistance. If we study the village rolls we notice their presence in towns like Kondakarla and Melupaka. In the lower delta lands where uncertainty reached its peak, their services were indispensable. It should be noted, however, that the contribution they paid to the agrarian economy depended on close supervision at the local level. They had to be present in the villages in order to perform their tasks. Their services could not be copied by a bureaucracy resident in the district headquarters. The danger inherent in the model of agrarian economy which was presented to the British administrators lay in its failure to highlight the long-term potential of the uncertain lands. If generally overassessed, they would be neglected and eventually run to waste; if properly assessed and managed, they were likely to contribute both to the revenues and to the general welfare of the region.

The sequel

When British collectors were made responsible for the revenues in the Northern Sarkars in the late eighteenth century, the Board of Revenue in Madras hoped to dispense with the services of the *racavaru*. The Board considered their commission exorbitant and favoured a settlement with the village headmen. In the Southern Division of Vizagapatam the collision between the *racavar* chiefs and the collector was particularly sharp. The *racavaru* were not prepared to collect the revenues on the conditions demanded by the collector, and as a result the Board of Revenue recommended that the raja should be expelled from his district along with the top *racavaru* who were refusing to co-operate. These events led to the battle of Padmanabham where the raja was killed together with a large number of warriors who remained true to their oath of fidelity. The collectors were left to deal with a power vacuum and a peasantry who reluctantly agreed to pay the revenues at a reduced scale. After a short trial with an unreliable *kachari* staff the collectors were eager to re-employ the *racavaru*, to the displeasure of the Board of Revenue. Brown was persuaded to continue a system of village settlements. In financial terms they spelled disappointment, while Alexander succeeded in raising the revenues when he reemployed the *racavar* renters during his short collectorship before the introduction of the permanent settlement. In general, however, there was little communication or understanding between the men who had served under the old regime and the Company servants. The British employees at the outports had relied on their *dubashes* and on their commercial agents and such habits were not easily broken. How vital these links were can be seen in the Southern Division. When the new estates created under the permanent settlement were eventually sold, the rich estates on the upper Sarada (the Munagapaka and Anakapalle estates) went to the Gode family, who acted as *dubashes* to the Chief and Council at Vizagapatam; the Kasimkota and Melupaka estates went to Alexander's *dubash* (the Karumanchi family); and the Dimila estate in the delta lands south of Melupaka went to the Brahman family (the

Baghavatulas) who acted as advisers and interpreters to Brown during the 1799 survey. This did not happen immediately, but within some ten years of the settlement.

That a model of the agrarian economy was created in accordance with British notions does not mean that it was immediately adopted. Old arrangements about cultivation rights and responsibilities could not be changed overnight. The first thing which happened was that all the estates were bought by the Raja of Vizianagram and sold by him when unfortunate harvests made it impossible for him to pay the revenues. This gave the small clique of Indians who knew English and acted as advisers to the British on all levels, an opportunity to watch the new system at work. What deals they made at local level when the estates were sold can only be guessed at; but when some of the estates reverted to British administration between 1831 and 1844 we can watch the new model in operation. Before we look at the revenue collections and the distribution of *inams*, the fate of estates which remained in Indian hands should be explored.

The Gode family managed extremely well due to their continued friendship and services to the British administrators. But their encroachments at local level were deeply resented. About 1830, the result was open rebellion, when they took over some of the estates in the southern extremity of the districts which had always been held by *racavaru*. The assistant collector reported that it was hatred of the Gode family which had sparked it off, while the collector, who had been away on a short visit to Madras, hurried to testify to the general good record of the Gode family as enlightened landlords. But the Gode family house in Anakapalle was burnt down and British troops had to defend their grain-pits, facts that obviously favour the first explanation. The Karumanchi family, who interpreted for Alexander, were threatened with bankruptcy in the late 1840s, but they were saved by the collector, who did not want to incur further responsibilities in the district. The estates already on hand gave trouble enough. The Bhagavatula family, who had lent support to Brown, was eventually forced out of the Dimila estate, although they had arranged for an English 'man of straw' to hold the estate for some years.

When the more vulnerable of the estates reverted to British administration, the renting system continued for a few years until collectors Freese and Arbuthnot began to experiment with a raiyatwari system.

They wanted to levy the revenues in money rather than shares, and they hoped to deal with raiyats individually and not through the *pedda kapus*. At some stage Arbuthnot even tried to impose the revenues directly on fields (a *khetwari* settlement), but he had no accurate information about their size, the quality of the land, and the means of cultivation available to the raiyats, and so he was unable to dispense with the advice of his *kachari* servants. The result of these efforts was described by collector Smollett in April 1842 in his *fasli* report for 1251. He explained that remissions were granted even in favourable years. The problem was that the *atukabadi*, or first settlement, was made by the taluk officers who distributed the land among the raiyats without reference to their means of cultivation. This led to heavy deductions when the final settlement was made. The raiyats could be divided into three groups: substantial, poor and extremely poor. The substantial raiyats paid only for the lands which they occupied. The poor raiyats, with too much land imposed upon them, were unable to pay their share of the revenue and an inquiry had to be instituted. This meant that the produce was measured. When the government share was worked out and converted into money, further remissions were granted as the commutation rate of paddy was well above ordinary selling price.

Other shortcomings were revealed in Arbuthnot's report for *fasli* 1252. No more revenue than that arranged at the early settlement was ever demanded. This meant that the substantial raiyats paid a limited amount, on the safe and secure fields which the *taluk* officers had allowed them to take up, while the poor raiyats were left with a high assessment on the fields which were visited by the uncertainties of the seasons. They could then apply for remissions, which were often granted merely to prevent their emigration. The *taluk* officers played an important part in the game, and even Arbuthnot realised that the government was being deceived,

> for notwithstanding all the precautions and efforts of a collector the tendency on part of the curnams and village servants will ever be to keep up the nominal revenue, by imposing too high an assessment on the lands in the occupation of the poorer ryots, so as to prevent a new distribution of the assessment, by which the assessment of lands in the occupation of the richer ryots would have to be realised.

In the last resort the assessment on the fields was very uneven, and was settled merely with reference to the amount fixed when the land was under zamindars or proprietors.[2]

This story was not confined to Vizagapatam. All over the Northern Sarkars collectors were up against difficulties. This was shortly to coincide with unfortunate seasons, and in Madras the Board of Revenue began to wonder whether something was seriously wrong. In the years from 1844 to 1855 some gifted servants on the Madras establishment began to examine the situation in earnest and work out fundamental measures of reform. Their endeavours were immediately thwarted by older members of the establishment, but they eventually prevailed and by 1855 the Madras council was dominated by the reformers. The reform measures then proposed are set out in some detail in the proceedings of the Madras government of 14 August 1855. A regular settlement and survey of the Presidency was recommended along with an inquiry into the *inam* tenures. What seems to have worried the councillors more than anything else was the escalating emigration of peasants and labourers to Burma, Ceylon and Mauritius, while good cultivable land in the Presidency was left out of cultivation. This did indicate that the burden of revenue was too high and that reduction and redistribution were overdue. Something had to be done to bring Madras out of its state of depression. The urgency with which the measures were recommended was closely linked with the administrative failures experienced in the Northern Sarkars.[3]

The reform policies suffered a heavy setback after the Sepoy Mutiny. It alarmed the authorities in London to such an extent that all reform measures were whittled down and occasionally turned into their opposites. This is particularly true about *inam* land, where titles were confirmed to the *sanad* holders and *karnams* without proper inquiry. The result can be studied in the final tables of the Inam Commission, which reported to government in 1859. They show an enormous amount of *sanad* land in all districts of the Northern Sarkars. The figures for Vizagapatam, Rajahmundry, and Masulipatam are shown in Table 3.

Vizagapatam heads the list for alienated lands while large areas had

[2] Madras Board of Revenue Proceedings (hereafter MBRP), 29 October 1840, and 12 December 1842. *Fasli* reports for *fasli* 1249 and *fasli* 1251.

[3] Madras Revenue Proceedings (hereafter MRP), 14 August 1855. Minutes by Lord Harris, 26 October 1854, Sir Henry Montgomery, 12 May 1855, and Walter Elliot, 8 June 1855.

also been granted in the districts of Bellary, Tanjavur, and Madurai (bearing an assessment between four and five lakhs).[4] A careful reading of the reports on the permanent settlement in the other districts of the Northern Sarkars suggests that the expansion of *inam* tenures was not confined to Vizagapatam. When collector Brown left the Southern Division, he went off to Ganjam to settle the revenues of that district under the permanent settlement. He made use of almost the same model, and the *sarkar* and *sanad* lands were settled in area measures. When the Inam Commission arrived half a century later they had to deal with an intricate problem. The Committee of Circuit, which tried to explore the resources of the Northern Sarkars from 1784 to 1788, began their researches in Vizagapatam and ended up in Ganjam, where they finally grasped the nature of the *kattubadi* paid by the *sanad* holders. The *kattubadi* related to a particular share of the harvest, while the holdings of the *sanadars* corresponded with a particular share of the village lands. The Inam Commission now had to make a choice between the figures presented by Brown and those of the Committee of Circuit. They chose the figures presented by Brown.

Brown's efforts to gain an insight into the resources of the districts on which he was reporting were equalled by those of collector Skinner, who was appointed to the second division of Masulipatam, and collector Read, who was responsible for the fourth division of Masulipatam. Skinner worked out the distribution of the revenues of his division under the permanent settlement according to principles similar to those adopted by Brown, and he was helped on his way by a former renter working for Sittaram Raju.[5] The research material presented by Read differs slightly. He showed the extent to which each estate was improvable. He gave the average valuation of the estates for five years preceding the famine and five years subsequent to the famine of 1790-2, but the size of the arable land was shown as that which had been cultivated in the most favourable seasons within 19 years (*faslis* 1194, 1195, 1197, 1198, 1204).[6] The model created in Vizagapatam was apparently

[4] India Office Library and Records (hereafter IOL), Official Publications, V/23/192, Report of the Madras Inam Commission, 1859, pp.344, 348.

[5] MRP, Proceedings of the Special Commission on the Permanent Settlement. Collector Samuel Skinner's researches are recorded in volumes 75 and 76.

[6] MRP, Proceedings of the Special Commission on the Permanent Settlement. Collector Read's researches are recorded in volume 73. His explanations

copied with minor deviations in the other districts of the Northern Sarkars. The game of musical chairs must, however, be looked upon as a continuing process. In 1844, when Montgomery wrote his report on the Rajamundry district he noticed how good *sarkar* land was continuously transferred to *sanad* land.[7]

The design presented above was not solely responsible for the poor utilisation of agrarian resources. Two more factors played a role. One was the strange and ineffective manner in which the irrigation system was run. To go into the details of this would require a separate paper, but two or three points should be mentioned. Although mid-Victorian civil engineering was greatly superior to any constructions which the *racavaru* could manage, it was badly supervised, slow in repair, and unconcerned with the actual utilisation of water resources. Local quarrels ended up in the courts, and decisions were taken with no concern for the interests of the peasants who cultivated at some distance, and the cultivators in the lower reaches of the irrigation system suffered accordingly. Water flows could not be measured, an argument accepted at the bench, whereas any *racavar* would have mounted his horse, accepted the evidence of his eyes, and deployed a working force for an immediate repair.

The third factor which affected the local economy was the elimination of the *racavaru* as grain dealers. The large pits of dry grain, which were kept as a reservoir for misfortunes like monsoon failures, are likely to have disappeared with the contraction of dry grain cultivation. The *racavaru* seem to have acted as a buffer between the producers and the market. They were always prepared to accept revenues in kind, and their calling was a mixture of production and exchange, while that of most merchants was solely exchange. When cultivators were exposed more directly to market fluctuations their position became more vulnerable.

Conclusion

Attention has been paid to the manner in which the Brahman interpreters and *kachari* servants manipulated the accounts to safeguard the interests of their community. The procedure adopted has been exam-

---

concerning the principles followed in his review of the resources of his districts are in the opening pages of his report, pp.3873-7.

[7] MRP, 28 May 1844. Report of Sir Henry Montgomery, 18 March 1844.

ined in detail. It looks sophisticated, but at the core there was a measure of simplicity. It was easy to move from a bumper harvest to a season of more average expectations. A new collector was readily persuaded of the mistakes made by his predecessor. The device used in Vizagapatam was copied in other parts of South India for similar reasons. By the time the survey figures were accepted and acted upon by the British administration and the law courts, they affected politics as well as economics. The position of the *racavaru* was undermined and a new elite emerged. In the Vizianagram zamindari the process did not go unobserved. In a report to the Madras Board of Revenue in May 1815 the collector of Vizagapatam expressed himself in no uncertain terms about the influence of the Indians who had served at the factory during the transfer of power. He wrote:

> The persons here referred to are men of opulence, their riches have accumulated at the expense of the chief families of the country, whose place they have usurped in society, and they retain that ascendancy and their influence by the practice of every artifice, which superior wealth and the slow course of justice, place so much in their power.[8]

Collector Smith can hardly be considered a neutral observer. He belonged to a faction among the Madras servants who believed in a different role for the Company servants than that which was favoured by the old families serving in the factories of the harbour towns in the Northern Sarkars. They had operated on a network of Indian allies who were not easily dispensed with. What is interesting about Smith's comprehensive report was the early recognition of a social transformation where the influence of the old warrior elite was replaced by that of the Indians who co-operated with the British merchants and Company servants at the outposts. Among other spoils, the *dubashes* came to possess the estates of Anakapalle, Munagapaka, and Melupaka, while Kondakarla retained a *racavar* family who had agreed to co-operate with the British. But the influence of the Brahman interpreters penetrated more deeply into the social fabric. A model of society in which ownership of land was to play an important role was drawn up and evidence created which could be presented in the newly established law courts to secure claims on land rather than shares in the produce.

It is one thing to create a model for the future, and another to deal with the demands of the agrarian economy at a particular time and

[8] MBRP, 8 June 1815. Report by Collector Smith, 26 May 1815.

place. The *racavaru* had participated in economic processes at several levels. They ensured the ploughing of lands in the early season by handing out *kauls* to the peasants. They offered *takavi* and *maramat* at the appropriate moment. They kept a keen eye on the state of the irrigation works and arranged for repairs. They paid the first instalment of revenues to the exchequer, took on leases for a series of years, and traded in grain. They offered supervision at close quarters and their services must be characterised as essential. How were they to be dispensed with? Some could be transferred to the *kachari*, but not all of them; and how would the *kachari* servants act with vigour and commitment? Their position in society was not linked with their performance as agrarian entrepreneurs. They could levy the revenue-demands, but in doing so they could easily run into a contradiction. If they wanted to uphold the ideas expressed in Alexander's survey it was necessary that uncertain and marginal land was kept under cultivation, even with a poor prospect of crops. If the land remained uncultivated, the *kachari* servants were threatened with a redistribution of the revenues. To avoid this, poor peasants were forced to take up the land and have their losses written off at the end of the season. Such practices can hardly be said to be rational. More serious was the exposing of peasants to market forces. They seldom possessed the capital necessary to deal with the fluctuations in prices which had been kept under control by the *racavaru* with their large grain stocks. Market forces do not serve consumers if they are also producers of consumable agrarian produce such as grain—now characterised as wage goods—and particularly not if the grain is subject to heavy fluctuations and no capital is available for intermediate relief. The neglect of the irrigation works, which has not been a primary concern in these pages, is nonetheless likely to have been the most important single factor responsible for the deterioration in resources.

Why could the duties performed by the *racavaru* for the benefit of local society not be replaced by new services performed by the *sanad* holders in the villages? The answer must be that the Brahman pagoda servants, administrators and *karnams* were agriculturalists neither by instinct nor by tradition. They relied on an age-old alliance with the *pedda kapus*, who were quite happy to cultivate their lands as they received a larger share on the *sanad* land than they were permitted on *sarkar* lands. The cultivators who were likely to suffer from the transformation were the ordinary *kadims* to whom *racavar* cultivation probably provided a safety valve against misfortunes such as the death of ploughmen or cattle. A *kadim* could leave his village and return without losing his cultivation rights, and so he could retrieve his fortune by working for a *racavar*. There is much to suggest that the peasants who emigrated to Burma were *kadims* who had been forced out of their overassessed holdings in coastal Andhra. They were not prepared to stay in their home villages as tenants to *karnams* or Brahman landholders who had deliberately undermined their position.

If we try to evaluate a political and economic system or regime according to the manner in which it mobilises and exploits the resources at its disposal, we must admit that no system or regime is able to utilise its resources fully. Like human beings they all carry their frailties. It is permissible, however, to place regimes within scales of resource utilisation and mobilisation. The old regime, reasonably expressed in the Vizianagram warrior state, suffered from many weaknesses such as perennial warfare, internal quarrels, and fights about succession. But those who participated in government tried to carry out their task according to well-established rules, and their honour and standing in society would suffer if they failed. They were not just warriors, they were entrepreneurs and agriculturalists who knew how much could be gained by the timely repair of the irrigation work, and by minor sums of *takavi* and *maramat*. They encouraged the peasants to take up cultivation by agreements on favourable terms, and left balances to accumulate over many years. They were closely in touch with the agrarian scene and, if they thought that the misfortunes of the cultivators were matters over which they had no influence, they were often ready to help. This was particularly true in years of famine. In peasant societies there is a kind of liberality which is founded on common experiences, and this was not lacking in the warrior state of Vizianagram.

The regime founded by the East India Company and later carried on by the Crown was by any measure a most awkward creation. It was backed up by military superiority, which in the case of Vizagapatam was clearly revealed on several occasions, most spectacularly in the battle of Padmanabham in 1794. Those who allied themselves with the Company servants were merchants and civil servants in the coastal trading towns. They seem to have been successful in improving their own status and that of their community, as the Brahman *sanad* holders in the villages improved their lot. They were able to educate their sons, who within a short period manned the courts and civil services of the district. But in terms of resource mobilisation the record of the British regime looks dismal. Good dry land went out of cultivation owing to high taxation, the delta lands became flooded and drought-ridden, and the community feeling within the villages was broken by an extreme polarisation between Brahmans and *pedda kapus* on one hand, and *kadim* tenants and labourers on the other.

Table 1.1: *Summary of land distribution in a fictive kasba town*

	Sarkar		*Sanad*		*Mirasi*	
	visams	gc	*visams*	gc	*visams*	gc
Safe						
1.0 to 0.833gc	315	315	105	105		
Per *visam*	90	75	30	25		
Uncertain						
0.75 to 0.667gc	133.33	100			40	30
Per *visam*	67.5	45			22.5	15
Marginal						
0.6 to 0.5gc	210	126			*Karnam*	
Per *visam*	60	30			30	15
Total	875.83	691	135	130	92.5	60

Table 1.2. *Collections of a kasba town*

Produce on	gc. x Rs.30 =	Rs.
Sarkar land, 390gc at two-thirds to the *sarkar*	260	7,800
Sanad land, 130gc at one-sixth to the *sarkar*	21.67	650
Sarkar land, 145gc at two-thirds to the *sarkar*	96.67	2,900
Mirasi land, 45gc at one quarter to the *sarkar*	11.25	337.5
Sarkar land, 126gc at three-fifths to the *sarkar*	75.6	2,268
Produce on *sarkar* land, 30gc at one half to the *sarkar*	15	450
Karnam cultivation, 15gc at a quarter to the *sarkar*	3.75	112.5
Total revenues to *sarkar*		14,518

Table 1.3. *Summary of comparisons of different harvests and surveys*

	Bumper harvest		Survey 1	
	visams	gc	*visams*	gc
Sarkar	875.83	691	840	691
Sanad	135	130	173.33	130
Mirasi	92.5	60	90	60
Total	11,03.33	881	1,103.33	881

	Good season		Survey 2 (Alexander's)	
	visams	gc	*visams*	gc
Sarkar	711.66	598.5	802.08	598.5
Sanad	135	130	210	210
Mirasi	62.5	45	91.25	72.5
Out of cultivation	194.17	107.5	(37.92)[1]	
total	1,103.33	881	1,103.33	881

[1] *Sarkar* land written off in survey 2.

Table 1.4: *Summary information on four villages*

ANAKAPALLE					
	Bumper harvest (*fasli* 1197)		Brown's survey (1)		
	visams	gc	*visams*	gc	
Sarkar	951.50	923.65	951.5	923.25	
Sanad	422.01	381.94	422	381.92	
Mirasi	142.76	99.51	142.75	99.92	
Total	1,516.27	1,405.10	1,516.25	1,405.09	
	Good year (*fasli* 1210)		Alexander's survey (2)		Difference
	visams	gc	*visams*	gc	gc
Sarkar	776.82	821.56	948	826.75	+ 5.09
Sanad	258.93	273.86	422	468.53	
Mirasi	64	48	142.75	108.04	
Out of cultivation	416.50	(261.68)	3.5 [1]	1.7	
Total	1516.25	1405.10	1516.25	1405.07	
MUNAGAPAKA					
	Bumper harvest (*fasli* 1197)		Brown's survey (1)		Difference
	visams	gc	*visams*	gc	gc
Sarkar	749.08	709.06	651.63	640.91	– 68.15
Sanad	88.44	103.62	173.01	153.12	+ 49.50
Mirasi	73.75	56.03	86.63	74.69	+ 18.66
Total	911.27	868.71	911.27	868.72	
	Good year (*fasli* 1210)		Alexander's survey (2)		
	visams	gc	*visams*	gc	
Sarkar	538.45	568.64	639.83	566.57	
Sanad	88.44	103.62	176.88	207.25	
Mirasi	66.75	52.3	94.57	94.89	
Uncultivated	217.63	144.14	(11.3) [1]		
Total	911.27	868.70	911.28	868.71	
KONDAKARLA					
	Bumper harvest (*fasli* 1197)		Brown's survey (1)		
	visams	gc	*visams*	gc	
Sarkar	361	362.58	342	362.33	
Sanad	73	82.25	84.25	70.21	
Mirasi	59	43.33	66.75	55.63	
Total	493	488.16	493	488.17	
	Good year (*fasli* 1210)		Alexander's survey (2)		
	visams	gc	*visams*	gc	
Sarkar	306.25	298.08	249	219.17	
Sanad	57	62.25	114	124.5	
Mirasi	54	40	66	64.5	
Flooded	64	80	64	80	
Uncultivated	11.75	7.83	(26)[1]	(3)[1]	

Table continues

Table 1.4 continued

MELUPAKA

	Bumper harvest (*fasli* 1197)		Brown's survey (1)		Difference
	visams	gc	*visams*	gc	gc
Sarkar	498.88	377.42	494	386.63	+ 9.21
Sanad	77.13	71.49	83	60.67	– 10.82
Mirasi	51.5	32.03	50.5	33.67	+ 1.62
Total	627.51	480.94	627.5	480.97	

	Good year (*fasli* 1210)		Alexander's survey (2)		Difference
	visams	gc	*visams*	gc	gc
Sarkar	325.53	251.59	432.5	256.80	– 5.22[3]
Sanad	57.13	51.49	83	80.17	
Mirasi	29.5	21.03	50.5	36.33	
Flooded	80	80	(60)	(60)[2]	
Uncultivated	135.35	76.84	61.5[1]	47.67[1]	
Total	627.51	480.95	627.50	480.97	

Notes: [1] Written off. In Kondakarla, all flooded land was added to *sarkar* land in Alexander's survey, 26 *visams* were written off, and 3 *visams* cannot be accounted for. The 29 *visams* were transferred to *mirasi* and *sanad* lands.
[2] Flooded produce was included in *sarkar* produce in Alexander's survey.
[3] Transferred from dry cultivation to wet.

Table 2.1: *Paddy production on sarkar, sanad, and mirasi land in a fictive kasba town* (Figure A)

Harvest measures (gc per *visam*)	*visam*	gc	*visam*	gc	Area measures (local *visams*)
SAFE	*sarkar*		*sanad*		
1.0	315	315	105	105	420
0.833	90	75	30	25	120
UNCERTAIN	*sarkar*		*mirasi*		
0.75	133.33	100	40	30	173.33
0.667	67.5	45	22.5	15	90
MARGINAL	*sarkar*		*karnam*		
0.6	210	126			210
0.5	60	30	30	15	90

Table 2.2: *Area and produce figures for two seasons and two surveys in a fictive kasba town*— I (bumper harvest)

Harvest measures (gc per *visam*)	Figure A (bumper harvest)		Figure C Survey (1)	Area measures (local *visams*)
	sarkar	*sanad*	*sarkar*	
1.0	315	105	420	420
0.833	75	25	100	120
		mirasi	*sanad*	
0.75	100	30	130	173.33
			mirasi	
0.667	45	15	60	90
		karnam	*sarkar*	
0.6	126		126	210
0.5	30	15	45	90

Note: All figures are in *garce* except where shown. The produce figures of figure A are used to back up the area claims of figure C. When added to one another, the produce figures for *sarkar*, *sanad* and *mirasi* land correspond, while the area figures of *sarkar*, *sanad* and *mirasi* land differ, and their situation has been changed in the two figures. The *sarkar-sanad* ratio on safe land is 1:3 in figure A.

Table 2.3: *Area and produce figures for two seasons and two surveys in a fictive kasba town*— II (good harvest)

Harvest measures (gc per *visam*)		Figure B (good but less abundant harvest)		Figure D Survey (2)		Area measures (local *visams*)
	sarkar	*sanad*	*sarkar*	*sanad*		
1.0	315	105	210	210	420	
	sanad			*mirasi*		
0.833		75	25	50	50	120
		mirasi				
0.75		100	30	115	15	173.33
0.667		45	15	52.5	7.5	90
0.6		63.5[1]		126		210
0.5		(30)[2]	(15)[2]	45		90

Notes: [1] 62.5gc would be out of cultivation. [2] A total of 45gc would be out of cultivation. All figures are in *garce* except where shown. The connection between the figures is that when added together the produce figures of *sarkar* land in figure B correspond to the produce figures of *sarkar* land in figure D. When added to one another the area figures of figure D correspond with the area figures of figure C with only 37.92 *visams* loss to the *sarkar*. The *sarkar-sanad* ratio on safe land is 1:1 in figure D.

Table 3: *Inam holdings in the districts of Vizagapatam, Rajamundry and Masulipatam, recorded in the tables of the Inam Commission*

	Number of title deeds	Extent in acres dry	garden	wet	Assessment Rs.
Vizagapatam	21,915	124,997	332	94,829	665,097
Rajahmundry	47,420	151,256	5,463	66,424	499,921
Masulipatam	12,669	199,883	138	54,061	261,338

Source: IOL, Official Publications, V/23/192, Report of the Madras Inam Commission, submitted 30 October 1869, pp.344, 348.

Table 4.1: *Area and produce on dry land in Brown's and in Alexander's surveys*

		Brown's survey (1) *visams*	gc	Alexander's survey (2) *visams*	gc
ANAKAPALLE					
Sarkar	written off			(31.25)	(31.25)
	safe 191.25	279.21	109	227.52	
	marginal	142	39.73	193	57.66
	uncultivated	4		(4)	
	total	337.25	318.94	302	285.18
Sanad	safe			37	37.33
	uncertain	65.5	32.75	65.5	32.75
	marginal	37	12.33		
Mirasi	uncertain			14	14
	marginal	14	5.60		
Total		453.75	369.62	453.75	369.26
MUNAGAPAKA					
Sarkar	safe	184	226.9	232.75	234.75
	uncertain	64	32		
	marginal	15.25	6.75	30.25	11.25
	written off			(0.25)	
Sanad	safe	6	3.53	6	3.53
Mirasi	safe			5	18.99
	uncertain			10	5
	marginal	15	4.50		
Total		284.25	273.68	284.25	273.52
KONDAKARIA					
Sarkar	safe	149.5	141.97	44.25	34.11
	empty			1.75	
	uncertain	160	80	160	80
	marginal	60	22	82	29.33
	uncultivated	1.75		44	(14.67)
	total	371.25	243.97	332	143.44
Sanad	safe			22	31.42
	marginal	22	7.33		
Mirasi	safe	44	14.67	44	44
	marginal	14	5.60		
	written off			39.25	(32.44)
Total		437.25	265.97	437.25	265.97

Table continues

Table 4.1 continued

MELUPAKA				
Sarkar safe	32	17.73		
marginal	40	9.87	51	12.04
empty			5	
written off			(24.75)	(7.1)
uncultivated	8.75			
total	72	27.60	56	12.04
Sanad safe			10	6.4
marginal	10	2		
Mirasi safe			5	5.33
marginal	5	1.25		
Total	95.75	30.85	95.75	30.87

Table 4.2: *Labour force and produce per plough* (in gc)

	Distribution of produce (Brown's survey)			Distribution of labour force (fasli 1209)		Assumed produce
	dry	wet	total	ploughs (no.)		
ANAKAPALLE						
safe	279.21	1,060.42	1,339.63	(1)	321	1,284
uncertain	32.75	112	144.75	(2)	35	140
marginal	57.66	232.68	290.34	(3)	26	104
Total	369.62	1,405.10	1,774.72		382	1,528
MUNAGAPAKA						
safe	230.43	599.94	830.37	(1)	192	768
uncertain	32	124.61	156.61	(2)	43	172
marginal	11.25	144.14	155.39	(3)	44	176
Total	273.68	868.69	1,142.37		279	1,116
KONDAKARLA						
safe	141.97	249	390.97	(1)	63	252
uncertain	80	125.83	205.83	(2)	17	68
marginal	44	33.33	77.33	(3)	68	272
flood danger		80	80			
Total	265.97	488.16	754.13		148	592
MELUPAKA						
safe	17.73	160.33	178.06	(1)	40	160
uncertain		94.33	94.33	(2)	20	80
marginal	13.12	146.30	159.42	(3)	15	60
flood danger		80	80			
Total	30.85	480.96	511.81		75	300

Key: (1) *kadims*; (2) *kadim paikaris*; (3) *paikari* strangers.
Notes: The figures for peasants and ploughs are shown in the statistical tables and quoted under individual villages. Generally they correspond but in the case of Anakapalle they differ and the figures shown in the village entry have been used.

Table 5.1: Anakapalle. *Produce and area figures for two seasons and two surveys (paddy lands)*

	Figure A: Produce in *fasli* 1197				Figure C: Collector Brown's survey						
gc/v	*sarkar*		*sanad*		*sarkar*		*sanad*		*mirasi*		*visams*
	v	gc	v	gc	v	gc	v	gc	v	gc	
1.25	180	225	60	75	160	200	80	100			240
1.167	141	164.5	47	54.83	176	205.33	4	4.67	8	9.33	188
1.133	60	68	20	22.67	80	90.67					80
1.067	67.5	72	22.5	24	90	96					90
1.0	117.38	117.38	39.13	39.13			136.5	136.5	20	20	156.5
0.833	178.13	148.43	59.38	49.48	205.5	171.25			32	26.66	237.5
0.8	60	48	20	16			80	64			80
0.75							64	48			64
0.667	24.24	16.16	143	95.33	240	160					240
0.533	76.75	40.93							76.75	40.93	76.75
0.5	46.5	23.25	11	5.5			57.5	28.75	6	3	63.5

Note:s: Additional to Figure A at 0.667 level are *mirasi* 72.76 *visams* and 48.51 gc; and at 0.5 level, *mirasi* 6 *visams* and 3gc. The levels shown ('gc/v') are *garce* per *visam*. The renter in 1197 was a Brahman. He is likely to have cultivated marginal land at his own expense corresponding with low-yielding *sanad*.

	Figure B: Good year, *fasli* 1210				Figure D: Collector Alexander's survey				
gc/v	*sarkar*		*sanad*		*sarkar*		*sanad*		*visams*
	v	gc	v	gc	v	gc	v	gc	
1.25	180	225	60	75	120	150	120	150	240
1.167	141	164.5	47	54.83	94	109.66	94	109.66	
1.133	60	68	20	22.67	40	45.33	40	45.33	80
1.067	67.5	72	22.5	24	45	48	45	48	90
1.0	117.38	117.38	39.13	39.13	78.25	78.25	78.25	78.25	156.5
0.833	178.13	148.43	59.38	49.48	118.75	98.95	44.75	37.29	237.5
0.8	32.81	26.25	10.92	8.75	40	32			80
0.75					64	48			64
0.667	Black paddy, out of cultivation				240	160			240
0.533	out of cultivation				76.75	40.93			76.75
0.5	out of cultivation				31.25	15.63			63.5

Notes: The levels shown ('gc/v') are *garce* per *visam*. Additional to figure B, at 0.75 level, are *mirasi* 64 *visams* and 48gc; and to figure D at 0.833 level, *mirasi* 74 *visams* and 61.67gc, at 0.8 level *mirasi* 40 *visams* and 321 gc, and at 0.5 level *mirasi* 28.75 *visams*. Out of cultivation, in figure B at 0.8 level, were *sarkar* 27.19 *visams* and 21.75gc, and *mirasi* 9.06 *visams* and 7.25gc. The breakdown in figure D at 0.5 level was *sarkar* 15.63gc and *mirasi* 14.38gc, and out of cultivation 3.5 *visams* (uncultivated 1.75gc).

Table 5.2: Anakapalle. *Area and produce figures for 'one good year' after due repairs and improvements (collector Brown's survey) extracted from the estate records (Anakapalle Handa).*

SARKAR

Paddy lands

visams	@ gc	= gc
160	1.25	200
176	1.167	205.33
80	1.133	90.67
90	1.067	96
205.5	0.833	171.25
240	0.667	160
951.5	(Total)	923.25
(4 *visams* uncultivated)		

Dry lands

visams (crop)	@ gc	= gc
16 *korralu*, *raghi*, and chillies	1.5	24
64 *gantelu*, and lamp oil seeds	1	64
32 *raghi*, and *gingeli* oil seeds	0.933	29.67
8.75 *ionalu*, and lamp oil seeds	0.867	7.58
8 *raghi*, gram and *gingeli* oil	0.769	6.07
16 *kandulu*	0.333	5.33
32 *allu* and *chamalu*	0.333	10.67
16 *ionalu*	0.333	5.33
224.75 (Total)		163.32

Garden lands

visams (crop):	maunds @	Rs.	= gc
6 tobacco	30	180	6
8 onion	240	320	10.67
4 *bringall*	200	200	6.67
12.5 turmeric	120	750	25
8 cotton	40	320	10.66
16 sugar	205	2,186.66	72.89
8 water pumpkins		480	16
46 mango[1]		232	7.73
108.5 (Total)			155.63

[1] Mango gardens and plaintains

SANAD

Paddy lands

	visams	@ gc	= gc
	80	1.25	100
	4	1.167	4.67
	136.5	1	136.5
	80	0.8	64
	64	0.75	48
	57.5	0.5	28.75
Total	422		381.92

Dry lands

	65.5	0.5	32.75
	37	0.333	12.33
Total	102.5		45.08

MIRASI

Paddy lands

visams	@ gc	= gc
8	1.167	9.33
20	1	20
32	0.833	26.67
76.75	0.533	40.93
6	0.5	3
142.75		99.93

Dry lands

14	0.4	5.60
14	0.4	5.60

Source: Madras Revenue, Proceedings of the Special Commission on the Permanent Settlement, IOL, P/283/vol.69, pp.2428-32.

Table 5.3: *Abstract of Kasba Anakapalle Kham, Jamma Wasil Baki for fasli 1197. Renter Jaggamahanti Nursimulu.*

	Rs.	(1) gc	(2) gc
Mattu (358.75 *visams*)	4,476.84	298.39	163.32[1]
Field assessments			155.62[2]
Sugar gardens	188.75	12.58	
Mixed land	47	3.13	
Total, dry and garden	(314.10)		(318.94)
Deficit		(– 4.84)	
Pallam (597 *visams*)			
Assessed	13,284.53	663.23	
Not measured; total produce[3]		72.49	
Black paddy (18.07gc)[4]		18.07	
To *sarkar* (54.37gc)	1,630.50		
Excused: *manyamdar*[5]	– 181		
Raiyats for their loss	– 119.75	10.03	
Total	(– 300.75)		
Malavati[6]	1,495.64	99.71	
Excused for *ChentalaTamudu*	– 8		
40 *visams*[7]	800	53.33	
Right by measuring[8]	312.25	10.41	
Total wet		(927.27)	923.25
Surplus		(4.02)	
Mirasi kattubadi	558.25		
(i)	(482.5)	64.33	99.93[9]
(ii)	(75.75)	10.1	5.6[1]
(mirasi deficit)		(– 31.1)	
Sanad kattubadi	1,822.13		
(i)	(1,351.88)	270.38	381.92[9]
(ii)	(470.25)	94.05	45.08[1]
(Sanad deficit)		(– 62.57)	
(perhaps incoporated in new *manyam* and *muqasa*)		(10.03)	
(53.33)			
Village god	35		
Takavi	34.5		
Total net *daul*	24,376.53		
(Total produce)		(1,680.23)	(1,774.72)
(Difference)[10]		(94.49)	
Wasil paid to renter	21,034.28		
Remitted to Pyakarao Juggiah[11]	800		
Remitted for *manyam dumbala*	494.38		
Sadirwarid and repairing	2,012.5		
Total	24,341.16		
Balance on raiyats	35.38		

Key: (1) Revenues converted to produce according to customary shares; (2) Corresponding figures in the area layout of Brown's survey. *Table continues*

Table 5.3 continued.

Notes: [1] Dry land. [2] Garden land. [3] *Sarkar* share (45.31gc). [4] *Sarkar* share (9.03gc). [5] On account of new *manyam* this year. [6] Superassessment on above amounts at different rates. [7] On account of Pyakarao Jagga *muqasa*. [8] On account of difference. [9] Wet land. [10] New *manyam* and *pyakarao muqasa* were apparently added to *sanad* land in Brown's survey with a *mirasi* claim on cultivation at the village border. [11] For her *muqasa*.

Source: P/283/69, p.2436.

Table 5.4: *Revenue distribution according to Alexander's survey*

	Assessment per *visam* *visams*	(Rs.)	Estimated value of produce (gc	per *visam*)	Full produce (gc)
Paddy lands					
	12*	30	1.5	12	18 (part dry)
	160	25	1.25	160	200
	240	22.5	1.125	240	270
	188	17.5	0.875	188	164.5
	16	14	0.7	16	11.2
	104	9.25	0.62	104	64.13
	240	7	0.47	240	112
Total	960				839.82
Dry lands					
	40	25	1.67	40	66.67
	120	16	1.067	120	128
	28	12	0.8	28	22.4
	72	9	0.6	72	43.2
	30	6	0.4	30	12
Total	290				272.27

Source: P/283/67, p.1375. *Dry in figure

Table 6.1: Munagapaka. *Area and produce figures for two seasons and two surveys (paddy lands)*

	Figure A				Figure C				
	Bumper harvest, *fasli* 1197				Collector Brown's survey				
gc/v	*sarkar*		*sanad*		*sarkar*		*sanad*		*visams*
	v	gc	v	gc	v	gc	v	gc	
1.2	120.19	148.23	40.06	49.41	160.25	197.63			160.25
1.125	120	135	40	45	160	180			160
1.1	25.12	27.64	8.38	9.21	33.5	36.85	4	4.32	37.5
1.0	181.14	181.14			57.13	57.13	70.38	70.38	181.14
0.833	92	76.67			16	13.33	76	63.33	92
0.75		*mirasi*:	62.75	47.98	62.75	47.98			62.75
0.667	210.63	140.42			162	107.99	22.63	15.09	210.63
0.533		*mirasi*:	7	3.73		*mirasi*:	7	3.73	7

Notes: The levels shown ('gc/v') are *garce* per *visam*. There is a range for some of the levels given (gc/v): 1.2 represents a range of 1.2 to 1.233; 1.1 a range of 1.08 to 1.1; and 0.75 a range of 0.7 to 0.8. Additional to figure C at 1.0 level is *mirasi* 53.63 *visams* and gc. The breakdown of the *mirasi* (figure A) and *sarkar* (figure C) at the 0.75 level was 31.25 plus 18.5 plus 13 *visam*. Additional in figure C at 0.667 level was *mirasi* 26 *visams* and 17.33gc. Note also the following details:

sanad	8.38 *visam*	9.21gc	@ 1.1 gc per *visam*			
mirasi	4	4.32	@ 1.08	*sanad*	4 *visam*	4.32gc
mirasi	31.25	25	@ 0.8	*sarkar*	31.25	25
mirasi	18.5	13.88	@ 0.75	*sarkar*	18.5	13.88
mirasi	13	9.1	@ 0.7	*sarkar*	13	9.1

There is a minute inaccuracy in figure C (survey 1) if compared with the estate accounts quoted in Table 6.2, as 0.5 *visams* of *sanad* land producing 0.78gc per *visam* (0.39gc) has been added to *sarkar* land at the 0.75gc per *visam* level. The moving has been done to simplify the accounts.

	Figure B				Figure D				
	Good harvest, *fasli* 1210				Collector Alexander's survey				
gc/v	*sarkar*		*sanad/mirasi*		*sarkar*		*sanad/mirasi*		*visams*
	v	gc	v	gc	v	gc	v	gc	
1.2	120.19	148.23	40.06	49.41	80.13	98.82	80.13	98.82	160.25
1.125	120	135	40	45	80	90	80	90	160
1.1	25.12	27.64	8.38	9.21	16.75	18.43	16.75	18.43	37.5
1.0	181.14	181.14			90.57	90.57	*90.57*	*90.57*	181.14
0.833	92	76.63			92	76.63			92
0.75			*62.75*	*47.98*	62.75	47.98			62.75
0.667	out of cultivation				210.62	140.41			210.62
0.533	out of cultivation				7	3.73			7

Note: See above for the range of some of the levels given (gc/v). In the columns headed '*sanad/mirasi*' the values for *mirasi* are italicised, at levels, 1.0 and 0.75. The breakdown of the *visam* totals given for *mirasi* (figure B) and for *sarkar* (figure D) at the 0.75 level are as shown in the table above. Additional to figures B and D at 1.08 level are *mirasi* 4 *visams* and 4.32 gc.

Table 6.2: Munagapaka. *Area and produce figures for 'one good year' after due repairs and improvements (collector Brown's survey) extracted from the Estate Records (Munagapaka Handa).*

SARKAR

Paddy lands

visams	@ gc	= gc
160	1.233	197.33
0.25	1.2	0.30
160	1.125	180.00
33.5	1.1	36.85
57.13	1	57.13
16	0.833	13.33
31.25	0.8	25.00
18	0.75	13.50
13	0.70	9.10
162	0.667	107.99
651.13		640.53 (Total)

Dry lands

visams (crops)	@ gc	= gc
12 *korralu* and black gram	0.8	9.60
4 *raghi* and *gingeli* oilseeds	0.733	2.93
124 *raghi*, *gingeli* oil, *pesalu*	0.633	78.53
64 *gantelu* and horsegram	0.5	32.00
4.25 *allu* and *chamalu*	0.5	2.25
3 chillies	0.5	1.50
8 *jonnalu* and black gram	0.375	3.00
219.25 (Total)		129.81

Garden lands

visams (crop)	@ Rs[1]	= gc
33 sugar cane[2]	3,760	125.33

Other garden

visams (crops)	@ mds.	(Rs.)	= gc
4 tobacco	20	(80)	1.67
4 cotton	30	(120)	4
2 turmeric	60	(120)	4
1 onion	200	(50)	1.67
Total[3] 11			12.33

SANAD

Paddy lands

visams	@ gc	= gc
4	1.08	4.32
70.38	1	70.38
76	0.83	63.33
0.5[4]	0.67	0.39
22.63	0.67	15.09
173.51		153.51 (Total)
Dry lands		
6	0.59	3.53

MIRASI

Paddy lands

visams	@ gc	= gc
53.63	1	53.63
26	0.67	17.33
7	0.53	3.73
86.63		74.69 (Total)
Dry lands		
15	0.30	4.50

Notes: [1] All rupee rates are totals. [2] Comprises 9 *visams* @ 200 maunds, and 24 *visams* @ 160 maunds, producing 5,640 maunds @ 1.5 maunds per rupee. [3] There is some uncertainty in reading the manuscript—the residual amount according to the accounts was 11.28gc. [4] Transferred to *sarkar* in model.

Source: P/283/69, pp.2328-32.

Table 6.3: *Abstract of Munagapaka Kham, Jamma Wasil Baki and Jamma Kurk for fasli 1197. Renter Vatchavoy Jagganad Raju*

		Rs.	(1) gc	(2) gc	notes
Mattu (205 *visams*)[1]	*bilmakta*	3,284.50	218.96	129.81	dry produce
				12.33	garden
Surplus			(76.82)		
Kurk pampakamu					
Divided in village		1,651.81	110.12	125.33	sugarcane
Deficit			(– 15.21)		shown in figure A
Pallam (241.63 *visams*)		5,962.88	298.14		top levels
Remitted to raiyats		– 208.61	– 10.43	wet land	
		(287.71)	283.23		
Surplus			(4.48)		
Medium level				27.64	
Pallam taken in shares		6,006.67			low level wet land
(i) One third to raiyats			177.13	181.14	
(ii) Two-fifths to raiyats			75.41	76.67	
(iii) One half to raiyats			79.8	140.42	
Total wet land[2]				(709.1)	
Mattu and *pallam**	*bilmakta*	1,162.50	77.5	listed in *pallam* land	
Surplus			(11.61)		(*87.75 *visams*)
Pyra crops		74.88	7.49		
Black paddy (Rs.20 per gc)		27.48	2.75		
Moturapha		165.75			
Mango gardens		7			
Manyam[3]		85.14	17.03		
Remitted to raiyats			10		
Wastage and cooly hire		– 47.70			
Sanad manyams		500.44	100.09	157.04	
Deficit			(56.95)		
Mirasi manyams		358.25	47.77	79.21	
Deficit			(31.44)		
Over collections		(1.25)			
Interest on bonds		61.36		(excluded from survey)	
Balance of last year		55.97		(– 68.62)	(– 1.18)
Total net *daul*		19,148.36			
Total produce			(1,212.82)		
Excluded			(69.80)		
Wasil or collected *eraisala* [4]		16,403.04			
Sadirwarid or village expenses		995.23			
Dastur baki		220.66		(balance on the collections)	
Sistu baki		1,529.42		(balance on the inhabitants)	
Total		19,148.36			

Key: (1) Revenues converted to produce according to customary shares; (2) Produce in the area lay out of Brown's survey. *Table continues.*

Table 6.3 continued

Notes: [1] 79.63 *visams* producing 95.41 gc, 116.5*visams* producing 116.52 gc, 9.75 *visams* producing 7.03 gc. Double cropping was likely on some of the dry land. [2] There are 709.1 gc in figure A and only 640.53 gc wet *sarkar* land in Brown's survey. The remaining wet land was claimed by *manyamdars*. [3] Inhabitants' cultivation; Rs.5 extra cess per gc. [4] Dues paid to Vatchavoy Jagganad Raju.

Source: P/283/69, pp.2326-27.

Table 6.4: *Revenue distribution according to Alexander's survey*

	visams	Assessment per *visam* (Rs.)	Estimated value of produce (gc per *visam*)	Full produce (gc)
Paddy lands	192	25	1.25	240
	160	16.5	0.825	132
	64	15	1	64
	64	12.5	0.833	53.33
	118	8.25	0.55	64.9
	40	5	0.33	13.33
Total	638			567.56
Dry lands	48	25	1.667	80
	96	16.5	1.1	105.6
	48	10	0.667	32
	71	6	0.4	28.4
Total	263			246

Source: P/283/67, p.1374.

Table 7.1: Kondakarla. *Area and produce figures for two seasons and two surveys (paddy lands)*

gc/v	Figure A: *fasli* 1197 *sarkar* v	gc	*sanad/mirasi* v	gc	Figure C: Brown's survey *sarkar/sanad* v	gc	*mirasi* v	gc	*visams*
1.25	120	150	40	50	160	200			160
1.0	84	84	28	28	112	112			112
0.867	15	12.75	5	4.25	20	17			20
0.833	127	105.83	*24*	*20*	*84.25*	*70.21*	66.75	55.63	62.75
0.667	15	10	*35*	*23.33*	50	33.33			50

Note: The levels shown ('gc/v') are *garce* per *visam*. In the column headed '*sanad/mirasi*' in figure A the values for *mirasi* are italicised, at levels 0.833 and 0.667; similarly in figure D the values for *sanad* are italicised, at level 0.833.

Table continues

Table 7.1 continued.

gc/v	Figure B: *fasli* 1210 *sarkar* v	gc	*sanad/mirasi* v	gc	Figure D: Alexander's survey *sarkar/mirasi* v	gc	*sanad* v	gc	*visams*
Flooded					64	80			64
1.25	72	90	24	30	48	60	48	60	96
1.0	84	84	28	28	*56*	*56*	56	56	112
0.867	15	12.75	5	4.25	*10*	*8.5*	10	8.5	20
0.833	127	105.83	*24*	*20*	151	125.84			151
0.667	8.25	5.5	*30*	*20*	50	33.33			50

Note: The levels shown ('gc/v') are *garce* per *visam*. The values for *mirasi* are italicised, in the column headed '*sanad/mirasi*' in figure B, at levels 0.833 and 0.667; and in the column headed '*sarkar/mirasi*' in figure D, at levels 1.0 and 0.867. Note that in figure B, 11.75 *visams*, 7.83gc, @ 0.667gc per *visam*, were uncultivated.

Table 7.2: Kondakarla. *Area and produce figures for 'one good year' after due repairs and improvements (collector Brown's survey) extracted from the Estate Records (Kondakarla Handa)*

Paddy lands

visams	@ gc	= gc
160	1.25	200
112	1	112
20	0.867	17
50	0.667	33.33
342 (total)		362.33

Dry lands

visams	(crop)	@ gc	= gc
46.5	*raghi*, *pesalu*, and *gingeli* oil	1	46.5
78	*gantelu*, *korralu*, and lamp oil	0.833	65
8	*gantelu*, and lamp oil seeds	0.8	6.4
160	*raghi*, *pesalu*, and black gram	0.5	80
60	*allu*, and *chamalu*	0.367	22
352.5 (total)			219.9

Garden lands

visams	(crop)	@ maunds	(Rs.)	= gc
4	sugar cane	80	(480)	16
10	cotton	20	(200)	6.67
3	tobacco	14	(42)	1.4
17 (total)				24.07

Other lands

	visams	@ gc	= gc
(1)	84.25	0.833	70.21
(2)	22	0.333	7.33
(3)	66.75	0.833	55.63
(4)	44	0.333	14.67

Key: (1) *sanad* paddy; (2) dry grain; (3) *mirasi* paddy; (4) dry lands.
Source: P/283/69, pp.2242-47.

Table 7.3: *Abstract of Kondakarla Kham, Jamma Wasil Baki and Jamma Kurk for fasli 1197.*

			Rs.	(1) gc	(2) gc
Mattu	(338 *visams*)	*sistu*	3,147.25	209.82	219.9
		malavati [1]	1,967.03	(131.14)	
Dry grain shares			20.38	0.71	
Garden cultivation					24.07
Mattu & pallam		*bilmakta*	1,479.53	73.97	
Pallam in shares			5,627.23	325.17	362.33
Increase on *anchana* etc.[2]				– 2.69	
Total *sarkar*				(606.98)	(606.30)
On account of manyams			34.5		
Manyamdar (paid in excess of his *kattubadi*)			30	– 4	
Hire for bullocks to carry grain to port			20.38		
Excessive collection			10.61		
Manyam kattubadi			940.19		
mirasi		(530.63)	70.75	70.30	
sanad		(409.56)	81.25	77.54	
Motarapha			35.75		
Palmyra gardens			45		
Balance of previous year			4		
Total net *daul*			13,341.47		
Total produce				(754.98)	(754.14)
Wasil					
Cash paid to the renter in money			8,805.44		
in grain					
delivered into the port @ Rs.29 per gc			1,995.00		
48.34 gc stored at various prices			1,352.97		
Sadirwarid, village expenses			447.27		
Dastur baki, balance on the collections			87.92		
Sistu baki, balance on the village			652.80		
Total			13,341.38		

Key: (1) Revenues converted to produce according to customary shares; (2) Produce in the area lay-out of Brown's survey.

Notes: [1] Comprising 158 *visams garuvulu* or enclosed fields @ Rs.7 per *visam,* 37.5 *visams ragada* or clay @ Rs.12 per *visam,* 95 *visams* wet *ragada* @ Rs. 15 per *visam*, and 47.5 *visams* of *konda garuvulu* @ Rs. 3.5 per *visam.* Double cropping was likely on *ragada* and wet *ragada varikatti* land. [2] Also overpaid collections—together 2-20-11 gc; withdrawn 1-10-5.5 gc.

Source: P/283/69, p.2,239.

Table 7.4: *Revenue distribution according to Alexander's survey*

	Assessment (Rs.)		Estimated value of produce		Full produce
	visams	per *visam*	gc	per *visam*	gc
Paddy lands	112	25	1.25	112	140
	160	25	0.8	160	128
	44	10.25	0.683	44	30.07
Total	316				298.07
Dry lands	76	12	0.8	76	60.8
	96	7.5	0.5	96	48
	160	3.5	0.217	160	34.67
Total	332				143.47

Source: P/283/vol.67, p.1373.

Table 8.1: Melupaka. *Produce and area figures for two seasons and two surveys (paddy lands)*

	Figure A Bumper harvest, *fasli* 1197				Figure C Brown's survey				
gc/v	*sarkar*		*sanad/mirasi*		*sarkar/mirasi*		*sanad*		*visams*
	v	gc	v	gc	v	gc	v	gc	
1.0	159	159	53	53	212	212			212
0.833	25.5	21.24	8.5	7.08	34	28.33			34
0.73	46.88	34.22	15.63	11.41			83	60.67	83
0.667	41.5	27.66	*9*	6	*50.5*	*33.67*			50.5
0.625	178.5	111.55			178.5	111.55			178.5
0.5	47.5	23.75			69.5	34.75			69.5

Note: The levels shown ('gc/v') are *garce* per *visam*. The values for *mirasi* are italicised, in the column headed '*sanad /mirasi*' in figure A, at level 0.667; and in the column headed '*sarkar/mirasi*' in figure C, at the same level. Additional to figure A are *mirasi* 20.50 *visams* and 15.03 gc at level 0.73, and *karmam* 22 *visam*, 11gc, at level 0.5. Note also for figure A: *sanad* 15.63 *visams* 11.41 gc @ 0.73 gc per *visam*, and *mirasi* 20.5 *visams* 15.03 gc @ 0.73 per *visam*.

	Figure B Good year, *fasli* 1210				Figure D Alexander's survey				
gc/v	*sarkar*		*sanad/mirasi*		*sarkar*		*sanad/mirasi*		*visams*
	v	gc	v	gc	v	gc	v	gc	
flooded	80	80			60	written off:	20	20	80
1.0	99	99	33	33	66	66	66	66	132
0.833	25.5	21.24	8.5	7.08	17	14.17	17	14.17	34
0.73	46.88	34.22	15.63	11.41	41.5	30.33	*41.5*	*30.33*	83
0.667	41.5	27.67	*9*	6			*9*	*6*	50.5
0.625	112.65	69.46			178.5	111.55			178.5
0.5	out of cultivation		69.5	34.75	69.5	34.75			69.5

Table continues

Table 8.1 continued.

Note: The levels shown ('gc/v') are *garce* per *visam*. The values for *mirasi* are italicised, in the column headed '*sanad /mirasi*' in figure B, at level 0.667; and in the column headed '*sanad/mirasi*' in figure D, at level 0.625; that level represents a range from 0.617 to 0.625. Additionally, out of cultivation, at that level, in figure B were 65.85 *visams* 42.09 gc, and at level 0.5 69 *visams* and 34.75 gc. Additional to figure B at level 0.73 are *mirasi* 15.03 gc; and in figure D at level 0.667 were 41.5 *visams* 27.67gc written off. Note also the following details for figure B: *sanad* 15.63 *visams* 11.41 gc @ 0.73 gc per *visam*, and *mirasi* 20.5 *visams* 15.03 gc @ 0.73 per *visam*.

Table 8.2: Melupaka. *Area and produce figures for 'one good year' after due repairs and improvements (collector Brown's survey) extracted from the Estate Records (Melupaka Handa).*

SARKAR *Paddy lands*

visams	@ gc	= gc
212	1	212
34	0.833	28.33
178.5	0.625	111.55
69.5	0.5	34.75
494	(total)	386.63

Dry lands

visams (crop)	@ gc	= gc
8 *raghi, pesalu* and black gram	0.667	5.33
12 *korralu* and *bobbarlu*	0.533	6.4
12 *gantelu* and horsegram	0.5	6
12 *allu* and *kandulu*	0.333	4
8 *kandulu*	0.267	2.13
4 *chamalu*	0.267	1.07
16 *allu*	0.167	2.67
72 (total)		27.6

Garden lands (not incorporated in fig. C)

visams (crop)	@ maunds	(Rs.) =	gc
3 sugar cane	180.5	(361)	12.03
2.5 betel	20	(250)	8.33
5.5 (total)			20.36
8.75 uncultivated			

Other lands: SANAD, MIRASI

	visams	@ gc	= gc
(1)	83	0.73	60.67
(2)	10	0.2	2
(3)	50.5	0.667	33.67
(4)	5	0.25	1.25

Key: (1) *sanad* paddy and (2) dry lands; (3) *mirasi* paddy and (4) dry lands.
Note: The separation of cultivated land in the Estate Accounts for Melupaka is not satisfactory as the garden lands have been omitted. At this stage of research it is not possible to place the sugar cultivation and the betel garden produce. Some reallocations can be expected in a final analysis. An approximation is made to illustrate the general theme. The garden lands have not been incorporated in the figures and tables shown in Tables 1.4, 4.1 and 8.1 They were placed at the medium level of land in the village called *varikatti*. In good years *varikatti* ground produced both a dry crop and a wet crop. It is clear from the renter's account for *fasli* 1197 that the *varikatti* land produced more than one crop, and most of the *varikatti* land is incorporated in the paddy lands in Brown's survey. *Source*: P/283/69, pp.2528-32.

Table 8.3. *Abstract of Kasba Melupak Kham, Jamma Wasil Baki and Jamma Kurk for fasli 1197*

			Rs.	(1) gc	(2) gc
Varikatti (123.5 *visams*)		*sistu*	1,768.89	117.93	dry land
	malavati	364.89	(24.32)	27.6	
excused for raiyats			– 24.18		
Mattu in shares			53.50	3.57	
visams (producing)	gc				
(1) 28.25	55				
3	5				
8	12				
39.25 (total)	72	paddy lands in figure C			28.33
(2) 3	2.4				
10	9.33				
13 (total)	11.43				
(3) 29.25	10.56				
4.5	1.73				
33.75 (total)	12.29	paddy lands in figure C			34.75
(4) 24.5	13.68				
11.5	5.75				
1.5	0.8				
-	3.57				
37.5 (total)	23.80	paddy lands in figure A			(23.75)
Malavati [1]					
Pallam [2]		in shares	6,127.75		
measured by grain *chuncham*				210.8	(214.46) fig.A
					212.00 fig.C
measured by country *chuncham*				114.57	111.55
Manyam kattubadi			798.75		
sanad			(361.25)	72.25	62.67
mirasi			(190.50)	25.4	34.92
surplus *manyam*			(247)	32.93	
Over plus collection			0.33		
Moturapha			50		
Gardens			6		
Takavi			164.09		
Baki, last year			304.52		
Net *daul*			9,614.56		
Total produce				(577.45)	(511.82)
difference [3]					(65.63)
Wasil or collection: paid to *sarkar*			6,918.58		
takavi			164.09		
sadirwarid			340.64		
dastur baki			5.70		
sistu baki, balance on raiyats			2,185.55		

Table continues

Table 8.3 continued.

Key: (1) Revenues converted to produce according to customary shares; (2) Produce in Brown's survey.

Notes: [1] Imposed at 4 annas per rupee on Rs.1,356.56 *sistu* on *kadims*, and 2 annas per rupee on Rs.206 *sistu* on *paikaries*. [2] *Sarkar* share of produce: 140.6 gc measured by grain *chuncham*, and 76.38 gc measured by country *chuncham*. [3] Understandable only with double cropping at medium level.

Source: P/283/69, pp.2523-4.

Table 8.4: *Revenue distribution according to Alexander's survey*

	visams	Assessment Rs. per *visam*	Estimated value of produce gc	per *visam*	Full produce gc
Paddy lands					
	112	12.5	0.625	112	70
	72	11.25	0.625	72	45
	168	8.75	0.583	168	98
	80	8	0.533	80	42.67
Total	432				255.67
Dry lands					
	16	5	0.333		5.33
	40	2.5	0.167		6.67
Total	56				12

Notes: There are minor discrepancies between the figures in Alexander's survey and the reconstructions made in figure D for the four *kasba* towns. They are small and unlikely to challenge the general finding which is that Alexander's survey expressed a new proportional division of the crops on wet land. The separation of the estate accounts for 'one good year' is made by starting on the *mirasi* and *sanad kattubadis* for 'one good year' entered under each village. The *sanad kattubadi* divided by 5 (paying Rs.5 or one-sixth of the produce per gc) gives the crop harvested on *sanad* land, multiplied with 4 we have an idea of the top level paddy lands within the village. When the model is refined and the details filled in the easiest way is to begin with the *sanad* and *mirasi* land in the estate records. They are fairly easily separated for the different villages, and show approximate levels of produce per gc on which the *sarkar* land can then be divided.

Source: P/283/67, p.1376.

Chapter 6

PADDY, GARDEN, CHENA, PLANTATION: WAS THERE A PEASANT STRATEGY IN THE KANDYAN REGIONS OF SRI LANKA BEFORE 1940?

Eric Meyer

All have not watered land enough for their needs, that is, such land as good rice requires to grow in; so that such are fain to sow on dry land, and till other men's fields for a subsistence. These persons are free from payment of taxes; only sometimes upon extraordinary occasions, they must give a hen or a mat or such like, to the King's use: for as much as they use the wood and water that is in his country. That is if find the duty to be heavy or too much for them, they may leaving their house and land be free from the King's service, as there is a multitude do, and in my judgement they live far more at ease after they have relinquished the King's service than when they had it.[1]

In the course of any study of agrarian change, a major issue is that of the existence of options open to the peasants, and of the nature and extent of non-technical constraints limiting their actual choice. In other words, is it possible to speak of a 'peasant strategy' when describing subsistence economies adapting themselves to increasing pressures under static ecological conditions? (See for example James Scott's analysis of South East Asian subsistence economies and Clifford Geertz's work on agricultural 'involution' in Indonesia.) Or is the term to be restricted to cases when peasants change their options in order to confront new external constraints, as in the case of Michael Adas's study of the lower Burma rice frontier?[2]

The interest of the Sri Lankan case lies in the dramatic change which took place in the Kandyan countryside with the opening of plantations. A determinist approach could lead to its analysis in terms of categories, 'subsistence to market', 'traditional to modern'. A systematic interpretation would insist on the establishment of dominant-depen-

[1] Robert Knox, *A Historical Relation of Island Ceylon* (London, 1681), p.43.

[2] J.C. Scott, *The Moral Economy of the Peasant. Rebellion and subsistence in Southeast Asia* (New Haven 1976); C. Geertz, *Agricultural Involution: the process of ecological change in Indonesia* (Berkeley 1963); M. Adas, *The Burma Delta. Economic development and social change on an Asian rice frontier* (Madison 1974).

dent relations between two sectors, estate and village, and deny the possibility for a 'peasant strategy' to arise in between.[3] But grass-roots analysis suggests a remarkable adaptability of the peasant to changing circumstances. Peasant strategies both before and after plantation development were based on diversification. 'Safety-first' principles meant most often distribution of risk. When the environment changed, the strategy remained the same—not to be tied by a single resource: this is precisely why the local villagers refused to become coolies, while accepting employment as casual workers in the plantations. The conspicuous absence of peasant rebellions in Sri Lanka is not only the result of a lack of peasant consciousness and of the land-distributing policy of the state, as shown by Mick Moore, but also of the ability of cultivators to adapt themselves to changing circumstances.[4]

The initial resources were a combination of permanent paddy subsistence cultivation, permanent garden cultivation (partly subsistence, partly for sale or barter), and temporary slash and burn cultivation. Each strategy fulfilled a different function and was practised by different people according to criteria of age, caste and gender, but at the global level of the village community they appear as complementary. Whether this complementarity was the result of a conscious 'production strategy' is unclear; at least economic rationality was not unknown among the people in charge of seigneurial villages, and especially temple villages.

Possible strategies were constrained by ecological factors, by tenurial systems, and by state policy, especially land legislation and taxation. How far did new external constraints brought about by colonial impact, at the level of state policy, market relations, and the development of a plantation economy, narrow or enlarge these options? State and individual policies were further dependent on knowledge and representations of the different agricultural activities, and influenced them: what was the impact of the colonial system and of the nationalist movement at that level?

This paper will focus on the history of *chena* (slash-and-burn) cultivation and of estate labour, and on the wet up- and mid-country areas where several types of resources were simultaneously available, rather

[3] A. Bandarage, *Colonialism in Sri Lanka. The political economy of the Kandyan highlands 1833-1886* (Berlin 1983).

[4] M. Moore, *The State and Peasant Politics in Sri Lanka* (Cambridge 1985).

than on the dry lowlands where the options were limited to irrigated paddy and dry millets on *chenas*. The major changes brought about by plantation development in the Kandyan regions will be assessed in the perspective of enlargement or restriction of subsistence options. An obvious difficulty is that, for want of direct historical information at grass-root level on peasants' farm-management, one has to reconstruct peasant strategies, if any, through qualitative and indirect sources, which are usually biased against certain forms of cultivation. For example, playing down or condemning *chena* cultivation was a constant theme in political and pseudo-scientific discourse: colonial Victorian moral representations re-enacted pre-colonial representations and were later re-employed by nationalist myths; they were enlisted to support repressive policies. Substantial attention will be devoted to the relations between the peasantry and the state. Whatever the actual choices in economic terms that cultivators might have made, the nature of the documentation will unavoidably lead us to emphasise the role played by the external constraints created by the state power, and the peasant strategies to overcome them. Finally, we shall have to examine the meanings and functions of the stereotyped representations of the rural—especially the 'rice bias'—still in current use today; these representations were constructed by nationalist ideologues on the basis of the 'hydraulic' civilisation of the dry zone (first millenium A.D.) without taking into account the actual working of the peasant economy in the densely populated areas of the wet zone.[5]

I. *Ecology, land tenure and cultivation: the pre-colonial regime*

In the Kandyan regions of Sri Lanka, before plantation development, the village territory was generally made up of three or four areas. The central part, watered by perennial streams or by artificial means, was exclusively devoted to the permanent cultivation of wet paddy. An inner ring around or on both sides of paddy fields was occupied by houses amid perennial gardens (*watte*) usually planted with food-producing trees. An outer and much wider ring (*chena*) was used for slash-and-burn shifting cultivation of grain, vegetables and roots; and when cultivation was over, for grazing and for wood (fuel, building, fencing) obtained from the re-growth of low secondary forest (*landu*). A third

[5] See also M. Moore, 'The ideological history of the Sri Lanka "peasantry"', *Modern Asian Studies* 32, 1 (1989), pp.179-207.

ring (not found everywhere and often difficult to distinguish from the former) was made up of primeval or secondary high forest (*mukulana*). These different areas provided complementary resources, but, according to local conditions, to weather variations, to land tenure systems or to socio-cultural factors, some of these sources were more extensively tapped than others.

Chena cultivation is commonly regarded as characteristic of the dry zone (centre-north and south-east of Sri Lanka); its technical, economic, and social features have been analysed by anthropologists doing field studies in that area, such as Leach, Yalman, Brow, Kloos and recently Spencer; most of them allude to the repressive policy pursued by the raj against *chena* (so vividly pictured by Leonard Woolf in *The Village in the Jungle*), and to the discourses and myths used to justify it, especially by contrasting *chena* with paddy cultivation.[6] But *chena* cultivation was no less prevalent in the wet and intermediate zones of the island, until the plantation economy reached its maximal development, eventually wiping it out.

In the system prevailing in the areas controlled by the Kotte and later the Kandyan kingdom, the availability of space and the diversity of resources made it possible for the peasant to minimise risks. Contrary to a common view (accepted by scholars such as Jayawardena and Roberts),[7], *chena* was not a minor or subsidiary resource in the wet and intermediate zones, but was central to the survival of large sectors of the peasantry. Their choice of *chena* cultivation was rational in socio-

[6] E.R. Leach, *Pul Eliya* (Cambridge 1961) esp. pp.289-91; N. Yalman, *Under the Bo Tree* (Berkeley 1967); J. Brow, *Vedda Villages of Anuradhapura* (Seattle 1978), esp. pp.100-4; P. Kloos, 'Agricultural policy in Sri Lanka; the views of a nineteenth-century colonial elite', *Focaal* (Zomer 1988), pp.56-79; J. Spencer, *A Sinhala Village in a Time of Trouble* (Delhi 1990). Spencer convincingly suggests (pp.105-6) that the general under-reporting in official and ethnographic literature of *chena* cultivation results from the convergence of colonial dislike, the cultural valuation of the sole paddy cultivation, and its accountability and visibility. See also the studies collected by J. Brow and J. Weeramunda, *Agrarian Change in Sri Lanka* (Delhi 1992), published after the writing of this paper.

[7] They argue that these resources 'were only of marginal utility to Kandyan village ecology'. By marginal, they seem to imply unnecessary; my view is that in a context of scarce, irregular and diversified resources, marginal means necessary. See M. Roberts, 'The impact of the waste lands legislation and the growth of plantations on the techniques of paddy cultivation in British Ceylon: a critique', *Modern Ceylon Studies* 1, 2 (1970), p.160.

economic terms in a context of limited population, extensive space and resource diversification. This is the kind of rationality which is branded the 'safety-first' approach by J.C. Scott in his analysis of the subsistence ethic of the Asian peasantry.[8]

The flexibility of the system resided in the ability of the peasantry (and especially the young people) to mobilise peripheral resources in case of need. But it does not mean that the village was self-sufficient: except in out-of-the-way areas, a certain amount of trade contributed to the balance of rural economy. In Sabaragamuwa for example, arecanuts, *jaggery* and various garden produce were extensively bartered for rice, fish, salt and textiles. Nor did the system always work smoothly: bad weather conditions or political troubles (especially the wars between the colonial powers and the Sinhala kingdoms) often led to village breakdown.

Village types

Climatic and relief characteristics combined with tenurial systems weighted peasant choices in favour of one form of cultivation or another. On the basis of nineteenth-century monographic information, it is possible to draw a typology of land use, but these are just ideal types and there was in fact a continuum between them.[9]

(1) Large rice villages (*batgam*) were found in the best watered valleys and flat lands. They were generally organised as surplus-producing estates for the benefit of the kings (lands then called *gabadagam*), or their courtiers (*nindagam*), or the temples (*viharegam* and *devalegam*). The tenurial system reinforced cultivation patterns through a systematic policy aiming at optimising paddy production. The *muttettu* (lord's demesne) was wholly devoted to rice production under the supervision of *vidanes*, and was gratuitously worked by low-caste people (generally Padu or Panna, then included in the larger category of Duraya, and now called Batgama). They had no options except to select species of paddy according to weather conditions.

The villagers subsisted on family tenures (*pangu*) where they prac-

[8] Scott, *Moral Economy*, p.15.

[9] Information mainly collected in the Kegalla kachcheri records, Sri Lanka National Archives (hereafter SLNA), lot 30, and the Land Settlement Department records. See also H.W. Codrington, *Ancient Land Tenure and Revenue in Ceylon* (Colombo 1938), and R. Pieris, *Sinhalese Social Organization: the Kandyan period* (Colombo 1956).

tised mixed farming on fields, gardens and *chenas*, over which the supervision of the *vidane* was less particular; but while some settled families had hereditary rights (*paraveni*), others were precarious tenants (*maruveni*) in the *vidane* hands. Their only option was to abscond if dissatisfied with their condition. There were constant attempts by the lord to deny *paraveni* rights, and by the *maruveni* to be recognised as hereditary tenants, which meant a secure subsistence: disruption of village life by wars or epidemics provided the best opportunities for such a change.

(2) The mixed farming village was the type most commonly found in the Kandyan hilly areas and in the lowlands of the Kurunegala district: there, *chena*s were associated with gardens and irrigated paddy in variable proportions according to the local ecological conditions. Tenurial systems differed: most of these villages (*koralegam*) were held by freehold peasants, generally of Goyigama (high) caste, while some were the property of temples or belonged to Radala families (of the upper subcaste of the Goyigama). In the latter cases, tenants performed services or furnished produce but production was not organised as strictly as in *batgam.*

The necessity of *chena* cultivation in combination with wet paddy was generally recognised, and the legal concept of appurtenance on a 'three-to-one' basis (a highland area three times the acreage of the irrigated area) was later constructed by the British to cover what was initially the acknowledgement of a technical complementarity. But a distinction must be made between highland areas close to the paddy fields (our 'inner ring')—which were necessary to wet paddy cultivation (such as threshing floors, *pillewa*), or for dwellings and growing vegetables (*watte*)—and the *chena* area itself which might have been much larger. According to the observations of a twentieth-century settlement officer, 'appurtenances' included only the former, and

> in a large number of cases the claimed *chena*s are claimed by people who are not paddy field owners, and calling these as appurtenant to paddy fields is to deny the claims of persons who are not field owners. One might as well say that *chenas* are appurtenant to gardens.[10]

It is mainly in these mixed-farming freehold villages that the possi-

[10] SLNA, Comments by Assistant Settlement Officer (hereafter ASO) Hunter on the third interim report of the land commission (1928); Ceylon Sessionnal Paper (hereafter SP) 22 of 1880 (Dry Grain Commission Report).

bility arose for cultivators to elaborate a land-use 'strategy'. At the beginning of a cultivation season, the peasants could choose between different species of paddy seeds according to the expected rainfall, or between different types of millets and/or yams (tubers) according to local conditions. If wet paddy crops failed owing to bad weather, pests or other troubles, *chena* cultivation of millets in the *yala* season (April to July) was the only means to tide over until the next *maha* season (September to February). An analysis of crop timing in the Kegalla district shows that a combination of paddy (hill or wet) and different millets and other *chena* crops could keep the peasant busy all the year round.[11] Hill paddy (*elvi*) was cultivated in *maha*, like irrigated paddy: it was therefore incompatible with wet paddy, and a substitute for it, while the millets (*amu* and *kurakkan*) were cultivated only in *yala* and were therefore compatible with and complementary to paddy.

A multiple-cropping pattern allowed the option of devoting more time either to paddy or to *chena*. The extent to which peasants did adapt themselves is impossible to assess. Data are available only after the British takeover, and even then 'Blue Book' general statistics on millets are pure guesswork, and paddy tax data on hill paddy must be used with care, tax evasion being much more frequent for that than for wet paddy cultivation. Taxation records suggest that hill paddy cultivation was extremely variable from year to year: for the period 1852-78 in the case of the Kegalla district, the data show peaks (in 1859, 1866, 1873 and 1876) which might be explained by weather conditions favouring *elvi* cultivation, and a trough for the period 1867-72, which is connected with the period of maximal *chena* repression as we shall see later. But it is only at the microcosmic level that one can reach conclusions. For example, a detailed analysis of the *wattoru* (tax receipts) for the village of Pannala (1858/9 to 1889/90) shows that while some large *chena* tracts were regularly cultivated with *elvi*, others appear only once or twice in 30 years: either the latter were cultivated the rest of the time with millets, which were never taxed; or they were cleared exceptionally when the need arose. In both cases, highland cultivation gave the system a flexibility lacking in paddy villages.[12]

(3) *Chena* villages were not found only in the dry zone, where climatic conditions made irrigated paddy cultivation highly unreliable.

[11] Ceylon Administration report (hereafter AR) Kegalla 1886, p.146A.
[12] AR Kegalla 1878, p.72.

They were also common in the wettest parts of the south-western hills, where marshy or narrow valleys could not be properly 'asweddumised' (converted into paddy fields). Sometimes *elvi* (hill paddy) was sown for two years in *owita* (unirrigated grassland at the bottom of valleys); afterwards millets and other fine grains were cultivated for a couple of years, and the *owita* was left fallow as grazing ground for three years.[13] More often, hill paddy was cultivated in highland areas where moisture was sufficient and the cultivation cycle could be as long as 10 to 20 years. *Elvi* was sown the first year (sometimes paddy seeds were mixed with millet or oil plant seeds); millets or yams were sown the second year; and then wood was allowed to grow for several years, often a decade or more.[14] This pattern was typical of the border between the Sabaragamuwa and the Western and Southern Provinces, and of the vast areas south and west of Sri Pada, in the Ratnapura and Kegalla districts (where it was called Kelerata—that is, forest country). In these sparsely populated villages, made up of small scattered hamlets, *chena* cultivation with hill paddy provided the only source of grain, but it was supplemented by hunting, food gathering (fruit and honey) and *kitul* (sugar palm) tapping.[15]

A large proportion of these villages was inhabited by members of the Vahumpura caste (often locally called *kande minissu*—people of the hills).[16] Under the service system imposed by the Kandyan kings, they were required to provide *jaggery* (palm sugar) and, more generally, cooked food, and occasionally to act as bearers and soldiers guarding the hill passes; although fully integrated in the caste system (in the upper strata of the lower castes) and now proud of their Buddhist

[13] U.A. Gunasekara, 'Land tenure in the Kandyan Provinces', B.Litt. thesis, Oxford, 1959, p.28; also SLNA, Diary of ASO Hughes, September 1924; *ovita* is considered either as paddy or as highland: 'They may be regarded in an old village as paddy land on the margin of cultivation'.

[14] R.E. Lewis, 'The rural economy of the Sinhalese, more particularly with reference to the district of Sabaragamuwa, with some account of their superstitions', *Journal of the the Royal Asiatic Society, Ceylon Branch* II, 4 (1848), pp.41-2.

[15] F. Lewis, *Sixty-four Years in Ceylon* (Colombo 1926), esp. pp.198 ff. (village of Bambarabotuwa); Diaries of the Agents of Kegalla and Ratnapura Districts and of settlement officers (hereafter SO), SLNA; also 'Ievers on *hen* cultivation', quoted in R. Pieris, *Sinhalese Social Organization*, p.86.

[16] Diary of SO J.P. Lewis, 18 December 1897: in the village of Narawelpita (Morawak) 'chena is the only means of subsistence for depressed Vahumpura villagers'.

identity, this community had some features similar to those of the Indian Adivasi and could possibly claim the same antiquity as the more celebrated Vedda. Other *chena* villages were inhabited by high-caste people (some of them locally called Vedda) and by members of lower castes, such as the Panna.

Chena villages were characterised as such by the peasants and distinguished from rice villages.[17] Among several typical cases in Ratnapura and Kegalla districts one can quote the large locality of Bambarabotuwa: 'an instance of what happens when there is a limited population and an unlimited land supply, which being out of sight was out of control', according to a settlement officer. The absence of state interference and inaccessibility were added factors: peasants were close to a food-gathering economy, settlement in hamlets was sparse and unstable, and the very few paddy fields 'in reality were only swamps roughly turned into paddy lands'.[18] Gilimale, a neighbouring village, had more diversified resources owing to the presence of small paddy fields, and to the trade linked to Adam's Peak pilgrimage: it was characterised by the presence of a seigneurial family. However when settlement was effected, the officer defined the village economy as follows: *chenas* are indispensable, there are no paddy fields, no estate employment, no money to pay for individual settlements; 'they seem to have no idea beyond *chena*...what they want is to be allowed to *chena* as heretofore.'[19]

Ecology, society and the state

Cultivation strategies have to be understood not only in economic but also in socio-political terms. In the case of Sri Lanka, the individual farming unit is not an entirely irrelevant concept, but it is encompassed within a larger unit. A strategy must be first conceptualised at the level of the village as a whole, or perhaps of kindred groups in which resource sharing or lending and work exchange are intense. Individual economic activity (such as *chena* cultivation, or later casual work on estates) is an option within that encompassing strategy. However, opt-

[17] Examples: Polatagama vs. Denawakawatta, Diaries of Assistant Government Agent (hereafter AGA) Kegalla, 12 August 1872; ASO Abhayaratne November 1933 on Moratihe Gabadagama, as compared with Moratihe Viharegama.

[18] Lewis, *Sixty-four Years in Ceylon*, p.198.

[19] Diaries SO, January and February 1931.

ing out of the *rajakariya* system was always possible, as underlined by Robert Knox in the epigraph of this paper. It gave the peasant society a mobility and a flexibility which contradicts the stereotype of the little republic.

Specific groups were compelled to rely more on garden and *chenas* than paddy. Some villages were mixed caste localities, where the lower castes had no *paraveni* rights to irrigated paddy land and depended more than the Goyigamas on highland cultivation. Besides, certain groups of any caste had little access to irrigated paddy: young men and women, when paddy fields were in full control of the elder generation, widows, or separated wives with children returning from their husbands' villages; also people who had lost their rights through indebtedness, family feuds and disputes with more powerful neighbours. These categories could rely on *chena* cultivation for a couple of seasons. Among these groups, women played a major role in garden and dry grain (especially *kurakkan*) cultivation, while paddy (including *elvi*) was a male occupation.[20] Millets acted as a survival crop: contrary to hill paddy, they were not taxed, they grew in a few months during *yala*, might be cultivated at shorter intervals than *elvi* and might follow it on the same field; they were the poor people's crop. *Chena* was like a safety valve, which prevented a breakdown in the village socio-economic balance, and eventually allowed poor peasants to postpone the decision to migrate. In that respect, a comparison with the villagers from Tamilnadu suggests that their greater propensity to migrate was related to the lack of flexibility of their cropping system.

But *chena* was not only the choice of impoverished groups. It could also be chosen by enterprising people intending to free themselves from the burden of royal service (*rajakariya*), or to avoid interference from family or usurers, especially when population growth led to subdivision of paddy land into infinitesimal (and usually mortgaged) shares. *Chena* was a temporary area which could be converted into permanent gardens—the first step in a village expansion process. In a sense, it could be argued that paradoxically it was more dynamic and evolutive—in a sense more modern—than paddy cultivation.

Chena cultivation represented a specific mode of social life, where

[20] As suggested by Joke Schrijvers in her *Mothers for Life: motherhood and marginalization in the North Central Province of Sri Lanka* (Delft 1986), and personal communication.

marginal groups could find a recognised space—a space of freedom. It was an unbound and unpermanent activity, in a temporary area of temporary dwellings, with no stable property and family relations, as opposed to the defined, bounded, policed paddy-cum-house area. It is extremely significant that *chena* cultivation was an unritualised activity: R.E. Lewis rightly emphasised in 1848 that, unlike paddy cultivation, *kurakkan* cultivation was effected without any rite. Colin Kirk in his recent study of a dry zone potter village confirms the resilience of that feature.[21] In the *chena* individualistic and temporary relations could take place. There young couples could be made and unmade; Kandyan unions were considered for some time as trial marriages, and such unions were described in a typical sentence: *geniyek henata geniyanava* (to take a woman to a *chena*).

Highlands, being exempt from *rajakariya*, were the usual refuge for people who tried to shake off the burden of taxes and services. It was unpoliced, beyond control of the higher authorities—the king and his superior headmen—and therefore the refuge of outlaws or of ordinary villagers in times of trouble. (This is well documented for 1817/18, and again in 1848, during the anti-British rebellions.) It is not clear whether or not the kings discouraged *chena* cultivation—except in *tahansi kele* (forests reserved for strategic purposes) where it was expressly prohibited; but it is likely that they viewed highlands as uncontrolled areas as opposed to the settled area of the paddy fields and gardens. Highlands were also places where the arbitrary power of local families could be asserted against the policy of the state and over the lesser peasant families, who had to bargain for undefined cultivation rights. In seigneurial and temple villages, access to highlands was at the pleasure of the lord, while in villages of the royal demesne it was controlled by the *vidane*. In freehold villages (*koralegam*), it could become a stake in power rivalries between local families, inasmuch as it was less defined than the paddy area, and therefore easier to control by arbitrary force.

[21] R.E. Lewis, 'The rural economy of the Sinhalese', pp.41-2; C. Kirk, 'Perceiving agrarian change: past and present in Ratnale, a Sinhalese potter village', in Brow and Weeramunda, *Agrarian Change*. In a sense, *chena* was a kind of forest cultivation and as such can be understood in the context of the traditional opposition in Indian thought and rituals between *grama* (village) and *aranya* (forest), bound and unbound, so nicely analysed by Charles Malamoud in *Cuire le monde: rite et pensée dans l'Inde ancienne* (Paris 1989), pp.93-114: 'Village et forêt dans l'Inde brahmanique'.

Colonial authors used to describe *chena* as a communal (and archaic) form of cultivation, because they could not discern rights of ownership but only rights of use. According to their European experience, individual enterprise was equated with property rights: they viewed *chena* areas, like pastures, as a local equivalent of the British commons.[22] On the other hand, many ethnographers have analysed *chena* cultivation as a rather individualistic activity as compared with paddy, a definitely social activity.[23] A possible explanation of the contradiction could be that in the course of time *chena* lost its communal character; another, that there were local variants. In most wet areas, it appears that highland cultivation was organised according to *pangu* just like paddy cultivation: particular areas were cultivated by particular lineages, and later claimed by them when the British undertook settlement operations.[24] However, in the North-Western Province, where the *panguwa* system was not so prevalent, the claims were more often asserted by nuclear families.

Chena was certainly more egalitarian than wet paddy in a context of availability of land and lack of outside interference: it was undertaken by the household, on land accessible to all, with minimal capital requirements: 'It provides few opportunities for the elevation of one villager above his fellows.'[25] But caste distinctions and power relations were reproduced in *chena* cultivation, and its precariousness (especially later in the colonial context) diverted a part of the resources obtained from it to the benefit of headmen and others. In a sense, it would be more appropriate to describe the 'peasant' strategy as a 'Kandyan chief' strategy.

22 R.E. Lewis, 'The rural economy of the Sinhalese', pp.40-1; Codrington, *Ancient Land Tenure*, pp. 4-5.

23 See especially Brow, *Vedda Villages*, p.100; Yalman, *Under the Bo Tree*, pp.100-1. On the other hand, Codrington, *Ancient Land Tenure*, and Leach, *Pul Eliya*, when describing 'wheel chenas' in the dry zone, insist on the communal organisation of the cultivation.

24 A detailed analysis of *chena* cultivation in a very large *batgama* of the Kegalla district, Dorawaka, where the original low-caste population had remained entirely on the spot, shows that there were two very large *henyaya* where a majority of the families cultivated, and which were in the 1880s in the process of partial conversion to gardens; and three smaller tracts where some of the families led 'parallel' cultivation. But in other villages the pattern seemed quite anarchical, at least in the eyes of outsiders who described *chena* as 'scattered all over the villages'.

25 Brow, *Vedda Villages*, pp. 103-04.

II. *New constraints and peasant responses under the colonial regime*

The imposition of colonial rule and the development of a plantation economy in Sri Lanka introduced a new set of rules and constraints which deeply affected cultivation practices.[26] The Dutch in the maritime provinces were the first to try to limit garden and *chena* cultivation in order to compel the peasants to preserve areas where cinnamon grew wild; at the same time they compelled the Salagama caste to deliver ever larger quantities of cinnamon. Their policy led to serious revolts during the second half of the eighteenth century.[27] When the British established the first coffee plantations on the Kandyan highlands, they similarly attempted to reserve the most convenient areas for their estates, and to restrict all other forms of soil utilisation, especially *chena*. In the dry zone, where the planters had no stake, protecting timber reserves and encouraging paddy cultivation were the arguments put forward to limit or forbid slash-and-burn cultivation. Repressive attempts were often successfully opposed by the peasantry, by passive or active acts of resistance. In other cases, some peasants found it in their interest to use the new opportunities created by the development of the plantation economy. The restriction of the options resulting from the disappearance of *chenas* was possibly offset by casual employment on estates, roads and other activities connected with the plantation economy.

Chena *repression and highland appropriation*

During the first decades of the British domination, the official policy was laissez-faire or even encouragement of *chena* cultivation for strategic purposes, so as to 'open' the country and destroy the forest belt used by kings and later by rebels as a cover for guerrilla attacks. Forest administration, which is always repressive, was non-existent. And revenue administration was not really concerned, although *elvi* was theoretically taxed after 1818 in the Kandyan provinces (millets were taxed only in the southern Matara district). Expropriation of surplus from *chena* was unremunerative and never insisted upon. At that time, 'waste land' was tacitly or expressly recognised as private property by most British officers, including those who later became advocates of *chena*

[26] On peasant agriculture under British rule, see generally K.M. de Silva, *A History of Sri Lanka* (London, Berkeley 1981), and Bandarage, *Colonialism in Sri Lanka*.

[27] De Silva, *History*, pp.163-5.

control.

There were no definite rules, and this very imprecision became an insuperable obstacle in the eyes of the would-be coffee planters in the 1830s. Hence the famous ordinance 12 of 1840 which established a legal presumption on the ownership of highland property in favour of the Crown, with a possibility to assert private title by producing Kandyan grants (*sannas*), British grants, or tax receipts (that is, proof of *elvi* cultivation since only paddy was taxed: this worked to the detriment of millet cultivators and may have led to an extension of highland paddy cultivation wherever possible). The purpose of the ordinance was to give clear titles to the planters, in the areas where they developed their estates.

However the ordinance was not applied systematically before the 1860s. The Revenue Officer for Matale wrote in his report for 1853:[28]

> Forest the natives know they have no right to fell. But every man had his *chena* and the natives cannot understand and never will understand that all chenas with respect to which the evidence required by ordinance 12 of 1840 cannot be adduced are the property of the Crown...While the law remains as it is, every agent must know that it is utterly impossible to protect the right of the Crown as it should be protected for every headman himself possesses tracts of such chena lands.

According to this officer's successor, Brodie, the legal claim of the Crown was not insisted upon: 'When it is proved that on many successive occasions the same land has been cultivated even with crops which pay no tax, then I believe government has invariably or almost invariably withdrawn all claim.'[29] But the arbitrary power of the *raj* and its local agents could be pushed very far on specific occasions. The agent in Kandy reported that 'in their desire to do their duty [that is, to implement the 1840 ordinance], I do not hesitate to say that not unfrequently a grave injustice is done to an individual by the Agent or his headmen'.[30] Some revenue officers became conscious that the new law was unjust and impractical. It pressed more on *kurakkan* cultivators—the poorest—than on *elvi* cultivators who could produce receipts, genuine or forged, or possibly pay for surveys. In a report of 1859, after pointing out the undefined legal nature of *chena* (rights to cultivate, not

[28] SLNA 10/177a.
[29] Letter AGA Matale to GA Central Province, 18 April 1859, in CO54/345.
[30] Templar (Report on the Central Province for 1864 in House of Commons Papers for that year).

to the soil) as compared with that of paddy ownership, Bailey characterised the British legal policy as follows: 'We appear at once to have done away with the subtle distinction between the right to cultivate and the title to the soil.' The first repressive outbursts, in the 1850s, were initiated by local officers connected with planters and were not fully supported by the higher authorities.[31] During the following decades, the working of the ordinance depended upon the land requirements of the planters, the personal predispositions of the successive governors, and the ability of planters' pressure groups to influence them. Most repressive were Hercules Robinson (1865-72) and Henry McCallum (1907-13). The impact of the ordinance also varied according to local conditions and to the idiosyncrasies of the different agents. For example, in Matale district, after a long period of rather lax enforcement of the law, there were repressive fits in the 1870s: but the permissive pattern was so well established that the agent lamented that:

> It appears to be the firm conviction of every Kandyan that all forest land belongs not to government but to the natives and that the fact that a man has cleared without the knowledge or with the connivance of the headmen a jungle once or twice at intervals of 7 to 10 years and sown *kurakkan* gives him a right to claim the land as his own.

Again in 1912 and 1913, the agent, Southorn, noted that the sudden repressive policy pursued by his predecessor (Tyrrell) had created serious tension—after years of laissez-faire. Tyrrell, acting on the orders of McCallum, had decided no longer to follow the practice of allowing highland 'appurtenances' to paddy fields, in order to check private land sales to outsiders.[32]

The mid-country districts were the scene of the most repressive out-

[31] See CO 57, Executive Council minutes, 4 September 1855, where Governor Ward censored the conduct of the agent at Batticaloa in the following terms: 'If the people of Batticaloa were not the most quiet, orderly and submissive people in the world, the conduct of the Agent must have produced and would almost have justified open resistance to his measures, which appear to me to combine the most characterized imprudence with the grossest injustice. For a nominal increase of land revenue Mr Morphew has sacrificed the character of government for good faith and plain dealing upon which depend all its relations with the people.... What one gentleman would scorn to do in dealing with another and could not do without vitiating his bargain, cannot be done by the Queen of England—and it is in her name that we are acting in dealing with a dependent people.'

[32] AR Matale 1872, p.61, SLNA PF 1954A, and AR Matale 1911/12, p.B18-19, and 1912/13 p.B22.

bursts, which coincided with a growing demand for highlands by planters. In every case, *chena* cultivation was depicted as a destructive and not a productive activity. The assistant agent at Kegalla wrote in his diary:[33]

It appeared to me strange that with so much fine timber forest, so easy of access, not one application for a permit to fell government timber had been made at the kachcheri since I had held office.... But as I penetrated further the reason was apparent. The jungles on all sides bore traces of having been from year to year ruthlessly destroyed. *Chenas* of all ages—of 30, 20, 10 years and of today are seen cut out of the centre of blocks of virgin forest presenting an appearance of wasteful destruction. To every enquiry I received the same reply, that forests and *chenas* were all regarded as native property.

He concluded: 'This form of cultivation must be put a stop to. Some new estates are being opened and are said to be doing well. I think all this land will soon sell as coffee land and I shall strongly recommend its being at once surveyed.'

In 1871, another systematic offensive was launched against *chena* cultivation in the Kegalla and Ratnapura districts.[34] In 1873, the agent at Ratnapura published a victory bulletin which was a piece of wishful thinking: 'I am glad to report that the rules proclaimed last year to prevent the destruction of forests have succeeded beyond my expectations.' He then described the lamentable state of things in a remote area after ages of *chena* cultivation, 'an unfruitful wilderness which barely supports the few people who exist in it, a sickly useless set like the land they inhabit'. In the same breath, he mentions the existence of a large untouched forest at the northern boundary of that district 'in every way suitable for coffee and tea cultivation, soil fair, climate pleasant.... At present it is the hunting ground of the villagers who collect large quantities of horn, honey and bees wax.'

In the Kegalla district, general dissatisfaction was caused by the sudden publication of these instructions and of another circular issued to headmen warning them to be very careful in the matter of *chena*, and that they would be held personally responsible for any illicit clearing.[35]

[33] SLNA 30/3, diary of the AGA Kegalla, 1-5 October and 15 September 1864.

[34] Ceylon Administration Reports, Kegalla 1872, pp.32-5; Ratnapura 1873, pp.111-12, 1874, pp.18-19.

[35] Ceylon Administration Reports, Kegalla 1873, p.117; diary Kegalla, 4 October 1872; report to Government Agent Colombo, 30 August 1872, in Sessional Paper no.15 of 1873.

The agent boasted that he had 'warned the people that they must take out licences for *chena* cultivation, and told them that [he] cannot admit their right to *chena* unless they can produce *sannas* [royal grants]'.

It was then decided to initiate a systematic and eventually unsuccessful policy of block settlement, later dubbed by its detractors '*chena* unsettlement', which I have studied elsewhere in detail.[36] In each village, one block of highlands was left to the villagers and the other reserved by the Crown. A major problem arose in villages where the settlement deprived certain lineages of the area where they used to cultivate *chenas*. Another form of settlement was tried under the Forest Ordinance after 1887; it created a right of user over large tracts of *chena* in certain areas, besides individual rights of full property recognised in response to claims. At that juncture, the repressive policy was found to be unrealistic :

Since 1815, the private title of villagers in highlands has been enjoyed without interruption and in every part of the Kandyan provinces *chena* lands have changed hands repeatedly on notarial deeds; it has been held by the Supreme Court that the payment of tax constitutes a good title against the Crown, and *wattorus* establishing proof of such payment could be produced for the greater part of the *chena* lands in these settlements. I believe myself that this individual right among members of each *pangu* existed in village life even under the Kandyan dynasty, and that it was recognized even in *gabadagam*. [However, communal lands should be settled on villagers] strictly prohibiting the alienation of such lands to other classes or for other purposes. I have repeatedly pointed out how greatly Kandyans have suffered from the failure of their coffee gardens; if in addition they lose their *chenas* they will certainly have no visible means of subsistence left and will become vagrants, and a serious burden on the general community, as soon as the small sums acquired by the sale of their *chena* lands have been consumed.[37]

After inconsistent successive settlements, the situation was an incredible mess in many villages:

A certain block of *chena* was declared Crown, another private; the *pangu* (lineage holding) whose lands were declared Crown were compensated by lands of other *pangu* out of the private block. When the people who had thus received lands had planted them, the original owners took a case against them and ejected them. The original owners of the Crown block then began to clear the Crown block.

[36] See E. Meyer, 'From land-grabbing to land hunger: highland appropriation in the plantation areas of Sri Lanka during the British period', *Modern Asian Studies* 26, 2 (1992), pp.321-61.

[37] SLNA, PF 173, Forest Settlement in the Kegalla district, report by GA Wace, 30 January 1889.

In other cases, two or even three successive settlements did not fit with each other: the villagers followed neither and no further settlement was possible![38] In 1905 the Agent, dealing with what remained of the 'old settlements' and with intractable interconnected land cases in the village of Paradenia, wryly concluded: 'The sooner the land is sold the better.'[39] Shortly afterwards, between 1900 and 1930, individual plots were purchased by share-hunters, while the areas set aside for *chena* cultivation were sold to planters with the help of headmen and the consent of government.

Troubles for paddy cultivation

The new conditions prevailing after the British take-over affected paddy cultivation in several ways.[40] First, the abolition of the elaborate mechanisms of the *rajakariya* after 1833 resulted in the neglect of the irrigation systems, especially in the dry zone.[41] The villagers themselves were generally keen on keeping these works in good repair but they lacked official encouragement. The most authentic testimony of their state of mind is to be found in the numerous petitions asking for government help.[42]

Secondly, the imposition of the paddy tax through renters discouraged paddy cultivation. It is generally recognised that agricultural taxation (one-tenth of the paddy produce) was rather light in Sri Lanka as compared with India. However paddy taxation led to very serious abuses. The evidence before the Grain Tax Commission of 1877 exposed the extortions practised by the paddy tax renters, and the use they made of the courts against the cultivator in case of default. Some officials maintained that paddy cultivation was actually paying, and that

38 Amitirigala (diary Kegalla, 3 January 1905); Maha Pallegama (ibid, 18 January 1905); Dorawaka (ibid., 3 March 1906).

39 Diary Kegalla, 15 May 1905.

40 Contra: M. Roberts, 'The impact of the waste lands legislation', who unconvincingly argues that plantation development had little impact on paddy cultivation.

41 SP 4 of 1867, Report on Irrigation, encl: Report by J. Capper on famine and mortality in the Seven Korales.

42 See, for example, the diary of the ASO F. Lewis for August 1908, reporting on a petition by inhabitants of the Kurunegala district asking for the return to the island of H. Parker, a very popular irrigation engineer (and an outstanding antiquarian and ethnographer), so that he could divert the waters of the Mahaveli river to the dry zone, a work effected 60 years later.

the very low yield figures provided by the headmen were greatly underestimated in order to evade the tax.[43] The conversion of paddy fields into gardens, especially in the Southern Province, was a more radical and effective device, as garden produce reached better prices than rice on the market. But even after the abolition of the tax (1892) the encouragement to paddy cultivation was quite limited, owing to the unequal repartition of paddy property:

In every village the poor possess a very small percentage of the paddy fields. The fields are owned by a few well-to-do people of the village who give no encouragement to the cultivators to produce better results. In many cases the fields are mortgaged and the crops given for payment of interest, in consequence the villager shows a preference for *chena* cultivation.

The general practice of share-cropping on conditions notoriously disadvantageous for the *ande* (half-share) cultivator, and the universal prevalence of indebtedness, were decisive factors in discouraging paddy cultivators as soon as other resources were at hand.

It was estimated in 1929 that a share-cropper family with two acres of paddy cultivated once a year, a garden, and a *chena* of two or three acres, had enough for its needs,[44]

but usually a tenant cultivator is obliged to supplement his earnings from time to time by going to work as a labourer on a neighbouring estate.... The cultivation of paddy is looked upon with increasing disfavour by the younger and more venturesome generations.... Interest in paddy cultivation is maintained more as a necessary insurance and by reason of the borrowing power of the crop than in the light of a remunerative undertaking.

In Uva where the *ande* system prevailed, an average tenant family was able to subsist for about two months of each year on the produce from his paddy holding, cultivation of which occupied him for about five months. The available highland for raising *chena* and garden crops was very inadequate, and consequently the majority of such cultivators, in order to provide sustenance for their families, were obliged to seek employment, usually on estates, for the remaining seven months of the year. In Negombo,

most of the cultivators are labourers on neighbouring estates where they find it more remunerative to work as daily paid labourers than to devote all of their time to the cultivation of their fields. Hence all the cultural operations are

[43] AR Southern Province for 1888, p.140A.

[44] SLNA A15/1929, Paddy Cultivation Subcommittee Reports, esp. Ratnapura District.

hurried through in the shortest possible time.

Even in Kandy district, 'paddy cultivation is by no means the sole occupation of the peasants'.

The import of cheap Indian rice, first for estate labourers, then for sale in the villages, was a major depressing factor. In Kegalla in 1929 rice cultivation was considered as a 'great' work (that is, rewarding in terms of status). It was not undertaken by calculating whether it paid or not. Imported rice being sold at 22 to 27 cents a measure, while the cost price of locally grown rice was around 40 cents.[45]

Finally, the opening of estates physically affected paddy cultivation (through the silting of fields, lack of grazing space for buffaloes). The fact was already recognised during the coffee period, before 1880:

> There cannot be a shadow of doubt that the extensive forest clearings made in Walapane and Hewaheta as elsewhere within the last 40 years or so, have very materially and prejudicially affected the productiveness of the paddy lands in these divisions, both by reason of the decrease in the regular flow of water in the streams on which the fields used to depend, and also because of the diminution in the supply of fertilising substances carried down in the low-lying lands from the adjacent forests during the rain.[46]

According to the Soil Erosion Commission of 1931, in every planting district of the island, except flat coconut areas, there were constant complaints of problems caused to paddy cultivation by erosion in uphill tea and rubber estates, and the subsequent silting of fields and irrigation channels. No statistics were kept at an all-Island level, and it is therefore impossible to quantify the phenomenon; but there is no reason to doubt the assertions of the commission (ignored by Roberts in his critical paper on the impact of plantations on paddy cultivation), and the repeated cases quoted in every report relating to the mid-country and the paddy-producing parts of the up-country. In most cases, fields had been silted by soil washed from estates: often whole valleys were affected for a couple of cultivation seasons, and some fields on the margins of the estates were abandoned for good.[47] In more serious cases, the

[45] Ibid, Kegalla, Uva, Negombo, Kandy districts.

[46] Report on the Walapane district quoted by Governor Gordon in his despatch to Knutsford, Secretary of State for the Colonies, dated 22 October 1889 (Eastern Pamphlets no.52, HMSO).

[47] Examples in Ceylon Administration Reports, Central Province, 1926; Matara district, 1927 and 1928; Diary of the District Agent Kegalla, April 1924; Diaries of the Land Settlement Officers, October 1933 (Kalutara district).

irrigation channels (*elas*) were choked, their banks breached by gravel and stones coming straight from estate drains, and water-flow to the fields was stopped.[48] In the worst cases (as in the Kotmale valley), earth-slips engulfed whole tracts of paddy fields.[49] The sources of water for irrigation were affected in several ways: even if one discounts the hypothesis of rainfall reduction, the springs included in the estates became less regular with the disappearance of trees and the thinning of the soil.[50] The rules for Crown sales usually reserved a wooded strip along the streams, but according to most reports, these reservations were ineffective in preventing erosion and silting: some officials remained quite strict, while others concluded that 'useless' stream reservations could be leased to planters, and planters themselves quite often encroached on them.[51] In the drier areas where tanks were used, their water-storage capacity was reduced by silt, or the reservoirs were even included in the estates and planted up: there were several such occurrences in the Nuwara Eliya district in the 1860s, and much later in the coconut belt (Kurunegala district). In the best-watered zones, water-powered factories sometimes tapped the streams to the detriment of paddy-field owners. And, at least during the coffee era, the streams passing through estates were regularly polluted by coffee berry processing operations; later, oiling of ponds and streams as an anti-malarial measure was also responsible for a large amount of pollution.[52]

In an indirect way, paddy cultivation is supposed to have been affected because the restriction of pasture land reduced the numbers and

[48] I have studied several cases at the grass-roots level in the villages of the Kegalla district, with the help of kachcheri records; the first detailed evidence relates to the village of Panawitiya in 1885.

[49] SP no.14 of 1890 (Report on the reassessment of Uda Hewaheta); SP 1951, Report of the Kandyan Peasantry Commission, pp.316-17.

[50] CO 54/517, Despatch 126, 29 March 1879, Governor of Ceylon to Secretary of State for the Colonies; AR Nuwara Eliya district, for 1884, pp.65-6A; Report of the Kandyan Peasantry Commission, pp.116-17.

[51] See SP no.43 of 1882 (D'A. Vincent report), § 110-114. Very frequent encroachments by rubber estates of the Ratnapura district: see SLNA, correspondence between the Government Agent at Ratnapura and the Colonial Secretary for 1912 and 1913; also AR for the Ratnapura district, 1926.

[52] SP no.2 of 1872/3 (Report of the Assistant Agent Nuwara Eliya on the paddy cultivation ordinance); Diaries of the Agent at Kurunegala for October 1922, and of the Assistant Settlement Officers in that district for August and September 1930; Diary of the Agent at Kegalla for November 1892, August 1899 and March 1918.

impaired the health of buffaloes. Roberts vehemently denies any such impact; according to him buffaloes were not regularly needed in paddy cultivation, and the actual reduction in their numbers was the result of epidemics brought from India rather than contraction of the available space. The question is somewhat outside the scope of the present contribution, but Roberts overstates his case when he maintains that buffaloes were not needed. Contrary to his assertions, labour supply was never abundant in the Kandyan regions in the nineteenth century, and it was only in the narrowest terraced paddy strips, and sometimes for the first turning of the sods in larger fields, that hoes were used instead of buffaloes. If Roberts is right in explaining the falling numbers of buffaloes by the occurrence of cattle murrain or rinderpest, the fact that buffaloes could no longer recoup their health in up-country areas must have exposed them more than before to the attacks. Cattle trespass on unfenced estates was from the beginning a major source of tension between planters and villagers: animals still needed the areas where they used to graze at large before the opening of plantations.[53]

The narrowing of subsistence options

The classical description of the process is given by J.C. Scott:[54]

> The narrowing of subsistence options forced the peasant family further out of production for his own use and further into the market.... The peasant might not be poorer but he was increasingly vulnerable to price fluctuations outside his control.... A life on the margin was more difficult. The only route to survival within the village led to a permanent dependence on those who hired labour.... Rents and taxes that had been tolerable only because this margin of subsidiary options remained open, became more intolerable once they were closed.

[53] M. Roberts, 'The impact of the waste lands legislation' pp.182-98: the author deconstructs the simplistic 'chain hypothesis' automatically attributing the decline of paddy cultivation to the development of estates, but he fails to delink the two phenomena. On the question of buffaloes and their necessity in paddy cultivation, there are innumerable references for every district, on which I am drawing for a forthcoming publication on 'Peasants, planters and buffaloes': among others, CO54/309, report of a Committee on Cattle Trespass, 5 March 1853, encl to desp.60, Governor to Secretary of State for Colonies, 26 October 1854; SP no.20 of 1869 (Report on cattle disease); SP no.6 of 1908 (Rice cultivation in Ceylon); paddy cultivation sub-committee reports, 1929 (SLNA A15/1929) in which the sub-committee of every district laid emphasis on the lack of buffaloes and silting from estates as the major troubles affecting paddy cultivation. Roberts does not appear to have used these files.

[54] Scott, *Moral Economy*, p.64.

Several testimonies concerning the period from 1860 to 1880, and also the depression of the 1930s, seem to bear out that process of increased vulnerability. The story of the Matara district, and especially of its northern border, the Morawak korale, is exemplary. Morris, the Government Agent of the Southern Province in 1868, had refused to allow *chena* cultivation in a situation of emergency—a drought affecting wet paddy cultivation. Elliott, his subordinate in Matara, was unwilling to carry out his orders, and described their results:[55]

Very great distress has been the consequence, especially in Kandaboda pattu where the people have not had a moderate harvest since 1862 and where in 1866-67 between 9 and 10 per cent of the population were carried off by fever and connected ailments, greatly aggravated by the want of proper food.... Dr Anthonisz (the Colonial Surgeon) has been shocked by the prevailing misery.... The fever is intensified and the death rate magnified by all the horrors of semi-famine.... Suppression of *chena* cultivation is a mistaken policy and calculated to do immense injury, and I submit that the present order for its suppression should be cancelled, the practice acknowledged.... I would repeat that *chena* cultivation is indispensably necessary for the well being of the people, as is shown by the undoubted misery caused in the neighbourhood by its indiscriminate suppression.

In spite of Elliott's arguments, four years later Morris carried on his policy:[56]

I am aware that Mr. Elliott entertains a strong opinion of the benefits of *chena* cultivation.... I would encourage the planting of gardens but in the Matara district I consider that fine grain cultivation ought to be put an entire stop to. There are abundant demands on the estates, the roads, the irrigation works and by private individuals for labour which has now to be imported from India, whilst our own available population is engaged in destroying the country.

When Elliott left Matara, his successor Fisher fell in line with Morris, recommending a total prohibition, but with little success:[57]

so wedded had the people become to the occupation of *chena* cultivation that I found the sale of 1/2 the crops had no effect on unlicensed clearing and I accordingly ordered the prosecution of all offenders.... These harsh measures have had the desired effect and entirely put a stop to the practice.

Later Elliott became agent of the province, and rules were relaxed from 1885 to 1892, since it was impossible to enforce the control owing

[55] AR Matara, 1868, pp.232-3 (Elliott).
[56] AR Southern Province 1872, p.149 ff.
[57] AR Matara 1875, p.19.

to the connivance of headmen.[58] Shortly afterwards another agent, Le Mesurier, adopted a more repressive policy, but he was expelled from the civil service for other reasons, and started perverting the system by collecting deeds for highlands in the very district he had earlier administered. Hence a new prohibition was aimed at that time at Le Mesurier but fell on the villagers as well:[59]

Soon after I assumed duties I obtained the sanction of government to interdict *chena* cultivation in the Morawak korale and if possible in the Hambantota district. There is such demand for labour throughout this district that there is really very little necessity for *chena* cultivation as a means of food supply.... The only division in which this prohibition is really felt is in the Morawak korale, where the extent of paddy land is quite insufficient for the requirements of the population. The people must however learn sooner or later to find a livelihood independently of Crown *chena* land, and the sooner they begin the better.... However it is desirable that all private claims to *chenas* be inquired into and dealt with in a liberal spirit.

The result was soon felt by the peasants:[60]

At Alapaladeniya, *chena* cultivation was allowed from 1885 to 1892 and in 1892 it was fully put a stop to; the people were then much better off than they are now, as there is not enough paddy land for them to cultivate. Men now go to Panagoda in Hinidum pattu to work in the plumbago pits...but all the people cannot get work at the pits. Nearly everywhere in the district I am told by the headmen that the people owing to the stopping of *chena* cultivation and the want of paddy lands do not get enough to eat. Yet the Administration report usually says that 'food was plentiful'.

In 1901, the settlement officer, Fraser, was again struck by the miserable appearance of the Morawak villagers, and, 'after considering the question without bias', found that '*chena* cultivation must be allowed'.

Peasant resistance

Chena repression has been analysed as 'the only source of discontent among the Kandyans' (Bailey in 1857). As early as 1840, the leading headmen of the Ratnapura district submitted a petition in which they defended *chena* cultivation threatened by the new law.[61] A few years

[58] AR SP 1888, p.155A; see also AR Matara 1886, p.91A.
[59] AR Southern Province 1897, p.E8, Matara, p.E20.
[60] Diary of SO J.P. Lewis, November 1900. Compare with AR Kegalla 1875, pp.65-6: 'It appears that there must have been a scarcity of food, but it was relieved I hear by emigration to the neighbouring coffee estates of much of the adult population during the year, and so avoided calamity. The cultivation of high lands in this division should I think be liberally regarded.'
[61] CO 54/345, Memorial of the inhabitants of Saffragam, 3 November 1840.

later in a *durbar*, Kandyan chiefs

> complained that Government claimed all uncultivated lands, but there were chiefs who had lands in different parts which they cultivated at intervals. Others had reserved the higher lands on purpose for timber, others did not cut down the timber in some parts of their lands for fear the rivulets and water courses should become dried up by exposure. If the government claims these as forest lands, what was to become of them all?[62]

And as late as 1928, H.R. Freeman, an ex-revenue officer, asked: 'How is it that these law abiding villagers become law breakers in the matter of their food supply?' Confrontation between villagers and foresters, which was common in British India, South East Asia or pre-industrial France,[63] in colonial Ceylon took the form of a covert or overt resistance of *chena* cultivators against the colonial authorities, and, to some degree, of the villagers against the planters and their agents.

The inroads of the plantation system had always been resented by the Kandyan villagers, who found their paths blocked, their cattle impounded or shot, their paddy fields silted by the estates. But peasant resistance was never an organised movement in Ceylon, and to the present day there has never been in the country a strong peasant consciousness.[64] There were a few episodes of overt resistance to the opening of plantations, but on the whole the villagers reacted by ignoring them and trying to bypass the rules established by the *raj*. One such episode is worth quoting as it may indicate the state of mind of the Kandyans in the early days (the 1840s)—and as it is could be regarded as a fascinating example of spontaneous *satyagraha*.[65]

[62] CO 54/238 Torrington to Secretary of State for the Colonies, 12 August 1847, encl. 2, Notes of the first conference held at Kandy, 29 July 1847, with a deputation of chiefs and priests of the Kandyan Provinces.

[63] R. Guha, *The Unquiet Woods. Ecological change and peasant resistance in the Himalaya* (Delhi 1989); Scott, *Moral Economy*; Corvol, *L'homme aux bois: histoire des relations de l'homme et de la forêt, XVIIe-XXe* (Paris 1987).

[64] This is analysed in Moore, *State and Peasant Politics.* See also E. Meyer, 'The plantation system and village structure in British Ceylon: involution or evolution?' in P. Robb (ed.), *Rural South Asia: linkages, change and development* (London 1983), pp.23-56; 'Rural society and the bourgeoisie in Sri Lanka, 1880-1940', *South Asia Research* 11, 1 (1991), pp.40-60; '"Enclave"' plantations, "hemmed-in" villages and dualistic representations in colonial Ceylon' in E.V. Daniel, H. Bernstein and T. Brass (eds.), *Plantations, Peasants and Proletarians in Colonial Asia* (London 1992), pp.199-228.

[65] P.D. Millie, *Thirty Years Ago, or the reminiscences of the early days of coffee-planting in Ceylon* (Colombo 1878), ch.XX.

A surveyor had pitched his tent on some of the grass lands below Harrow Estate, the whole of the existing estates being then of course forest. He had had a weary toil as there were neither roads nor bridle paths. When he passed on, troops of natives came up in his rear and surrounded him on all sides. They however offered no violence, but talked and bewailed and lamented as only natives can do. They would not go away, but surrounded his tent after it was pitched. The next morning he commenced with theodolite and chain, but the natives stood in front of the former, and threw themselves down on the ground before the course of the latter, saying: 'Pass over our bodies, our dead bodies, before you measure and sell the hunting grounds of our forefathers'. So, without any actual violence being used, the work was stopped, and the surveyor had to depart and report the state of affairs at headquarters.

Resistance generally took passive or covert forms, with the connivance of headmen: common were non-reporting of *chena* clearings, pleading destitution, and using forged tax receipts. Hiring lawyers, generally with the help of interested parties such as low-country speculators, and refusing to obey orders, became general practice, especially in the Kurunegala district: but at that stage it had little to do with actual needs of *chena* cultivators.[66] Another method was to play the game by laying claim to huge areas. It is likely that settlement operations led the villagers to lay claims which would not have occurred to them, by a process of reification. Such was the case in the Matara district:[67]

Formerly the villagers used to cultivate large extents on permits from the Crown, paying share to the Crown. During that period the title of the Crown was never questioned. Then the policy was adopted of restricting *chena* cultivation in the hope that the people would devote themselves to permanent products, especially paddy; this may be good policy where paddy land is available in a village, and the people have the means to acquire it. But if the people have no money, they cannot buy it and can only work as coolies or go to a big irrigation work. This *chena* question has greatly strained the relations between the Revenue officers and the villagers.

The revenue officer fined the villagers who disregarded his orders (five rupees per acre instead of a permit at one rupee per acre), and when he found that they continued, he forbad cultivation and sequestered the crops.

This, the villagers say, drove them to desperation for they must live, and they began to clear, claiming the land on any frivolous title. It is in this way that a great many claims have originated. Now the villagers refuse to acknowledge Crown title. The villagers spoke quite sensibly and said *they did not want the*

[66] SLNA, H.R. Freeman, Evidence before the Land Commission, 1928.

[67] Le Mesurier, Memorial to the Secretary of State for the Colonies, 12 June 1896, in SLNA, PF 570.

land but only their immemorial right to chena.[68]

Resistance, combined with the inner contradictions of the repressive policy, paralysed its working. According to D'A. Vincent, Forest Commissioner in 1882:[69]

For 10 or 15 years it has been held that in compliance with an order from Government, *chena* cultivation could be stopped, but the impossibility of stopping the cultivation of grain on which the poorer people depend as much or possibly even more than on paddy, and the inability of any administration to carry out [an] order in opposition to the mass of the people and likely to inflict hardship, unless supported by the law, have not been considered. The issue of the order and its occasional repetition appears to have satisfied the conscience, especially when it has met with an occasional assurance that the order was being carried out and that *chenas* were stopped.... It has been frequently pointed out that if the wants of the people are considered and *chenas* are allowed in moderation, the people are easy to manage; if *chenas* are forbidden absolutely, they get exasperated and when owing to some excuse of famine, the orders of government are suddenly relaxed, the forests are devastated wholesale.

The government had eventually to take into account the reactions of the peasantry. In the North-Western Province, during McCallum's governorship in 1911, a fresh Government Agent decided that 'permits for *chenas* should be insisted upon'; he proposed to proclaim in the villages by tom-tom beat 'what the government view of the title to land is'. The then Controller of Revenue, Fraser, in a private letter, warned the Colonial Secretary:

Unless great care is exercised by government in restraining Thorpe's well-meant but almost *fanatical* zeal, you will steadily have a recrudescence of the whole land question in Ceylon. I understand the people of Weudawili are contemplating organized action.... I see no object in exasperating the people by fines and prosecutions.[70]

The development of new options—gardens and small plantations

To end *chena* cultivation by converting it to garden cropping was the aim of the modernists. Ludovici was the first to point out that vegetable cultivation on highlands was very profitable, and that *chena* products were more marketable than paddy. Later pro-native administrators such as Davidson, probably the most intelligent civil servant of his time,

[68] Diary SO Fraser, 14 June 1902.

[69] Report on the Conservation and Administration of Crown Forest in Ceylon, Sessional Paper 43 of 1882, ch.6.

[70] SLNA, pending file 2059; Fraser to Bowes, 31 July 1911 and 22 August 1911.

wanted to sell land from the Crown block to enterprising villagers to help them develop gardens. Obviously *chena* was incompatible in the long run with market gardening; but could the authorities compel the peasant to take up garden planting by preventing him from cultivating *chena*? Settling titles to the highland seemed to be a better course: this was the new policy, and it did work. However Davidson recognised that garden development was difficult as long as there was no substitute for *chena*-produced grain:

> *Chena* cultivation is the result of the communal system of holding highlands in vogue in the Kandyan provinces, and as a method of agriculture it decreases and disappears before the growth of wealth and the security of a personal title in land. I can illustrate this by means of a table in which are placed the *korales* of this district in order of wealth and population, showing how in the more favoured places the population has grown up to the capacity of the paddy fields to feed it, and the leisure to turn the highlands into gardens instead of being compelled to cultivate them with grain.... As soon as the peasantry has a margin between it and hunger, it will improve its lands permanently, and *chena* cultivation will tend to decrease.[71]

But the success of plantation products led to the abandonment of paddy fields and their planting with rubber or coconut in certain areas after 1900. This was attributed to the much higher profits which could be derived from plantation products, rather than to environmental factors affecting paddy cultivation. The comparative profits of the different forms of cultivation were thus summarised by the Government Agent of the Central Province in 1908:[72]

> Paddy is far less remunerative to native land owners than tea and coconut, for while paddy yields only 60 Rs per acre, coconuts are estimated to yield 187 Rs per acre, and native tea 125 Rs per acre. When the labour necessary for the cultivation of these three products and the chances of the failure of the paddy crops are taken into consideration, it is obvious that as an investment, paddy cultivation, while actually not unremunerative, cannot compare with coconut and tea, and possibly it may be added, with rubber and cacao. The fact is that paddy cultivation cannot altogether be regarded from the point of view of profit and loss; it is not grown primarily to sell, but to use.

However the conversion from paddy to plantation remained very limited: the resilience of paddy cultivation and its technical stagnation were clearly linked to its subsistence character.

71 AR Kegalla 1896, p.J17: see also report of 15 May 1895, SLNA 45/1868 PF 1187, Kandupita Forest Settlement.

72 SP 6 of 1908, Rice cultivation in Ceylon, encl: Report of the GA Kandy, 6 March 1908.

The dramatic increase in garden cultivation and concomitant development of rural markets in the whole of the island at the turn of the century was noticed by every observer.[73] Some areas eventually specialised in market products, such as Kinigoda plantains, or Rakwana and Uva vegetables. As early as the 1890s, in Kegalla, it was estimated that:

> 209,959 bushels of rice were locally grown, and adopting the corrected census figures for 1891, the population being about 129,000 we have 1.5 bushel for each person, manifestly insufficient for the needs of the people. The very large extent of gardens, estimated at 20,000 acres, furnishes means whereby the population was enabled to import a sufficient supply of food, combined of course with dry grain cultivation on *chena* lands.[74]

But most *chenas* had been alienated for estate plantation development, and the reports for Kegalla in the 1920s all notice that '*chena* cultivation seems to be dying out, the villagers preferring to work for wages on the estates and buy imported rice in the bazaars'. This trend is documented by statistics collected by Patrick Peebles: between 1921-24 and 1946, areas cultivated under the *chena* system in the whole of the island were reduced by 42 per cent. The reduction was only 32 per cent in the North Central Province, in the dry zone; in provinces including dry and intermediate or wet areas it reached 37 per cent (Central Province), 40 per cent (North-Western Province), or 42 per cent (Uva); in the wettest areas, it was as high as 75 per cent (Sabaragamuwa).[75]

New options—estate employment

The increasing and mostly casual employment of Kandyan villagers on estates is an important chapter in the history of rural Sri Lanka, one which has too often been neglected.[76] After the restriction in resources resulting from the contraction of *chenas*, it represented a decisive

[73] E.B. Denham, *Ceylon at the Census of 1911* (Colombo 1912).

[74] Diary Kegalla, July 1893, commenting on the Administration Report for 1892.

[75] Calculated from P. Peebles, *Sri Lanka. A handbook of historical statistics* (Boston 1982), table VII, pp 118-19.

[76] E. Meyer, 'Between village and plantation: Sinhalese estate labour in British Ceylon', in M. Gaborieau and A. Thorner (eds.), *Asie du Sud: traditions et changements* (Paris 1979), pp.459-68. In 1946, 15 per cent of the population residing on estates was Sinhalese, and Sinhalese labour including non-residents was much higher; in 1939, on larger estates, about one-third of the Sinhalese employed were resident, one-half regular non-resident, and the rest employed on contracts.

enlargement of subsistence options. Its impact on the standard of life of the villagers who lived near enough has been hotly debated. Further quantitative research is required to reach a global and definite conclusion as to whether such employment balanced the loss of traditional options.[77] At the least one can argue that resources derived from the presence of plantations became indispensable to the economic balance of the village, and that economic fluctuations which affected plantation products had beneficial or devastating consequences on the villages. It was estate work which enabled the more enterprising peasants to turn to garden cultivation, by replacing *chena* work with more rewarding activities. Whenever residence was not imposed, Kandyan villagers eagerly sought employment on estates, especially when they were led to seek money wages, because of the disappearance of their *chenas*, troubles to paddy cultivation, collapse of native coffee and grain tax crisis, land hunger increased by population growth, and monetary needs induced by new consumption models: estates were then the only source of cash within reach. The attraction of estate and other wages became quite obvious during the first decade of the twentieth century especially in the mid-country. In the Ratnapura district, the very serious decline of paddy cultivation was attributed to gem mining and moreover to estate employment: 'It is often said that the owner of a small field will generally leave estate work to cultivate his share. But as wages rise, the temptation to stay on increases and the villagers are beginning to realize that coolie work pays better than rice cultivation.'[78]

Labour flows between villages and tea estates were regulated by the varying needs of the planters, who employed villagers as extra labour in rush periods or on contract for specific tasks such as weeding, while keeping their regular Tamil labour nucleus; it also depended on the villager's agricultural and festive seasons.[79] The nature of work in rub-

[77] L.R.U. Jayawardena, 'The supply of Sinhalese labour to Ceylon plantations, 1830-1930, a study of imperial policy in a peasant society', Ph.D. thesis, Cambridge, 1963; Meyer, 'The plantation system'.

[78] AR Sabaragamuwa 1907, pp.12ff.

[79] According to the reports of 1929 on paddy cultivation in the plantation district of Nuwara Eliya, agriculturists 'are not seriously indebted as estate work is available when paddy cultivation is finished' (SLNA A15/1929, Paddy Cultivation Subcommittee Reports, Nuwara Eliya district). In his recent monograph on the Badulla district ('Spring Valley: a social, anthropological and historical inquiry into the impact of the tea estates upon a Sinhalese village in the Uva Highlands of Sri Lanka', Ph.D. thesis, University of Sussex, 1982, p.145)

ber and in coconut cultivation made it much easier for villagers to work more regularly, but part time, as tappers, or to pick and transport the nuts, while keeping an eye on their own 'farm'. There were other forms of employment indirectly connected with the plantations, such as sawing timber, and selling garden produce or drinks to estate coolies. Peasant strategies in the new economic context somehow reproduced the traditional pattern of resource diversification described above: women and young people were employed as tappers or coir workers, and paddy cultivators often tapped rubber trees in the morning before working their fields, except during the sowing and harvest periods.

It is well known that the local Kandyan villagers refused to reside on the estates and chose casual labour instead. They never became a plantation proletariat. Two series of connected factors are usually put forward to account for that situation. The first was the choice of the planters, who preferred to employ immigrant Tamil labour resident on the estates, tied by their indebtedness and personal allegiance to the Tamil recruiter and overseer (*kangani*): a pattern typical of the plantation system all over the world, based as it was on the exploitation of cheap labour under strict supervision. The second factor was the pride of the villagers, and notably the caste pride of the Kandyan Goyigamas, leading to their refusal of 'slave labour' at the beck and call of a foreign master (this implies that the lower Kandyan castes would have been less particular in the matter).

But more basic considerations were at work: the villagers wanted to retain their ability to shift between different resources according to circumstances. The experience of coffee collapse in the 1880s made them wiser for a time. Those who became exceedingly dependent on rubber and coconut again paid a tremendous price during the Great Depression, but even then the majority of villagers had not become entirely dependent on plantation products so that the country escaped a major catastrophe. There is no peasant strategy if there is no freedom of

Kemp shows that 'whilst most of the productive work of tea cultivation was indeed left to the regular Tamil workforce, the work required by maintenance and improvements, rush periods or fluctuations in the Tamil population was given to local Sinhalese. In the Uva highlands, the physical juxtaposition of village and estate was more than a case of mutual avoidance': for the period 1908-50, the estate regularly employed between 180 and 320 village workers, under 12 to 18 *kanganis*; in the 1970s, there were about 575 villagers on the books, against 883 adult Tamils and 665 young Tamils.

choice: the Kandyans by refusing regular estate work felt they safeguarded that freedom. Here we have a clear case of 'strategic choice' which involves a set of factors, some related and others unrelated to mere economic rationality. My point is that the Kandyans evolved a subtle compromise by working as casual labourers on estates wherever possible, while keeping residence in the village. Residence was what mattered in a caste-bound society, while work relations on the estate, governed by economic considerations, were unimportant to them so long as they did not impinge on private life. Economic rationality stopped short of considerations of honour and private life. There were cases of people who abandoned their houses to work for the cultivation season, especially but not exclusively among men of the lower castes. But this was invariably done with a view to coming back to the village. The aim of every peasant of whatever caste was and still is to preserve and enhance his *paraveni* (hereditary) rights in village lands, and to keep his house in good repair or build a new one. For that they were prepared to work on estates or in towns, just as they are now prepared to seek employment in the Middle East.

III. *Elite representations of cultivation systems*

The policies pursued by the authorities were largely dependent on the views held by the most influential pressure groups during the successive stages of the colonial period.[80] Roughly speaking, the position of the planters and their allies was predominant at least up to the beginning of the twentieth century, but was increasingly challenged by the views held by the 'proto-ecologists', and by the different 'pro-native' groups. With the rise of the nationalist movement, the defence of the villager became a dominant theme, which led to a reversal of the government policy in the late 1920s, but the new standard representation of the rural was a reconstruction bearing little resemblance to the actual state of things in the countryside.[81]

The enterprising planter and the lazy native

In the colonial discourse, *chena* cultivation typified and epitomised everything the Victorians considered as evil: unrestricted freedom as

[80] For this section, see also my 'Plantations as "enclaves"'.

[81] Spencer, *Sinhala Village*, p.108: 'Chena cultivation, drifting across the landscape, often escaped the classifications of power'.

opposed to responsibility; desultory, temporary and nomadic way of life as opposed to settled activity; laziness as opposed to exertion; imprecision as opposed to visibility. It was characterised as the lazy man's cultivation, the Kandyan cultivation, the antithesis of European-cum-Tamil estate production. Later it came to be opposed to paddy cultivation in similar terms. The myth of the lazy native was general throughout the colonial milieux of Asia and Africa.[82] In 1866, Governor Robinson expressed it in typical terms:

> The wants of the native population of the island are few and easily supplied by an occasional day's work in their own gardens or paddy fields; their philosophy, their love of ease and indolence, or their limited ideas—whichever may be the real cause, render them perfectly content with what they already possess, and it is therefore futile to expect that they could by any mere outlay of government money, be ever brought to supply with food an immigrant population [the Tamil coolies], whose presence in the island is mainly rendered necessary by their own apathy.

The racial component of the myth and its power implications were denounced by Ludovici: 'the acknowledged value of [the planter's] enterprise to the country gives him such an importance in the eyes of the government as to qualify him to assert his claims with an amount of self-confidence unattainable by a native.'[83] He suggested that the myth was connected with the experience of the European himself and his fears:

> Even the European in this country soon yields to the enervating influence of the climate...and he soon enough takes to the fashion of damning the climate and speaking of the natives as a lazy apathetic race.... This belief in native apathy has become fashionable, it has had a prescriptive existence of above half a century and the members of government seem to cling to it as a venerable article of faith in their code of political science.

Although field observers were well aware of the arduousness of *chena* cultivation, the myth of laziness was so deeply rooted that it survived the failure of the policy it tried to justify. In 1884, the agent at

[82] S.H. Alatas, *The Myth of the Lazy Native* (London 1977).

[83] L.Ludovici, *Rice Cultivation, its past history and present condition* (Colombo 1867), quoting the opening address to the Legislative Council, 3 October 1866. In a letter of the Colonial Secretary to GA Colombo, 21 January 1873, Ludovici is mentioned as a private surveyor of great experience who could be entrusted with the task of effecting the *chena* survey. For a similar attitude in south India, see H. Cleghorn, *Forests and Gardens in South India* (London 1861).

Puttalam wrote: '*Chena* is a lazy kind of cultivation in which women and children [*sic*] take part, it introduces lazy habits, the work is easy, the crops are large and the result is a supply of poor indigestible food'.[84] It was also viewed in truly Victorian terms as an immoral activity undertaken by dangerous vagrants: 'The love of a wandering life which it engenders is one of the greatest evils and we shall never improve the character of the people till we induce them to settle on their legitimate property and devote their energies to a more permanent cultivation in gardens and paddy fields.'[85]

The duty of the severe and benevolent colonial master was therefore either to train the natives to industrial discipline, or failing that to direct their energies to manly pursuits such as paddy cultivation. The administration report for 1872 by the Government Agent of the Southern Province (Morris) epitomised the colonial discourse:

In an oriental and uncivilized country, it is incumbent on the rulers to think for the people and to exercise in some sort a patriarchal sway over them. In no way is the necessity for this shown more than in interposing to protect for them their native land and to prevent its being, to use the word of a report I have quoted [written by his father] 'reduced to a desert'. The objection to chena cultivation may be thus summarized: it is wasteful; it engenders a nomadic and improvident habit among the people and retards civilization; it is ruinous to the general welfare of the country; and it is in reality but a poor means of support, for though the crop produces from 20 to 40 or 80 fold, the grain is so extremely small that the produce is of small bulk.[86]

As late as 1911, the Governor, McCallum, who made efforts to help the planters obtain Sinhalese labour, minuted as follows on the proposal by Leonard Woolf (then a member of the Ceylon Civil Service) that *chena* cultivation should be allowed in Hambantota district:[87]

It is the proper policy of government to discourage the *chenas*, which has only been the means of bringing up our people in an atmosphere of idleness and dependence of the government, while the ruin to the country has been incalculable.... If *chenaing* is prevented everywhere and in every province including Uva, there will be no need of villagers leaving one district for another except to find work; and if they only work instead of spending idle life, depending on government and damage to Crown land, it is immaterial whether they are in one

84 AR Puttalam 1884, p.100A.
85 AR of the Agent at Ratnapura for 1864 (in House of Commons Parliamentary Papers p.139); note by Fowler dated 8 September 1906 in PF 2226.
86 AR Galle 1872. The objections are set out in list form in the original.
87 SLNA, MP 04261/1910, minute of Governor McCallum, 29 May 1911, quoted by Jayawardena, 'The supply'.

district more than another.

Chena cultivation was regarded as leisure not work. It was also an untaxed activity which brought no money to the exchequer. Finally, it was considered as a collective activity based on communal property, and the duty of the colonial government was to universalise individual property rights in the island:

> The fixed policy of government is in favour of the extension of permanent garden cultivation. With this end in view, endeavours are being made to limit the area held under the *panguwa* system of land tenure and now wholly devoted to the destructive methods of *chena* cultivation. The government hopes thereby to be able to sell small allotments of land to villagers under Crown grants, thus facilitating the transfer of land from the 'family' to the 'individual'.[88]

The proto-ecological discourse

When in the 1870s, the preoccupation with the preservation of the environment became popular among scientists, *chena* cultivation was accused of causing of ecological disruption. This was an easy means of shifting attention from the responsibility in that process of the growth of plantations. In Ceylon, forest administration was weak because the interests of the all-powerful planters clashed with those of forest conservancy. The ecological discourse was used by the state to repress *chena* cultivation in order to foster the interests of the planters: rather than a concern for forest conservancy, it was that alliance which accounted for the repeated fits of *chena* repression during the nineteenth and early twentieth centuries. This repressive discourse was given romantic colours borrowed from the spectacular features of slash-and-burn cultivation: a criminal activity, a form of arson, carried on in the heart of forests, it was interpreted as a kind of desecration of nature. This kind of proto-ecological discourse was common in nineteenth-century romantic as well as positivist literature: for romantics, the forest was sacred. *Chena*, destructive of forests, was supposedly disruptive of ecology, and especially was the cause of repeated droughts: the equation of drought with fire was obvious.[89]

According to Morris, who held sway over the Matara and Kurunegala districts in the late 1860s and 1870s, where his repressive policy

[88] Kelani Valley Railway Commission Report, SP XXII of 1895, p.46.

[89] See Corvol, *L'Homme aux bois*, esp. p.280, where she quotes the French writer Blanqui giving an apocalyptic description of slash-and-burn (*brûlis*) cultivation compared with a volcanic eruption.

contributed to near famine conditions, there was 'a steady and rapid desertisation of the country by the destruction of forest and jungle for *chenas*, and the consequent diminution of the rainfall'. The food shortage in 1874 in the North-Western Province was, according to him, the result of the sequence in which *chena* cultivation led to the disappearance of forest then to drought, and then to the failure of paddy cultivation. He inferred from this theory that *chena* cultivation should be stopped at once—as if that would restore rainfall at a stroke.[90] Paddy and *chena* cultivation were considered to be antagonistic and mutually exclusive: suppress the latter and you will develop the first by diverting peasant exertion. In the North-Western Province, the abandonment of many villages was attributed to the neglect and breach of tanks, and the concomitant development of *chena* cultivation: the question of whether slash-and-burn cultivation was the cause or the effect was not seriously considered.[91] But, according to a chief headman, Mahavalatenne, the choice was not between locally-produced rice and *chena* products: if *chena* was forbidden, the people would have to subsist on imported rice.

In 1873, C.H.K. Thwaites, director of the Peradeniya Botanic Gardens, brought to the notice of his colleague, J.D. Hooker, director of the Royal Botanic Gardens at Kew, the fact that the indiscriminate felling of forests for coffee and tea plantations created grave troubles for the environment. He prophesied that the damage would increase with the adoption of new products such as tea. Hooker, who was recognised as an authority in tropical affairs, sent a letter to the Colonial Office in May 1873, pointing out that the disappearance of forests in tropical colonies had very detrimental results:

> The mischief done in other colonies is especially to be guarded against in this case. The planting of tea and coffee has suddenly become an object of active and to some extent of almost speculative enterprise.... There is reason to fear that districts may have been hastily cleared of forests under this inducement, which it may prove impracticable to permanently occupy when the fertility of the soil has been lowered by a few years' cultivation.

Gregory, the Governor of Ceylon, after extorting from Thwaites the assurance that rainfall was not actually affected, attempted in a some-

[90] AR North Western province, 1874, pp.125-7: he based his theory on the work of Col. Corbett from India; SP no.16 of 1877, Grain Tax Commission report, appendix, answers to questions, J.G. Jayatileke, Kurunegala.

[91] Diary of ASO Hunter, January 1928.

what muddled and uneasy reply to minimise the responsibility of the the plantations for any environmental problems by raising the *chena* question. He added:

The clearing for coffee cultivation is so infinitesimal with regard to what has been cleared and what is likely to be cleared, though doubtless large in acreage, compared with the enormous extent of forest above the height at which it is supposed coffee can be cultivated, that no apprehension whatever need be felt as to any injury to the climate arising on this account.

He went on, with an affected magnanimity, that the interests of the native population were to be protected by keeping for village use patches of forests—'In these patches natives obtain their free sticks, wood for fuel and for building, and shelter for their cattle, and it is almost tantamount to expulsion to clear them'—and that *chena* cultivation when actually necessary was to be allowed under strict supervision. The Colonial Office passed on this report to Hooker. He commented that, regarding the extent of forest destruction, he 'might have been misinformed by some of the most intelligent of the planters who probably included the *chena* cultivation in the same category with the coffee', but he maintained that the probable development of tea 'to an immense extent' would 'inevitably lead to the destruction of the forests in those regions where it is of paramount importance that they should be preserved'.[92]

Rehabilitating chena cultivation

A minority of close observers of *chena* cultivation defended the practice on technical and, more often, on economic grounds, and denounced the crass ignorance of their colleagues. Some questioned the reality of its ecological ill-effects and noticed that *chena* cultivators were careful to protect water supply:

The *chena* system is undoubtedly responsible for a large amount of preventable erosion, but it is possible that the damage is not so great as is often thought. The *chena* cultivator does not introduce drainage systems into his land, but on the other hand he does not keep the surface of the soil so scrupulously clean as the estate owner, and the very presence of weeds assists in preventing soil wash.[93]

[92] SP no.29 of 1873 and CO54/487, Governor of Ceylon to Secretary of State for the Colonies, 31 July 1873. For a detailed study of the ecological discourse in Sri Lanka, see my 'Forests, *chena* cultivation, plantations and the colonial state in Sri Lanka (ca.1840-1940)', in R. Grove and S. Sangwan (eds.), *History and the Environment in South and South-East Asia* (Delhi, forthcoming).

[93] In SP 3 of 1931, ch.3; also Ellis (SLNA, PF 1075, Land settlement work in

Others, without defending *chena*, insisted that it was well-adapted to a traditional society near the subsistence level, on account of its being specific to a 'tribal' way of life, or accommodating to the local ecology.[94] In this view, *chena* products were the poor man's, not the lazy man's crops. The condition of 'feudal village' cultivators was, accordingly, much better than that of others because they had unlimited access to *chena*.[95]

If a comparison be made between service villages and any adjoining village held by peasant proprietors and not liable to service, the results will be in favour of the former.... This is attributable in a great measure to the greater security the service tenants possess from the settlement of village boundaries in reference to the *chena* which, though cultivated in the most rude and primitive way, are of the utmost value to the poorer classes.... To refuse to allow the cultivation of *chena* altogether is often to condemn a whole village or district to starvation. And yet this has been attempted at times, in complete ignorance of the real state of the case. On one occasion, the temporary administrator of government publicly told a deputation that the government was determined to put down *chena* cultivation, as the *chena* plant was a most obnoxious weed which exhausted the soil and so on. Fortunately the deputation did not understand English, and the interpreter failed to explain the reasons of the temporary head of the government.

The most penetrating observers saw the strategic complementarity of *chena* and paddy cultivation:[96]

---

the NWP, memorandum dated 10 February 1904): 'The representations made by scientists and others as to the injury done to the soil by this species of cultivation must be considerably exaggerated'. See contra: evidence of planters before the Commission on Soil Denudation in the Kelani Valley (SP 42 of 1905), which blames 'the wasteful and pernicious system' and pretends that 'in mountain regions the rivers have a tendency to become naturally silted'!

[94] See in this respect the brilliant defence of 'swidden' cultivation in Indonesia by Clifford Geertz in the second chapter of his classic *Agricultural Involution*. According to Geertz, 'swidden' farming is integrated into and maintains the general structure of the pre-existing natural ecosystem; contrary to the man-made and highly artificial wet paddy cultivation, its rationale is to reproduce the multiplicity of plants and resources found in the original tropical forest (which is itself the ecosystem with the highest diversity), eliminating the useless plants and adapting itself to a shallow soil which can only be regenerated with ashes and subsequent growth of trees; but intensification is impossible in such a system, and this is precisely why it has been condemned from the start by outside observers and revenue collectors.

[95] Report on service tenures, AR 1870, p.286 (by Dickson); contra, A.C. Lawrie, *A Gazetteer of the Central Province of Ceylon* (Colombo 1896), vol.I, pp.42-3, noting the harassment of *chena* cultivators by temple authorities.

[96] Ludovici, *Rice Cultivation*, ch.1.

Enthusiastic Government Agents with visions of imaginary forests sprung from the *chena* looming in the future, have prohibited *chena* cultivation as to this they attribute the decline of paddy cultivation. We fail to see in this connection the cause and the effect.... If ever a paddy field is abandoned it is from causes other than choice.... Wet and dry cultivation are not carried on simultaneously.... *Chena* cultivation far from interfering with paddy cultivation, is only intended to supplement the produce of the former.

For the modernists, *chenas* came to be seen as a necessary but provisional evil with a view to a successful transition to permanent garden cultivation:[97]

The policy of government has of late been to restrict *chena* cultivation; with this every right minded person will agree. But have we substituted anything for it? Have we made it easy for the villager to take up permanent cultivation in place of *chena* cultivation? I think not. To ask the average villager leading a hand to mouth existence, to purchase land at 10 Rs an acre and fees is to demand of him an impossibility. Even if it were possible for him to raise this amount, it would mean that he would have to borrow it or that in paying it he would exhaust his capital.... The result in one case will be that the moneylender will reap the benefit of the villager's labours by selling him up, and in the other case he will have no capital to enable him to develop his purchase.

Gradually the population is increasing, gradually more land will come under permanent cultivation and the spread of coconut plantations will no doubt in time diminish the amount of *chenas* required in the less arid parts of the district. But this gradual growth will be at once checked if a reasonable amount of *chena* cultivation is not allowed, at all events for some years to come.... A liberal *chena* policy is necessary for the preservation of health and even existence. The undue restriction of *chena* cultivation, until the villager has learnt to go in for garden cultivation, can only result in the impoverishment and gradual extinction of villages. The substitution of garden cultivation is coming with the spread of population from the South and the example of estates.

The settlement officer, Fraser, in a report to the Colonial Secretary in Colombo on 'the chaotic state of land matters' (later censored on the order of government from his annual administration report), put it bluntly:[98]

The English government having upset or allowed to fall into disuse the land policies existing in his eastern possessions at the time of annexation, cannot escape the responsibility of substituting for these the most perfect form of land policy which our western civilization can devise.

[97] Diary of SO Fraser, 31 October 1903; Diary of SO Wait, 14 March 1909. Also Tyrrell, GA NWP, in his AR for 1924, p.F2.

[98] SLNA, PF 1684A, 'Formation of a Land Settlement Department', SO to Colonial Secretary, 9 January 1907.

In his diaries, he developed his views as follows:[99]

> The ethical objection to *chena* cultivation is that it is detrimental to character in making the people lazy and improvident. But for my part, I cannot see how enterprise and exertion upon *chenaing* can be half as demoralizing as periods of enforced idleness. For it is not as if gardens and fields were being neglected for *chenas*. No, they are cultivated, in many cases over cultivated, planted with trees until production is impeded by overcrowding. Even if given new areas to convert into gardens, the people, destitute of money, must have a means of livelihood until the new gardens come into bearing. And this means can only be found in *chena* cultivation.... There is no doubt that these lands formerly produced an immense quantity of Indian corn and fine grain, whereas they now produce next to nothing. The 'waste' all seems to be on the side of the present system if the question is regarded from the economic point of view.... To ask villagers to desist from *chena* cultivation is much the same as asking them to starve to please government.

These arguments were necessary in view of the growing obsession with a release of vagrant population due to landlessness: according to a number of officials, everything had to be done to avoid the prospect of an uprooted rural proletariat. The object of most planters was therefore antagonistic with the views of government, which tried to avoid what it called a silent revolution. The only means to prevent it was to keep the peasant on the land, to admit any form of cultivation including *chena*, and then to encourage conversion to garden-cultivtion or small-scale plantations. The alternatives were not between *chena* and estate employment, nor between *chena* and paddy, but between *chena* and garden.

However, these views never became popular, even among the Ceylonese elite, who inherited the prejudices of the colonial administration, and chose rather to attack the plantation system for the sake of the paddy cultivators.[100] They did defend the *chena* rights of the villagers against what they called the encroachments of the Crown and its settlement officers, so far as cultivation rights could be converted into property rights for the emergent middle-class elite; but they never attempted to defend *chena* cultivation, which in the south-west and central areas had become quite periphery in the village economy, due to the rapid development of gardens and small plantations, and the employ-

[99] Diary of SO Fraser, November 1901 and June 1902.

[100] The only exception was a memorial on the Kurunegala land question dated 29 October 1907, in which influential members of the Kandyan aristocracy and low-country new-rich asserted the necessity of *chena* in the dry zone and for some groups, such as the Vahumpuras.

ment of Sinhalese on the larger estates, between 1900 and 1930.

The defence of paddy cultivators against planters and the 'rice bias'

Ward, Governor in the 1850s, was the first official genuinely to attempt to revive paddy cultivation, especially in the dry zone. After his departure, a section of the Ceylonese press, opposed to the influence of the planting lobby, undertook to defend the peasantry: Ludovici, who fought against the image of the lazy native, insisted that if paddy cultivation was in a bad state it was the result of the lack of help and funding from the colonial government, in contrast with the kings of the past, praised for their involvement in irrigation works, and the Dutch who had encouraged the conversion of virgin land into paddy fields.[101] The ordinance of 1840 was blamed for restricting the village territory by making it impossible to develop new paddy land on the slopes of the hills.

In the 1880s, when the collapse of the coffee economy created a serious crisis in the Kandyan countryside and compelled the peasants to withdraw to subsistence production, the theme was revived by colonial administrators encouraged by the traditionalist Governor Gordon (later Lord Stanmore); some of them, such as Le Mesurier, successfully lobbied for the abolition of the paddy tax, so that after 1892 Ceylon was the only country in Asia where subsistence agriculture went untaxed.

But land, more than tax, was the sore point, every facility still being given to its development into plantations. One result was that when the nationalist movement took roots among the Ceylonese bourgeoisie after the turn of the century, one of its major planks was the defence of the paddy producer against the planters, although this position apparently impinged on the interests of most of the elite members as plantation owners.[102] The Corea and Senanayake families, and later S.W.R.D. Bandaranaike, played a major part in this political game. The early nationalists were members of the legal profession, and most of them had a close knowledge of the legal aspects of land matters. In the first decade of the twentieth century, the best representative of this class was C.E. Corea, the founder of the Chilaw Association, connected with the

[101] Ludovici, *Rice Cultivation.*

[102] V. Samaraweera, 'Land, labor, capital and sectional interests in the national politics of Sri Lanka', *Modern Asian Studies* 15, 1 (1981).

Ceylonese coconut planters of the North-Western Province.[103] They insisted on the rehabilitation of paddy cultivation, and on the preservation of communal rights, which, in their ethno-national view (largely borrowed from British authors such as Maine and Phear), were characteristic of the original 'Aryan-Sinhala' culture. They paid little or no attention to the technical side of the land question, and when they took or pretended to take up the defence of the Kandyan villagers, they never failed to condemn large-scale land expropriation for the benefit of European planters but hardly considered *chena* cultivation or garden development as a matter worthy of political action. Still less did they understand 'peasant strategies'.

From the 1920s, with the rapid development of the small and medium Ceylonese rubber and coconut plantations, the growing land hunger in the villages, and the increasing assertiveness of a political class backed by the press, the land question became the major political issue in Ceylon. Clifford, Governor of Ceylon in the late 1920s, tried to cut the ground from under the feet of the nationalists by setting up a Land Commission to review the past and present land policy and to recommend a new approach. The voluminous evidence collected by the Land Commission, and the specific cases it had to examine (including a few developments on the questions of pasture lands and the silting of paddy fields), were found to be compelling reasons to bar any further alienation of land to European tea companies.[104] Land hunger was then being felt in Kandyan districts, and a section of the civil service, led by C.V. Brayne, proclaimed its sympathy for the 'plight of the paddy cultivator' and the need to stop indiscriminate estate development. The 'poor peasant' had to be shielded from the influence of plantations, and 'rehabilitated'. The myth of a 'disintegrating village' and of a peasantry in need of 'moral rehabilitation' had its roots in the idealised image of the pre-plantation village, but the enquiries on rural society undertaken after 1935 gave it a scientific backing. During the period of the Great Depression, colonials and nationalists alike wondered at the resilience of the 'indigenous economy', supposedly better able than the 'new eco-

[103] C.E. Corea, *Communal Rights* (Dehiwela 1916).

[104] Minutes of evidence before the Land Commission, SLNA (especially October 1928 sittings); see also H. Clifford, *Some Reflections on the Ceylon Land Question* (Colombo 1927); Interim and final reports of the Land Commission, SP nos.18 and 33 of 1927, 2, 5, 16, 35 of 1928, 1, 17 and 18 of 1929.

nomy' to overcome the crisis. The Dutch economist, Boeke, who first developed the dualistic thesis, insisted on 'the absorbing capacity of the Javanese rural communities' and 'the advantages of a little developed capitalist system which has not yet created an influential proletariat'. Analysing the economy in dualistic terms and pretending that the peasant sector was safe from economic fluctuations generated in the modern sector, while the crisis provided full evidence to the contrary, was a piece of self-deception: those in the colonial milieux were anxious to shift responsibility and to make others and themselves believe that they had left the traditional structure untouched. What is more intriguing is that the same course was adopted in the nationalist milieux, although it was given a pessimistic twist: the colonial plantation system did affect the village in a negative way, and it was the nationalists' duty to come to the rescue of the 'poor peasant' and help him reconstruct a mythical golden age. Such a view could reconcile traditionalists with the emerging elite, and contribute to offsetting the responsibility of the latter in the spoliation of the peasantry.

Following Brayne's recommendations, attempts at 'rural reconstruction' were made by D.S. Senanayake, the Minister of Agriculture and Lands. When 'mapping out' the villages after settlement, the settlement officers were ordered to give full priority to the needs of the community. Wherever possible, 'village expansion schemes' were set up to give house and garden sites to an expanding population: sometimes uncultivated areas belonging to estates were acquired for the purpose. In the most congested districts, a systematic policy of estate acquisition for redistribution to peasants was undertaken in the 1940s. It led in a few cases to the expulsion of Tamil coolies. But the main answer was to focus on the large undeveloped areas in the dry zone, where colonisation schemes could be started and (hopefully) a strong paddy-producing peasantry recreated, unhampered by the proximity of any plantation. What M. Moore calls 'rice bias', taking for granted that the duty of the state was to devote massive investments to the paddy sector, became the motto at least until the 1980s. These areas were the heart of the ancient irrigation-based Rajarata kingdoms, and their 'reconquest' became the major project of D.S. Senanayake and his successors, up to the present day (the Mahaveli development scheme being its crowning achievement). The unexpected results were the development of what may be called a 'non-peasant' society of rural colonists, and mounting tensions with the nearby Tamil settlements in the Northern and Eastern Provinces, eventually leading to the present ethnic crisis.

After independence, the official report of the Kandyan Peasantry Commission of 1951, and the academic enquiry conducted by S. J. Tambiah and N. K. Sarkar, popularised the thesis of the disintegration of an hitherto 'well integrated and self sufficient social and economic system, well adapted to certain ecological factors'.[105] Politicians used an increasingly strong phraseology: rural reconstruction policy was officially aimed at 'curing the rural ills by making a direct and concentrated attack on five fronts, namely social, physical, mental, economic and spiritual'. Verbal inflation had more results in minds than in fields. Through the development of mass education and the growing influence of the press, these myths gained wide currency, and the sons and daughters of the Kandyan villagers learnt that they had to be saved from the malefic influence of the plantations, while their parents were actually increasingly dependent on them. Estate and village were represented as incompatible entities, with the result that any radical attempt at village reconstruction could imply estate destruction. This view was and possibly still is lingering in the minds of the young J.V.P. rebels who played a leading part in the 1971 and the 1989 revolts against the established order: the Maoist strategy of conquest of the town from the countryside being adapted and converted into the conquest of the estate enclave from the villages, with its possible outcome the destruction of the conquered entity and a falling back on subsistence agriculture.

These views influenced the representations the villagers had of themselves, and as a result, their own choices. A 'peasant strategy' is not to be analysed in strictly rational economic terms. It has to take into account ideological motivations, which are more easily manipulated, and may run counter to the apparent economic interests of those who adopt them. Paddy cultivation was already perceived as the noble activity *par excellence* in the caste ideology prevailing since the Kandyan

105 N.K. Sarkar and S.J. Tambiah, *The Disintegrating Village* (Colombo 1957), p.XI. Another volume was published two decades later with the same title, but with a question mark added in the introduction; however the deconstruction of the myth was limited to that of idyllic image of the pre-plantation village and to the absolute dualistic model, and the idea of a 'traditional paddy-centred communal village', disintegrating under modern impact, was retained: B. Morisson *et al.*, *The Disintegrating Village. Social change in rural Sri Lanka* (Colombo 1979), pp.7, 15-17.

times, while artisanal work, hired labour and even garden cultivation were despised. When this Goyigama-dominant view was upheld by the state and its agencies (among the policy-makers it took the form of the 'rice bias'), it led the very groups who depended on other resources to conform to it in words or even in practice.[106] Utopias like 'village rehabilitation schemes', 'village uplift programmes', or 'rural reawakening movements' which flourish in modern Sri Lanka, incidentally with the blessings of foreign donors, are modern products of the 'paddy cultivator' ideology. They are readily accepted—at least superficially—by the rural population, because they provide it with a recognised peasant identity and with state or N.G.O. subsidies.[107] But at the same time the actual strategy still relies on plantation products. A telling example is that of the Southern Province. Although the hotbed of the J.V.P. insurgency, it witnessed in the 1980s a dramatic development of tea smallholdings worked by local villagers, while the continuing neglect of the up-country state plantation sector during the same period, a result of the rice bias, led to its becoming the most unprofitable sector in the economy.

The peasant strategy has always been not to put all one's eggs in the same basket, using new opportunities without entirely depending on them, and acting with guile in one's relations with the state and local powers: the Kandyan peasantry appears to have fared rather better than many of its Asian counterparts. Among the reasons for the difference may be its ability to manage the transition from *chena* to garden and small plantation cultivation, the relative lightness of the fiscal pressure, the relative inefficiency of the repressive policy, possibly the salvaging

[106] Moore, 'Ideological history'; Kirk, 'Perceiving agrarian change'. In his research on a potter village in the dry zone, Colin Kirk insists that those villagers who depend extensively on pot-making are keen to declare themselves as paddy cultivators, while the main cultivation practised by them is *chena*. This is interpreted by him as a case of false consciousness linked with the general stereotype of the Sinhala peasant as paddy cultivator (the tank, the temple and the field), which is the model of moral and social order supposedly disturbed by present-day land hunger and marketisation. The peasants are mystified by a Goyigama discourse largely diffused through the media and the rural education system, which masks the actual state of things: paddy cultivation is highly unegalitarian, but community-based, while pot-making and *chena* cultivation are entrepreneurial and potentially more equalitarian.

[107] See the contributions by J. Brow and M. Woost to J. Spencer (ed.), *Sri Lanka. History and the roots of conflict* (London 1990).

by the settlement department of uncultivated land for sale to the villagers, and the existence of a pioneer front in the dry zone to relieve population pressure. The fact that the peasant society was not highly stratified certainly helped to preserve a degree of flexibility in peasant strategies. The manner in which the plantations burst in on a peasant society could have created the conditions for a major breakdown or upheaval. This is at least how outsiders, especially politicians and intellectuals, later imagined the process. Instead the Kandyan peasantry, while resenting the presence of the plantation economy, and seriously deprived of land for its expansion and of the flexibility provided by peripheral resources, eventually managed to get something out of the changes on its own initiative.

Chapter 7

AGRARIAN CHANGE AND THE LABOUR PROCESS

Ravi S. Srivastava

Introduction

The nature of agrarian change in colonial and post-colonial India has been the subject of a protracted debate. The purpose of this paper is not to join this debate here, but to throw light on the nature of an uneven agrarian transition on the basis of concrete village-level experience of changes in the labour process.[1]

This transition, in our view, is not teleological. Instead, changes need to be viewed in terms of their impact on the reproduction strategies of different classes. Major 'shifts' in these strategies occur when these classes are no longer able to sustain them in pre-existing ways.[2] The paper shows that, to the extent that such shifts occur, they are associated with subtle changes in extant labour processes (rather than simply a movement towards wage labour) which reflect the diverse objectives of rural classes under uneven transition.[3]

The analysis here is based on changes in the labour process in two villages situated in the eastern and western regions of Uttar Pradesh, and focuses on dominant landholders and their relations with tenants and labourers.[4] The two regions encapsulate vastly diverse agrarian ex-

[1] Agrarian transition is considered here as a movement from a conceptually somewhat ambiguous pre-capitalist agriculture to one in which 'wage labour' and expanded reproduction by capitalists may become the *sine qua non* for agricultural growth. Pre-capitalist agriculture is defined as one in which those with superior rights in land or its produce exercise considerable extra-economic clout, and appropriate surplus labour from direct producers as rental incomes or from hereditary or otherwise dependent labourers, and have little interest in productivity-raising investment for which they must bear the risk. Note that this concept allows scope for raising absolute surpluses or for the formal subsumption of labour under capital. (Patnaik, 1986, has outlined the contours of pre-capitalist Indian agriculture.)

[2] Srivastava (forthcoming a & b).

[3] A process which Byres (1981) refers to as 'partial proletarianisation'.

[4] 'Dominant' landholders are those who, according to Bharadwaj (1979), exercise a preponderant influence in setting the terms of exchange. In terms of our classification (for details, see Srivastava 1989a) these are landlords, rich farm-

periences. By all accounts, the eastern region showed few outward signs of change till recent years. Thus our interest here is as much in inter-village differences as in within-villages changes over time.

Of the two villages studied,[5] Chaukra is in Muzaffarnagar district of western Uttar Pradesh, which agriculturally is one of the most dynamic in the country. The dominant agriculturalists in the village are Tyagis—both Hindu and Muslim, but other castes, notably the Jhheenvar, are also considerably internally differentiated, and several among them have middle-sized holdings. The land-holding structure is oriented towards large and middle peasant holdings. Agricultural land in the village has been irrigated by canal, a public tubewell (since 1985/6), and private tubewells which have steadily grown in number. The number of tractors, too, has grown from seven in 1977/8 to over 30 at present. Sugarcane is the principal crop. Cropping intensity in the village has also been increasing, and most good land is double or triple-cropped. Cane-processing industries (*kolhus*, crushers and mills) provide major avenues for non-agricultural employment during the cane-cutting season (November to April).

Alipurjeeta lies in Allahabad district in south-eastern Uttar Pradesh. The dominant agriculturalists in the village are Thakurs, who had the bulk of the zamindari rights. Following them are Muslims and Brahmans. The labouring and poor peasant castes are mainly drawn from the scheduled castes and the depressed backward castes. Alipurjeeta is a former zamindari village, in which, till the 1950s, the dominant landowners were either rentiers, or cultivated with the help of traditional farm servants.[6] Up to the advent of canal irrigation in 1974, village lands had some well irrigation and gram with barley and arhar with bajra were the principal crops. The main crop cycle is now wheat fol-

---

ers and (to some extent) rich peasants.

[5] Of these villages, Chaukra was surveyed by this author in 1977-8, 1985-7 and 1990-1. Alipurjeeta was first surveyed by the Agro-Economic Centre, Allahabad, in 1961-2. It was resurveyed by this author in 1985-7 and 1990-1. The surveys in 1990-1 were carried out jointly by the author and N.C. Saxena. In the 1985-6 survey an attempt was made to reconstruct labour histories from the early 1950s. The data from the 1990-1 survey have still not been fully analysed and, except where stated, the results reported refer to the 1985-6 reference year.

[6] It could be argued that, at the beginning of the study period, the agrarian organisation of Alipurjeeta was 'pre-capitalist' in the sense outlined earlier (also cf. Bhaduri's characterisation of Birbhum; 1973). By the same token, such relations were already in an advanced state of dissolution in Chaukra.

lowed by paddy. The level of mechanisation in the village is comparatively low (with one tractor). However, the number of private tubewells in the village (three in 1985/6) is now slowly on the increase.

During the study period, the pre-existing modes of agrarian organisation—involving both the dominant landholders' own participation in the labour process, and the modes of appropriation of surplus labour—have been subject to different kinds of change in the two villages. In Alipurjeeta, these changes are principally an outcome of the landholders' inability to sustain pre-existing modes of surplus appropriation in the face of a shrinking production base and implicit resistance from the workforce. The changes are facilitated by an environment in which increased investments in agriculture have become both more feasible and necessary. In Chaukra, the continued intensification of agriculture and the growth of a non-agricultural sector have created pressures on dominant landholders to reduce labour costs and minimise supervision costs. In the latest phase, this has implied a limited switch to labour-saving land-use patterns.

Characteristics of the agrarian labour market

Traditional labour control mechanisms have been under pressure in Alipurjeeta leading to diverse outcomes: a loosening of the labour market on the one hand, and an intensification of direct extra-economic coercive pressure on the other. In Chaukra, high labour absorption in agriculture and high levels of non-agricultural employment (both particularly during the rabi season) have led to significant changes in the labour payment systems.

The two main labouring castes in Alipurjeeta are the Chamar and the Pasi. The Chamars are integrated in the traditional scheme of agricultural production as the main labouring caste, even though they do perform certain other *jajmani* functions. The Pasis are newer entrants to the labour force and have the reputation of being recalcitrant. Chamars, therefore, are more likely to be subject to traditional, hereditary dependence-relationships than Pasis. These relationships have, however, been weakening over the years. On the other hand, the dominant surplus appropriators themselves are divided into different castes, and individual members of one caste are more likely to follow one strategy of surplus appropriation than another. In particular, members of the Thakur caste are more likely to adopt directly coercive tactics than are members of other castes.

Compounded with the traditional distinctions among labouring castes is their residential segregation, with one of their localities completely encircled by the houses of Thakur landowners on one side, and by fields belonging mainly to the same landowners on the other. It is members of this group of labourers who are subject to the most blatant coercive tactics and threats, leading to forced exaction of labour, and underpayment or non-payment of wages.[7] Thus crises in pre-existing labour relations in Alipurjeeta have led to different responses and strategies by surplus appropriators. One response from a section of dominant landholders has been to *intensify* the coercion exercised upon the most vulnerable section of labourers. It is this more blatant use of coercive power which sometimes leads to the non-payment of wages; underpayment results from coercion as well as monopsonistic power by a larger body of big employers in Alipurjeeta. Unwillingness to live with this coercion (particularly as farm servants) was repeatedly given as a reason by labourers for emigrating and undertaking the extremely arduous tasks associated with migrant labour. Thus, in Alipurjeeta, extra-economic factors continue to play a very important role in determining the dynamics of the labour market.

There is considerable evidence of monopsony in the labour market in Chaukra; large employers here attempt to keep down the level of wages they must pay to their labourers. Employers also collusively exert social (extra-economic) pressure to see that contractual obligations, arising out of indebtedness and so on, are respected. This is important since the legal framework does not always provide sufficient support to them. Hence, large employers in Chaukra use their considerable economic power to enter into favourable contracts, and their social and economic power to enforce them. However, unlike in Alipurjeeta, in Chaukra extra-economic forces do not intervene directly in setting the terms of these contracts.

Wages in Chaukra have been under pressure due to factors operating on the demand side both within and outside agriculture. Prevalent time-rate (daily) wages in Chaukra are Rs.8-10 for most operations, twice the levels prevailing in Alipurjeeta. Since large employers in Chaukra are also supervision-constrained, they are increasingly resorting to piece-

[7] The labourers must travel close to the houses or the fields of the landowners. This forms the basis of most threats.

rated employment, which reduces supervision and labour costs while raising the normal intensity of labour.[8] Consequently, most major operations are now also piece-rated. The nominal wage rates prevalent in Alipurjeeta were Rs.4-5 with or without a component of variable value paid in kind. Except for harvesting operations, paid as a share of the crop, and in certain cases (for example, potato-digging) where a fixed amount in (3 kg.) is paid in kind, time-rates are widely prevalent in Alipurjeeta.

Tenancy

The higher classes account for a very significant proportion of households, as well as land rented out, in both villages.[9] But in contrast to Chaukra, where resident cultivating-lessors predominate, rentier landowners (mainly resident) are still important in the lease market in Alipurjeeta.[10] While Chaukra a higher proportion of households let out land, Alipurjeeta has a higher proportion of the owned area leased out (30 per cent, inclusive of seasonal leases, compared to 24 per cent for Chaukra). The workers, the poor and middle peasantry dominate among

[8] Marx considered piece-rates more favourable to the development of capitalism than time-rates (Marx 1976:693). According to him, the former method of payment is associated with several qualities. Firstly, the quality of labour is controlled by the work itself. Second, piece-wages provide an exact measure of the intensity of labour. Thirdly, the supervision of labour becomes, to some extent, superfluous. Fourthly, the method gives rise to a hierarchical system of exploitation. Fifthly, it enables the capitalist to raise the normal intensity of labour. Finally, while wage earnings may become more unequal, the average wages and wage costs may themselves fall. Some of these aspects have also been the subject of recent neo-classical literature.

[9] For details of the methodology of classification of households see Srivastava (1989a). Rich farmers account for 75 per cent of total rented-out land (and also 96 per cent of land which is leased out seasonally). Rich farmers and rich peasants together rent out 84 per cent of land annually. In Alipurjeeta, landlords alone rent out 86 per cent of land, with rich farmers and rich peasants accounting for another 13 per cent. Landlords, rich farmers and rich peasants maintain their preponderance as lessors, even in numerical terms. The latter two classes account for 69 per cent of all lessors in Chaukra. The three classes together comprise 87 per cent of all lessors in Alipurjeeta.

[10] The percentage of owned land leased out annually in Chaukra varies from 14 for workers and poor peasants, to 10 for lower middle peasants, 9 for upper middle peasants, 21 for rich peasants, and 13 for rich farmers (13 for all households). In Alipurjeeta, the percentage of owned area leased out is nil for lower middle and upper middle peasants. It is 12 for workers and poor peasants; 3 for rich peasants; 8 for rich farmers, and 67 for landlords.

lessees.[11] But there are differences in the importance of these classes in terms of land rented *between* villages. In Alipurjeeta, workers and poor peasants lease in most of the land (54 per cent). The order is reversed in Chaukra where middle peasants lease in 61 per cent of the land. In the last case, while middle peasants predominate in the *annual* market for leasing in, workers and poor peasants (in that order) predominate in the *seasonal* letting markets. Tenants are also less differentiated in Alipurjeeta than in Chaukra.[12]

The total area involved in the rental market is considerable: 35 per cent of owned land in Alipurjeeta and 16 per cent in Chaukra. If land received for cultivation for payment for services rendered and seasonally-leased land are also included, the percentages change to 36 in Alipurjeeta and 20 in Chaukra.

Rental modes and practices differ considerably between the villages. These differences constitute the matter of this subsection. First, there are marked variations in the period of leases and hence in tenurial security in the villages studied. In Alipurjeeta 33 per cent of all holdings have been held on lease for five or more years. The corresponding figure for Chaukra is three per cent. At the other extreme, 60 per cent of leased holdings are held under seasonal leasing arrangements in Chaukra. The corresponding figure for Alipurjeeta is two per cent. While these figures differ dramatically between the villages, it is apparent that the highest incidence of tenurial insecurity in both villages is among the poorer classes. Conversely, as we move upwards, lessees belonging to the better-off classes generally have a greater proportion of relatively stable tenurial contracts. Tenurial contracts in Chaukra are made for periods varying from one season to two or three years (this latter period marks the full period of the sugarcane cycle). In Alipurjeeta, contracts are in principle renewable at the beginning of each kharif season, and a fifth of contracts are actually renewed each year. But landlords allow a large number of contracts to continue undisturbed. However, while the

[11] In Alipurjeeta, they account for all the land leased in. In Chaukra, these households account for all but 4 per cent of land leased in. The same conclusion also holds for the market for seasonal leases.

[12] In Alipurjeeta, 50 per cent of tenant households are poor peasants, the remaining (29 and 21 per cent) being lower middle peasants and upper middle peasants. In Chaukra, 48 per cent of households belong to the first two classes, and 32, 16 and 4 per cent belong to lower middle peasant, upper middle peasant and rich farmer households.

duration of the actual lease does reflect a change in the environment in which decision-making takes place, it does not lend security to the tenants in any of the villages. The tenants enjoy no legal rights, and can be evicted at will (except in cash-rent contracts which run for specified periods).

Crop-sharing is the dominant form of rent in both the villages. Indeed it is the only form in Alipurjeeta, except in cases of land allotted to farm servants. In Chaukra, the situation is changing to one where cash rents are assuming greater significance. During 1985-86, onlyfive per cent of leased holdings were on cash rents.[13] But a number of contracts, involving both residential and non-residential landlords, had been renegotiated for the following year. These would have roughly trebled the area under cash rent. The cash-rent contracts are negotiated for two- or three-year periods, and the contractual amounts are generally payable in advance. Hence, they are restricted to those peasants who can raise the necessary resources—through owned surplus or borrowing.

Crop-sharing arrangements also vary in Alipurjeeta, in the division of the crop and its by-products, but the predominant division (in 95 per cent of cases) is a 50:50 share between landlord and tenant in the crop and (in the case of wheat but not paddy) in its by-product. However, higher shares for the landlord (two-thirds and three-quarters) also occur in a small number of transactions. These higher shares are confined to rental contracts with poor and lower middle peasants. Threshing is generally carried out by tenants at their own cost, but this practice is not universal and in certain cases landlords share in these costs, so that rent is determined by the net final quantity of produce. The model crop share in Chaukra is now three-quarters for the landlord (62 per cent of cases), the traditional half-and-half share having been relegated to second place (34 per cent of cases). The other shares observed were three-eighths and two-thirds (each in 1.8 per cent of cases). By-products are shared in proportion to the crop share.

Here again, there is a systematic (inverse) relationship between rental shares and class status. Workers, for instance, pay three-quarters of output as rent in all cases. The percentage declines to 72 and 44 for poor peasants and lower middle peasants. On the other hand, no upper

[13] During 1977-8, when we carried out our first survey in the area, only one small plot of village land was on a cash rent.

middle peasant or rich farmer pays more than 50 per cent of the produce rent.[14] There is a clustering of higher rental shares (typically the three-quarters share) around shorter duration leases in Chaukra.

Cost-sharing arrangements also need to be considered. Traditionally in Alipurjeeta they involved '*beej-khad*' (seed and fertiliser). The tenant bore the responsibility of supplying owned and other inputs. With the introduction of canal irrigation, the incidence of canal rates fell on the landlords. In due course, they shifted half the irrigation charges on to the tenant. Now the input costs generally shared (in 83 per cent of cases) are seed, fertiliser and irrigation charges. In two cases landlords also shared the cost of manure. In two other cases the cost of hired labour was shared as well. Some tenants have pooled resources to lease land. Joint cultivation of leased land by family members is a feature seen elsewhere, but in Alipurjeeta such contracts are also entered into by two non-related parties. In a singular instance, a self-cultivating landowner (cultivating with the help of farm servants) has leased out his land to a share-cropper on a quarterly basis. All costs, including hired labour charges, are shared.

In Chaukra, cost-sharing has involved fertiliser and irrigation costs. But, with changes in rental arrangements (the three-quarters share), landlords have tended to provide a corresponding proportion of other input costs, such as seeds or sapling, and for ploughing the fields with the assistance of the share-cropper. As with new output-sharing arrangements, traditional cost-sharing arrangements have been overridden by a new scheme, with the landlord contributing three-quarters of the costs in over 60 per cent of cases.

There is evidence that tenancy is in a state of flux. Traditional cost shares have adjusted to include new inputs. This has facilitated productivity gains, and has also helped landlords shift the incidence of canal rates on to tenants. But the major change is towards not only sharing the cost of *more* inputs, but towards a higher share for landlords in both input costs and output. Both the nature and the direction of the changes appear to be dictated by the bargaining strength of the land-

[14] Patnaik (1983) expects a situation to prevail where large and small tenants would pay a similar proportion of output as rent because of competition, even though the former would still pay a smaller portion of surplus produce because of higher productivity. In the situation here different classes of cultivators appear to face a differentiated lease market.

lords. Thus tenants in Alipurjeeta noted with resentment the success of landlords in shifting the burden of half the canal rates on to them, since in some adjoining villages landlords still paid the entire rates. Similarly, in Chaukra, tenants have protested about the high landlord share of the output on seasonal leases (three-quarters as against two-thirds in several other villages in the area), and about their share in production costs (paid entirely by landlords in adjoining areas).

Landlords' participation in decision-making varies between villages. In Chaukra, landlords help decide on key material inputs in 89 per cent of cases.[15] They also participate in decisions regarding cropping pattern in 79 per cent of cases. In fact, in 74 per cent of cases, they take such decisions unilaterally. The equivalent figure for material inputs is 50 per cent—the extent of their involvement being strongly associated with the degree of cost-sharing. This is also true of decision-making regarding cropping patterns, though less so. Landlords' participation in decision-making about labour inputs is less common (16 per cent of transactions). A similarly high level of landlord participation is also found in Alipurjeeta (in 79 per cent of cases for cropping patterns and 88 per cent for material inputs). Landlords unilaterally decide cropping patterns in 55 per cent of material inputs, and 12 per cent of cases. But there is no significant association between cost shares and decision-making on material inputs.[16] As regards labour inputs, landlords participate in only 1.5 per cent of cases.

The following pattern emerges. In Chaukra, landlords mostly share in agricultural decisions, and in a majority of cases landlords actually take unilateral decisions regarding the crops to be grown and application of inputs. Direct supervision, or consultation regarding labour use is much less common. In Alipurjeeta, too, mutual consultations—in which, no doubt, landlords have a greater say—are becoming prevalent, and several landlords are beginning to supervise share-cropped land closely, but, unlike in Chaukra, actual dictation of material inputs or cropping patterns is less common.

What are the implication of changes in the tenancy situation? The

[15] This includes cases in which landlords decide the cropping pattern unilaterally, as well as those in which such decisions take place through joint consultation between landlord and tenant.

[16] Tenants decide on material inputs in 19.5 per cent of cases involving cost sharing. On the other hand, joint decisions are taken regarding input use in 60 per cent of cases involving no cost share.

foregoing empirical discussion suggests that transactions differ from village to village, and that the outcome of such transactions also differs. But there is a certain systematic basis to the inter-village differences. Despite the existence of widespread leasing-out by resident as well as non-resident landlords, there is considerable heterogeneity in the lessor-lessee relationships in Alipurjeeta. A large number of landlords share in the cost of key material inputs. In several cases, landlords closely watch the production on leased holdings. Landlords do not actually supervise the labour process, but do keep an eye on different stages of production. Leasing contracts, in which landlords provide a greater proportion of the costs and receive a proportionately greater share of output, are also no longer unknown. These are now developing in the form of seasonal lease agreements for labour-intensive crops (mainly potato and vegetables). Such arrangements are still rare in Alipurjeeta. The main change has been a shift—which is not yet generalised—towards closer supervision by landlords.

Changes which are incipient in Alipurjeeta have taken root firmly in Chaukra. Here modal cost and crop shares have shifted upwards, and this has taken place alongside increasing landlord control over production on leased holdings. Tenant households participating in such lease arrangements are proletarianised households, and are no longer required to own any means of production (other than basic tools), for these are supplied by the landlord. The only requirement for leasing is the existence of sufficient family labour.

Thus, a new set of leasing arrangements has become predominant: these involve high costs and output shares (and higher risk) for the landlord, and a high degree of control by the landlord over the processes of production. These leasing arrangements take place chiefly for labour-intensive crops (paddy, potato, onion, *urad*, vegetables and occasionally sugarcane). The arrangements are mainly confined to the kharif season, when non-agricultural employment is low, and are so pervasive that they form the main source of labourers' subsistence for that season. (This is true not only for the village, but for the region as a whole.) At the margin, such arrangements tend to reduce tenants to piece-rated workers with the added proviso that they must share the risk. The existence of high rental shares means that net daily returns to share-croppers are sometimes well below established daily wages. Hence, not only do landlords save on supervision, but they are also able to side-step pressure for increased money wages, and to enlarge net returns to themselves.

The evidence suggests that between our villages there is a shift from leasing-out as a means of rental appropriation, to leasing-out as a form of control over labour, and the appropriation of its produce. This feature, as we shall see later, is highlighted further when we consider interlinked transactions. On the other hand, a class of better-off tenants is emerging, who also pay rent at a lower rate.

Characteristics of farm servants

This section discusses the employment and working conditions of farm servants. To begin with, changes in the general conditions of employment of farm servants over time are briefly described. Current aspects of their employment are discussed later.

In Alipurjeeta, there are three distinct categories of farm servants. *Jeura* is the name given to traditional hereditary farm servants generally belonging to the Chamar caste.[17] Formally their contracts are renewable during the festival of Holi (generally in March, that is, *before* the harvesting of the major rabi crop, wheat). The *jeuras* are paid a share of the crop. They are also paid a daily wage in kind of about 1.75 kg. of grain for certain operations. A very complex set of rights and obligations mediates the relationship between the *jeura* and his wife, and the employer, but these are no longer fully observed in many cases.[18] Another type of farm servant is the *halwah* whose primary responsibility is to plough the fields using the master's animals. The *halwah* is given a plot of land, and is paid a daily wage in kind (about 1.75 kg of grain) for certain operations. As for the *jeura*—though in a less complex way—a gamut of rights and obligations define

[17] The word *jeura* also refers to other customary, hereditary services, such as those of washermen, barbers, messengers, and so on.

[18] The *jeura* must report for duty at the employer's residence every morning and evening. After completing the morning's work, he is allowed to seek alternative employment, or attend to his own work, if the master has no daily tasks available for him. The detailed obligations of the *jeura* are traditionally defined, as also his remuneration at various post-harvest stages. He is additionally paid a wage in kind for certain work. His wife is expected to participate in some daily chores (connected with domestic work or livestock care) in addition to work connected with the harvesting and processing of produce. Again, some traditional remuneration is set out for this work. Employers have, in particular, been cutting back on the share of grain due to the *jeura.*

relations between the employers and these farm servants. Though his is not a traditional service relationship, the *halwah*'s position displays certain traditional features. Finally, a new type of farm servant—the *mahindar* —is emerging. As the name denotes (*mahina* means 'month'), the *mahindars* are paid a monthly salary and employed for definite contractual periods.

Until about three decades ago, the cultivation carried out by dominant land-holders rested primarily on the two types of traditional farm servants, the *jeura* and the *halwah*, with the former being numerically preponderant. The casual labour force played a relatively more insignificant role. But the *jeura* system is no longer the *predominant* system of labour relations in Alipurjeeta. Moreover, there have been changes within the *jeura* system itself. Elements of employer-employee relations in Alipurjeeta reflect both continuity and change which figures alone cannot unravel. Customary, hereditary dependence (the patron-client relationship) provides the backdrop against which extra-economic factors continue to play an important role both in sustaining appropriation of surplus and in providing subsistence. At the same time, changing production-relations have provided workers with new alternatives and new forms of resistance, and employers with new modes of control including those involving debt and interlinkage. The transitional relationship poses certain problems for analysis since certain modes of control (for example, interlinkage) are consistent with both traditional and new strategies. Again, labour turnover may, in a specific instance, be less of a qualitative change than it appears. Within the hereditary context, it is sometimes a form of worker resistance, which may peter out and involve a return to the *same* employer after a period.

Chaukra, a generation ago, at the core of the labour force on the self-cultivated holdings of the dominant holders in Chaukra were two types of farm servants—the *hali* and the *naukar*. The *hali*, of whom there were only seven in 1951/2, were inter-generational farm servants. They also performed certain minor supervisory functions because of their long-standing relationships with their employers. They were paid a share of the produce (one in seven maunds) and given certain other perquisites. The *naukar* (of whom there were around 33 in 1951/2) were more contractual farm servants, and were paid a monthly salary (37 kg. of grain or Rs.5 per month in 1951/2). Their contracts were renewable each July, and hence in principle they were free to leave at the end of the period. Casual labourers formed a third component of the labour force.

The *hali* system was reported to have disappeared in the 1950s. Farm servants in Chaukra are now employed by the month, season or year. With one or two exceptions, they are paid a monthly cash salary and a bonus in kind for retention of their services during the wheat harvest. Hereditary or customary elements may cause some continuity in relationships over generations, but in general these relationships are far more contractual than those in Alipurjeeta, and continuity is more often a product of indebtedness, which generates some immobility or other interlinkages. Again, wage differentials in Chaukra in several cases reflect skill and productivity differentials among farm servants.

Some general issues arise in relation to long-duration labour contracts. Attachment and labour-bonding have formed the subject of several contributions by Bardhan (1979, 1983, 1984). The essential prerequisites for 'attached' labour in his models are peak season configurations (recruitment costs or spot wage-rates) which rise with the level of development. In the field-work villages, in Alipurjeeta four out of eight rich farmers, and one out of two rich peasants, employ farm servants, while in Chaukra all eleven rich farmers employ farm servants. This is superficially consistent with Bardhan's results. However, the analysis of changes in long-duration labour relationships over time does not confirm an increase in the importance of attached labour.

Moreover, the evidence does not confirm the sole importance of peak-season labour supply and demand configurations as an explanation of attachment. Other factors emerge as important determinants, such as employers' social status, or their own or family participation in farm labour and supervision; cropping patterns, or the time-distribution of labour requirements over the working *day* (particularly necessitated by the care of livestock and draught animals). Some of these factors decline with the extent of capitalist penetration.[19] On the other hand, given the articulation of 'backward' and 'developed' areas in Indian agriculture through migration of the rural labour force, attachment—often with strong pre-capitalist features—could be a strong feature of

[19] Breman's careful study of the Dublas in Gujarat (1974) shows how the change in cropping pattern reduced the demand for farm servants. The study shows why the decline in traditional hereditary dependence took place, and how in certain areas it came to be replaced by a different form of 'attachment'.

those areas in which the nature of agricultural growth and other processes (for example, demographic change) have induced shortages of labour.[20]

Interlinkages between labour and other transactions

Recent debates, within both neo-classical and classical doctrines, emphasise different implications of interlinked modes of exploitation. In one explanation, which basically focuses on static issues, interlinked markets are an efficient, albeit second-best response to information asymmetry and uncertainty. In the other explanation, interlinked modes of surplus appropriation serve mainly to reinforce the scope of pre-existing exploitation, and lead to regressive changes which include the underdevelopment of markets and the consequent slowing down of the transition to capitalism. The first approach ignores the limits set by pre-existing strategies, while the other minimises the ways in which surplus appropriators can make use of the *existing* relationships in the context of transition. On the other hand, we have argued that the interlocked relationships arise out of prior set of production relationships and the strategies of subsistence or accumulation of different classes, which may themselves change under certain conditions.[21]

Below we summarise some of the main results from our study to highlight the nature and change in interlocked relationships in the context of transition, and to evaluate the role of extra-economic factors in fostering such relationships. The first issue is transactions linking lease,s labour and credit. For the two villages, the percentage of tenants performing labour for their landlord employer ranges from 36 in Alipurjeeta to 60 in Chaukra. The percentage of cases in which tenants perform labour for their landlord-employers goes down in all the three villages as we proceed up the class ladder. In Chaukra, 82 per cent of tenants among workers or poor peasants perform labour for their landlords. The corresponding figure for lower middle peasants is 43 per cent. In Alipurjeeta, the percentages are 51 for workers or poor pea-

20 Singh and Iyer (1985) show the results of this articulation between migrant labourers from north Bihar and the developed district of Ludhiana; as also between migrant labourers from the tribal districts of Chhotanagpur and the less-developed district of Hoshiarpur (necessitated, at least by out-migration from this district), the latter having more pre-capitalist features. Bhalla (1976, 1983) also discusses the outcome of attachment in Haryana's regions.

21 Srivastava (1989b).

sants, 21 for lower middle peasants, and 10 for upper middle peasants.

Tenants involved in credit transactions with their landlords are more likely to work for their landlords. In Chaukra, 72 per cent of debtor tenants are employed by their landlords. Of such tenants, 33 per cent work for their employers on a priority basis, and a similar proportion work for them as farm servants.[22] Thus, only 33 per cent of those tenants who are employed by their landlords work for them on a strictly casual basis. Further, all tenants employed on a priority basis are workers or poor peasants. Finally, 29 per cent of those who are employed on any basis are also underpaid.[23] In Alipurjeeta, a smaller proportion (56 per cent) of debtor tenants work for their landlords. But among them there is a distinctly higher percentage of priority labour (70), and also a higher percentage of underpayment (60). These results confirm that there is some association between debt work status of tenants in Chaukra and Alipurjeeta.[24] They also suggest a strong association between committed labour and adverse terms of employment (low payment or abnormal working hours).[25] But the committed-labour status of tenant-labourers is independent of their status in respect of loans.[26]

Thus our evidence on land-labour-credit interlinkages leads to the following main conclusions. First, the employment of tenants is positively associated with their loan status. Further, a much higher proportion of tenants work for their landlords in Chaukra than in Alipurjeeta. Second, however, the more adverse forms of the labour relationship, involving priority labour and/or underpayment, are more prevalent in Alipurjeeta among tenants who are employed by their land-

[22] 'Committed' labour or 'priority' labour refers to the 'beck-and-call' relationship cited by Thorner and Thorner (1962). The priority labourer must first offer himself for employment to a particular employer before seeking employment elsewhere.

[23] 'Underpayment' here refers either to abnormal working conditions (long working hours), or low wages (relative to prevailing rates)

[24] Chi-square values are significant at 10 per cent level of probability in both villages.

[25] The chi-square values are significant at 5 per cent level of significance for both villages.

[26] In Chaukra, 46 and 50 per cent of transactions of working tenants (excluding farm servants) involved priority labour in both the two categories (non-debtor and debtor respectively). In Alipurjeeta, 85 per cent and 78 per cent of transactions with working tenants involved priority labour in the case of non-debtor and debtor tenants.

lords, than in Chaukra. Third, adverse working conditions are not systematically correlated with the credit relationship. These conditions could, therefore, result from the lease relationship or from other forms of economic or extra-economic dependence. In Alipurjeeta, for instance, they have more to do with the persistence of *traditional* lease-labour relations, than with the development of new forms of attachment as a response to recruitment costs or peak season shortage. There are two conditions traditionally associated with the relationship between weak lessee and strong lessor in Alipurjeeta. The first is the requirement to work on the landlord's farm on a priority basis, which may or may nor involve underpayment, and the second is the obligation of performing *begar* or unpaid labour for the landlord for a few days in the year.[27]

This brings us to the interlinkages between credit and labour. In Chaukra, 20 credit transactions are interlocked with product or labour markets. Of these, 60 per cent involve repayment through deductions in agricultural produce or other commodities, while 35 per cent involve repayment through labour services. Twenty per cent of all such transactions involve underpricing of agricultural produce, and 40 per cent involve undervaluing labour. As many as 33 credit transactions in Alipurjeeta generate interlinkages with product exchange or labour services. Of these, 76 per cent involve repayment through deductions in agricultural produce, and 21 per cent involve repayment through labour services. Of all such transactions 15 per cent involve undervaluing labour, and 3 per cent involve undervaluing agricultural produce.

Thus, 40 per cent of credit transactions in Chaukra and 15 per cent in Alipurjeeta involving kind repayments also involve extraction of some implicit interest through underpaid labour. In general, credit-generated interlinkages are higher in Chaukra. Most of such transactions involve repayment by means of produce and no implicit interest, but employers as a group do extract such interest in a significant proportion of cases. The proportion of all credit transactions involving both implicit interest payment and repayment through labour is relatively small. However, the highest proportion of credit-labour linkages is in Chaukra.

27 'Begar' traditionally takes the form of either ploughing the landlord's fields for the arhar or jowar crop in kharif, or of work on repairs and so on around the landlord's house or cattle shed. It is possible even for non-cultivating landlords to extract the latter form of 'begar' from the tenants.

In Alipurjeeta, the traditional (*jajmani* or service) ties still determine the core relationships. In addition, the dominant landholders can draw upon the pool of agricultural labourers and poor peasants, and can make contracts which have traditional features. Dominant landholders enter into two primary sustained relationships with the poor—as tenants or as farm servants. They are entitled to extract *begar* from the former, and also priority labour services if they are themselves self-cultivating. In the case of farm servants, contracts place explicit obligations on their wives and household members. Relations of debt (not advances, which are deductible from annual wages) can and do lead to labour-tying and bondage in certain cases. Thus it is not correct to maintain that interlocking leads to personalisation of relations, since the converse is often true, as is illustrated by our evidence from Alipurjeeta in particular.[28]

The poor, for their part, seek to diversify their subsistence strategies by leasing in land from more than one source, by entering into partnerships with other family members (brothers, for instance), or by seasonal emigration. Farm servants seek to acquire assets such as livestock, or even to lease in land from other landowners on which they can use their surplus family labour-time. The joint outcome of the strategies of the poor and the rich is a partially constrained one for the former, arising primarily because of their economic weakness and lack of viability.

But in Chaukra, while the relationship may be derived from historical attachments of the labourer or his kith and kin, it is increasingly being mediated by new forms of interlocking based on lease and credit relationships. Here, too, interlinkages operate mainly between a dominant landholder and a few agricultural labourers or poor peasants. Some of these may have loose hereditary ties, as farm servant or tenant, with the household of the dominant landholder. The pool of tenants constitutes an important source of occasional labour for any dominant landholder—all the more so if there is a continuing credit relationship between them. But even in the absence of a credit relationship, subsistence requirements in the context of uncertainty of employment often generate dependence and adverse terms of employment. Significantly, interlocked relationships are a more important source of labour mobilisation in Chaukra than in Alipurjeeta. In particular, this serves to underscore the different implications arising out of inter-

[28] Cf. Bardhan (1980), and Basu (1983).

locking.

In Chaukra, interlocked relationships, arising out of more recent strategies of surplus appropriators, have led to fragmentation and labour-market segmentation. Compared to Alipurjeeta, in Chaukra the growth of the labour market has been more dynamic, and the constraints there on the mobility of labour have more to do with direct economic processes. The effect of interlocking on wages or the functioning of the labour market in Chaukra is to place a *relative* constraint on the mobility of labourers—farm servants can and do leave while opting to work as labourers on a priority basis; labourers can choose to settle accounts or reproduce themselves as cane-crushers, or through raising livestock, and so on.

In the ultimate analysis, as Bharadwaj (1985) and other commentators have argued forcefully—and given the lack of direct sanction by the state—the persistence of certain forms of surplus appropriation, arising out of both extra-economic and economic factors, results from lack of alternative subsistence strategies for very large numbers of the poor, and may lead to the more adverse forms of interlocked relationships. If there are alternatives available, our evidence shows that the poor *preferentially* withdraw from relations leading to personalised dominance and unequal arrangements. This evidence controverts the schema outlined by Bardhan (1980), which implies an opposite movement.

Changes in the modes of labour appropriation have also been accompanied by changes in the nature and extent of participation in the labour process by dominant landholders themselves. In Alipurjeeta, the dominant landholders (Thakurs, Brahmans, Kayasthas and Pathans) did not traditionally participate in main or even auxiliary on-farm labour. Some amongst them were ritually barred from using the plough. But with the decline in traditional labour relations, many were compelled to perform auxiliary labour, and some even to plough the fields themselves. A Brahman smallholder gave the following account of the change: proprietors were increasingly having trouble in finding a *halwah*; faced with these difficulties, several Brahman proprietors held a meeting in the mid 1970s and decided that they would plough the land themselves. Further, they decided that if any person from their community raised the question of taboo, they would ostracise that person!

In Chaukra there was no ban on the Tyagis holding a plough, though in the 1950s few would use the plough themselves. However, the participation of Tyagi farmers in auxiliary and, in some cases, in main on-farm labour seems to have grown. Greater mechanisation in Chaukra implies that the plough was never a dominant concern, though most rich farmers still keep bullocks, for levelling the fields, for light ploughing, or for transporting cane to the crushers. However, with increasing investment in urban activities such as trade, business and education, there seems to have been some 'depeasantisation' of Tyagi youth, and that appears to be aggravating problems of agricultural supervision in the latest phase.

From crops to trees?

In recent years supervision-constrained as well as non-resident lessor households in Chaukra appear to be turning to an option which is labour-saving as well. In the last decade there has been a spurt in tree-planting in Chaukra. According to the survey in 1990/1, 25 per cent of all landowning households in Chaukra had planted trees since 1982, 94 per cent of them eucalyptus. A breakdown shows that in Chaukra, 23 per cent of resident landowners and 82 per cent of non-resident landowners planted trees. In Alipurjeeta, tree-planting has been a less important activity, and only 16 per cent of all landowning households had planted trees. Fewer of these (55 per cent) were eucalyptus. Moreover, the number of trees planted per reporting household was lower. On the other hand, again, a larger proportion of non-resident than resident households planted trees (29 per cent compared to 15 per cent).

Apart from considerations of profitability (during the 1980s eucalyptus-planting was promoted as a highly profitable venture), were there any other reasons for the growth of eucalyptus? Data show that eucalyptus requires less labour and supervision than field crops. In our field-study, eucalyptus required 50 to 60 man-days per hectare annually, compared to 375 man-days per hectare for paddy and wheat, and 220 man-days for sugarcane. The requirement for watch-and-ward by landowners was further lessened by the practice of leasing out woodlots in the initial years for crop inter-culture. Tree-planting has thus appeared as an attractive option for lessors and supervision- and labour-constrained cultivators.[29] Between 1985/6 and 1990/1, several non-

[29] Eucalyptus is planted on boundaries and as woodlots. Strictly speaking, the latter enables saving on labour and supervision. However, boundary plant-

resident lessors in Chaukra planted trees as woodlots. Further, our study of resident planters shows that (in the medium-large land-size group), while planters as a group were significantly more supervision-constrained than non-planters, among planters, woodlot planters in particular were considerably more supervision-constrained than other planters.[30] The increase in agriculture thus may underscore a trend towards labour-extensive farming among more supervision-constrained farmers in Chaukra.

Conclusion

This study focuses on a certain set of issues which are relevant to an understanding of the characteristics of labour relations during transition. The unevenness in agrarian relations across regions is reflected in the nature and extent to which the appropriators of surplus labour are themselves dissociated from the process of production. During transition pre-existing forms of labour appropriation undergo subtle changes reflecting diverse objectives and changes in the strategies of rural classes. While several factors affect these changes, this paper emphasises that the relationship of the labouring classes to the dominant classes is crucial The case studies presented here reflect changes from two contrasted initial situations. The differences between the villages, as well as the uneven change within them, serve to underscore some of the implications of the transitional situation.

Despite the variety of changes to be found, traditional features reinforced by extra-economic coercion occupy a significant place in any explanation of the dynamics of labour relations in Alipurjeeta. The evidence provides details of the role of extra-economic factors in determining both the characteristics of, and the payments to, agricultural labourers. In the case of Alipurjeeta a section of rich farmers use threats and other means of extra-economic coercion directly to enforce adverse payment conditions. In both villages, large employers also use monopsonistic power and other devices to keep down the level of wages. The evidence suggests that the wages paid by such employers are *lower* than the wages paid by smaller employers.

In Chaukra, more radical changes in production organisation have

ing may also be seen as an intermediate stage in which the farmers gain experience with trees.

30 The supervision constraint has been computed as an index of net sown area person available for agricultural work or supervision in the household.

significantly modified the labour process and the requirements for supervision and so on. This has implied far-reaching changes in labour contracts and wage-payment systems. The differences in wage-payment systems reflect the problems of control and supervision faced by large employers in Chaukra, who are rapidly moving towards piece-rated wages in most operations. Changes in tenancy in Chaukra have enabled landlords to sustain increases in the intensity of production, while avoiding increases in money wages. The changes also help in keeping supervision costs low, while exploiting tenants' family labour. Concomitantly, they enable cultivating landlords to mobilise on-farm labour. Such changes are still peripheral in Alipurjeeta, but at the same time landlords (and this includes the sizeable pure rentier class) have moved towards greater participation in decision-making and greater supervision of leased holdings, effecting some productivity gains and reflecting their concern over the enforceability of contracts under weakening traditional relationships.

Changes in labour relations have modified the nature and extent of participation by members of dominant proprietory castes in the labour process. Among them, smaller upper-caste landholders in Alipurjeeta have broken the caste taboo and wielded the plough. The extent of participation in main and auxiliary labour by dominant landholders in both villages has increased. But in recent years there is some evidence of withdrawal from participation in agricultural labour and supervision by newer entrants to the workforce in Chaukra, aggravating supervision problems. In the most recent phase, supervision-constrained farmers in Chaukra have taken to eucalyptus-planting as a means of reducing supervision and labour costs. Except for isolated lessors, eucalyptus-planting in woodlots has not emerged as an important option for dominant landholders in Alipurjeeta.

Tradition plays an important role in the (condition of) employment of farm servants in Alipurjeeta. Such employment is now far more contractual in Chaukra. The employment of farm servants does not appear to be strongly related to the specific high points of *seasonal* labour requirement or recruitment costs. There is, further, no evidence to suggest that the proportion of farm servants in the total labour force has grown over the years. If anything, the contrary appears to be the case.

As stressed above, during transition, with shifts in the orientation of the strategies of surplus appropriators, interlinked transactions begin to reflect these shifts and acquire 'two-sided' characteristics. The systematic differences between Alipurjeeta and Chaukra demonstrate how the *nature* of such relationships is affected by the orientation of the dominant classes (given the balance of class forces). All the major forms of interlinkages show that in Chaukra such relationships are increasingly being geared to the mobilisation of on-farm labour, reflecting a great stress by employers on particular ways of minimising the problems related to labour control, supply and supervision. In Alipurjeeta, while the general strategies of participant classes—in particular, the alternatives available to the poor—play a part, the specific nature of many forms of interlinkage is still determined by tradition. They are therefore determined in a prior sense by production relationships, rather than through individually-negotiated contracts.

The evidence shows that if alternatives are available, the poor preferentially withdraw from interlocked relationships. The shift in orientation in interlinked transactions is also closely associated with the nature of shifts in the separate transactions, and this is indicative of the penetration of capitalism. At the same time, however, the differences in the nature of such transactions in the three villages still underscore the fact that surplus appropriators may make use of existing modes of surplus appropriation in very different ways, leading to different outcomes.

References

Bardhan, P.K., 1980. 'Interlocking factor markets and agrarian development: a review of issues', *Oxford Economic Papers* (March), pp.82-98.

—, 1984. *Land, Labour and Rural Poverty: essays in development economics* (Delhi).

Basu, K., 1983. 'The emergence and isolation and interlinkage in rural markets', *Oxford Economic Paper* (July), pp.262-80.

Bhaduri, A., 1973. 'An analysis of semi-feudalism in east India', *Frontier* (Autumn).

Bhalla, S., 1976. 'New relations of production in Haryana agriculture', *Economic and Political Weekly* (March), pp.A23-A30.

—, 1983. 'Tenancy today: new factors in the determination of mode and level of rent payments for agricultural land', *Economic and Political Weekly* (annual number, March), pp.832-54.

Bharadwaj, K., 1979. 'Towards a macro-economic framework for a developing economy: the Indian case', *The Manchester School* (September), pp.270-302.

—, 1985. 'A view on commercialisation in Indian agriculture and the development of capitalism', *Journal of Peasant Studies* 12, 4 (July), pp.7-26.

Breman, J., 1984. *Patronage and Exploitation: changing agrarian relations in South Gujarat* (Berkeley).

Byres, T.J., 1981. 'The new technology, class formation and class action in the Indian countryside', *Journal of Peasant Studies* (July), pp.405-54.

Marx, K., 1976. *Capital: a critique of political economy*, vol.1, intro. by Ernest Mandel (London).

Patnaik, U., 1983. 'Classical theory of rent and its application to India: some preliminary propositions with some thoughts to share-cropping', *Journal of Peasant Studies* 6, 4, pp.375-42.

—, 1986. 'The agrarian question and the development of capitalism in India', *Economic and Political Weekly* (3 May), pp.781-93.

Singh, M. and K.G. Iyer, 'Migrant labourers in Punjab', U. Patnaik and M. Dingwaney (eds.),

Srivastava, Ravi S., 1989a. 'Tenancy contracts during transition: a study based on field in Uttar Pradesh', *Journal of Peasant Studies* (April).

—, 1989b. 'Interlinked modes of exploitation in Indian agriculture: a case study', *Journal of Peasant Studies* (July).

—, forthcoming a. *Capital Accumulation by Rural Households in India* (Delhi).

—, forthcoming b. 'Some issues of labour and supervision in tree planting. Evidence from U.P. villages', in Vishnu Ballabh and N.C. Saxena (eds.), *Socio-Economic Aspects of Tree Crops on Farm Lands* (Delhi).

Thorner, D. and A. Thorner, 1962. *Land and Labour in India* (New Delhi).

Chapter 8

TIME AND MONEY: THE MEANING AND MEASUREMENT OF LABOUR IN INDIAN AGRICULTURE[1]

Sumit Guha

In the course of a recent attempt to measure changes in labour intensity in Indian agriculture over the nineteenth and twentieth centuries,[2] I encountered serious problems in defining and measuring labour, and even greater ones in measuring labour-time. Upon a close examination, agricultural work begins to dissolve into widely different sets of actions, and even to include long periods of inaction (crop-watching, animal-tending), demanding very different levels of skill, strength and endurance, and performed by workers differentiated according to age, sex and status, among other things. This was recognised in contractual arrangements from an early period. When Thomas Coats studied the Deccan village of Loni in 1819, he found peasants who cultivated in partnership with others. He wrote: 'Partnerships are commonly formed of four men, who each supply a woman and two bullocks, which are sufficient to do the work of one plough'.[3] Mention of the sexual division of labour, of course, brings in the almost insurmountable difficulty of separating household and agricultural labour. The intimate dependence of the two appears in a study by P.C. Patil. The farm household contained two women, and in 1928/9 they put in nearly as many days of farm work as the men; the next year the farm was partitioned, and the remaining woman could spare little time from domestic tasks and 'worked' for only 33 days, while 159 woman-days of labour had to be hired in. Needless to say, the woman of the household was not 'unem-

[1] I am indebted to the participants in the Centre for South Asian Studies' seminar, where this paper was first read, for their suggestions, and to Peter Robb for additional comments during its revision. I have also drawn upon Indrani Chatterjee's ideas at several places in the course of the paper. None of the above is responsible for any errors that may yet remain.

[2] Preliminary, and somewhat inexact results, were published in S. Guha, 'Labour intensity in Indian agriculture 1880-1970: some findings', Review of Agriculture, *Economic and Political Weekly* (29 September 1990).

[3] Thomas Coats, 'Account of the Present State of the Township of Lony', in *Transactions of the Literary Society of Bombay* III (1823), p.223.

ployed' for 332 days of that year.[4]

The problem of measuring agricultural (or indeed other) work is not merely that of the social scientist: the ruler, the chief, the employer or the *paterfamilias* all had to gauge the level of effort that their dependants were putting forth in their several functions, in order to check dereliction of duty. Viewed in this context, the measurement of agricultural work is but an aspect of the far wider problem of social control inherent in (perhaps) every society—too large an issue to tackle here. I shall confine myself to considering the issue of the management of agricultural work in a complex agrarian society—Western India from the turn of the eighteenth to the twentieth century. Social and economic conditions changed substantially over the period, and the typical forms of labour relations changed as well. I shall not attempt to chronicle these changes, but rather will look at the various forms of labour relations as examples of different strategies of control. Social change would manifest itself in the changing mix of these various relations in agrarian society as a whole, and perhaps even in the disappearance of some forms (chattel slavery?). My object at present, however, is merely the elucidation of the internal logic of the relations themselves, largely ignoring changes in the social matrix in which they were embedded.

Assessing the quantity and quality of work done presents peculiar problems in agriculture: the long span of time over which many separate operations are performed, and the impossibility of distinguishing their effects from those of the accidental forces of nature, make supervision difficult and laborious. Many agricultural proverbs enjoin the master to be ever-present in the field—[5] but that counsel of perfection was not easy to follow, and many would find it uncomfortable, inconvenient, or degrading. A number of social arrangements can be seen in terms of reducing either the amount, or the need, for supervision.

Perhaps the most widespread was incorporation of labour into the household, or domestication. One method of recruitment was by marriage. Govind Krishna was a 'well-to-do Thakur' tribal of the Thana district. He had 'three wives but no children...and a servant on wages of Rs.5 a year and his food'. He had probably paid bride-price, but

[4] P.C. Patil, *Principles and Practice of Farm Costing with Farm Studies* (Bombay 1933), pp.598-605 and 624-9.

[5] A. Manwaring, *Marathi Proverbs* (Oxford 1899, Delhi reprint 1991), nos.63-5.

thereafter his wives cost him nothing more than their food, estimated at two pounds of *nagli* (a coarse grain) each per day, or less than nine rupees a year at that time (1883-84).[6] While most wives worked, both indoors and out, marriage was probably rarely viewed as being purely a labour relation. The labouring dependant was most frequently servant or slave, and such persons might also be given familiar but subordinate status with the object of getting them to identify their interests with those of the larger household: a strategy widely used, both in India and elsewhere.[7] Thomas Coats, a close observer, recorded of the slaves found in Loni, that they:

are inmates with their masters; their treatment is very good: they are fed and clad in the same way as the members of the family, scarcely any difference being observable, except their taking their meals apart. They get presents of pocket-money on holidays, if they behave well, and their masters are at the expense of their marriages.... The men labour in the fields.[8]

Incentives by themselves might not be adequate for control, and subordinate kin and slaves both might require chastisement on occasion. Independent fiscal interests could operate within the shell of the household. Coats noted that married women:

all keep their private purse, and carefully conceal its contents from their husbands, which is supplied by the wages of extra labour, presents from their relations, and sometimes when he is good-humoured, by a donation from the husband. If the wants of the lady are urgent, or she is less scrupulous, she pilfers grain and sells it to recruit her funds. This money is employed in procuring ornaments for her person, or favourite child; and if these should be so expensive as to excite the suspicion of her husband, she is obliged to attempt some satisfactory explanation.[9]

What might happen if the explanation was not satisfactory may be

[6] *Annual Report on Crop Experiments in the Bombay Presidency, 1883-84: Northern Division*, pp.44-6.

[7] Among the Swat Pathans, the 'tie between master and servant is usually an extremely close and intimate one—much more so than those between brothers, friends, or even father and son.... Theirs is a symbiotic relationship: they are unequal and complementary, and the fate and career of the one depends to a very great extent on the actions of the other'; F. Barth, 'Political leadership among the Swat Pathans', in S.W. Schmidt *et al.* (eds.), *Friends, Followers and Factions: a reader in political clientelism* (Berkeley 1977).

[8] Coats, 'Account of Lony', pp.225-6.

[9] Ibid., p.217. One is reminded of Hobbes' asking his reader as to what opinion he had of his children and servants when he locked his chests; T. Hobbes, *Leviathan* (Harmondsworth 1978), p.187.

inferred from the observation of a report on slavery in the Deccan prepared in the 1820s: 'The power of punishment rests with the master, but it is not to be exerted to a greater extent than would be *used towards one of his own family*.'[10] Of course, all controls might fail on occasion, as in the case of Yesa, a slave who behaved in such a headstrong fashion that his master freed him in 1786/7.[11] Coats also mentions that slaves might be manumitted if they grew burdensome, and their familial status inevitably made them costlier to maintain than other workers. More common, therefore, were various types of tied, or permanently or semi-permanently attached labourers, who were also sought to be incorporated into the household. Of such hired servants, Coats wrote:

> they are all grown-up men and have families in remote villages: their wages are from 25 to 30 rupees a year, which is made to consist of 13 months; together with their food, clothes and lodging. They live under the same roof with their master, and their fare and clothing scarcely differ from his.... The servant sometimes calls on his master to advance him a sum of money for his marriage; and if it is complied with, it is generally on the condition of the servant binding himself to serve for a certain number of years for his clothes, board and lodging....

Arrangements of this type were very common, and W.H. Sykes found many examples in his surveys of the Deccan.[12] Rural officials and gentry did much of their farming through such men, and, when the first British Collector of Khandesh prepared an estimate of costs and returns in farming in 1821, his statement assumed that the 'Ryot' possessed eight bullocks and employed four full-time labourers costing him Rs.50 a year each.[13] H.D. Robertson, Collector of Poona, also made a similar calculation, putting the total cost of four men at Rs.184, but added that 'we may on most occasions suppose that the subsistence and the wages of two of the four persons charged for, go to the father and one son of the family'. The peasant's wife and younger children were expected to earn 'by selling cow-dung, by the sale of a cow's milk

[10] R.D. Choksey (ed.), *Period of Transition 1818-1826* (Poona 1945), p.232 (emphasis added).

[11] G.C. Vad *et al.* (eds.), *Selections from the Satara Raja's and Peshwa's Diaries* (Poona 1991), VIII, pt.3, p.99.

[12] Coats, 'Account of Lony', pp.225-7; W.H. Sykes, 'Special Report on the Statistics of the Dukhun', in *Reports of the Seventh Meeting of the British Association for the Advancement of Science, 1837*.

[13] *Selections from the Papers at the East India House...under the Company's Governments in India* (4 vols. London 1820-26), vol.IV, pp.691-2.

or ghee about 30-0-0', in addition, of course, to unpaid domestic or field labour, which passed unmentioned.[14] Reporting from Nagpur in 1827, Richard Jenkins noted that the children of the poorer cultivators,

> and all the lower orders,begin to work about the age of eight or nine years, when they are usually employed in herding or looking after the cattle.... Women continue to work in the fields and perform hard labour until fifty or fifty-five, after which they are employed in spinning and work within doors.[15]

Labourers hired by the year were not necessarily incorporated into the household: quite often they lived separately, and were paid part of their wage in grain which they prepared and cooked at home. This might have been especially common on the tribal fringes of Maharashtra—areas such as Nagpur and Khandesh—where caste considerations might prevent the inclusion of the labourer within household.[16] Their condition could nonetheless approximate to outright servitude, tempered, however, by the possibility of flight, at least in 1825 when the situation was described by G.A. Rigby, Bhil Agent in Khandesh:

> The caste employers hire Bheel labourers at wages on which it is utterly impossible that they can procure the means of supporting existence, unless they purloin the produce of the Lands they are employed to cultivate, or make up the deficiency from some other and generally more objectionable source... [robbery, perhaps]. On advances made such extortionate usance is imposed, as absorbs the wages, and not unfrequently leads the Bheel to abandon his Hut in despair.[17]

The employer's strategy in this case seems to have been to reduce the Bhil to debt-bondage, with the latter retorting with the weapons of the weak—pilferage or flight. Thus domestication was not the only strategy used in long-term labour relationships. Half a century later, with the British legal and administrative system in place to enforce contracts of servitude, the scales tipped further against the Bhil. In the year 1880/1, E. Lawrence, an Assistant Collector in western Khandesh, noted ten cases of enforcement of contracts under Act XIII of 1859, and

[14] Ibid., IV, p.577.

[15] R. Jenkins, *Report on the Territories of the Raja of Nagpur* (reprint Nagpur 1923), p.26.

[16] An informant in Dharwar in the early 1820s said, 'Not many Dhers and Mangs are slaves. The reason is that people of these castes cannot be employed in the houses of the high caste'; G.S. Sardesai (ed.), *Selections from the Peshwa Daftar* (Bombay 1931-34), vol.42, p.32; translation mine).

[17] Report, 30 June 1825, in Pune Archives, Deccan Commissioners' Records, vol.202, no.1782, p.11.

commented that the

> continuation of a penal Act for the enforcement of civil obligations appears somewhat an anomaly in these days of increasing civilisation and order, and puts an extra means of power into the hands of an already powerful enough class, i.e., the agriculturist who has servants under him who are very much in the position of serfs.[18]

Labourers on long-term contracts continued to be employed by gentry farmers in the Pune area. When the Sarvajanik Sabha was collecting evidence on British land taxation, an 'Inamdar and Zemindar in the Indapoor Talooka' calculated the cost of cultivation on the basis of a holding of 96 acres, employing 16 bullocks and five permanent farm servants, four men and a boy. Another informant, the Kulkarni of Kedgaon village, Bhimthadi taluka, claimed to employ twelve annual servants and work alongside them.[19] Sixty years later the system was still in place, and its advantages were described by Jagalpure and Kale in their study of the village of Sarola Kasar. The labour of annual or monthly servants 'is more efficient than that of daily wage earners as they take more interest in their work and have a greater sense of responsibility'.[20] The long period of contract, and the more-or-less successful incorporation into the household, or domestication, provided one way around the problem of monitoring and supervision: to internalise the supervisor, and make wage-employment morally equivalent to self-employment. If this was achieved, then little supervision was needed, and the problem of measuring labour-time would largely be by-passed. Moral suasion would also be supplemented by the other concomitant of co-residence: the possibility of near-continuous surveillance. This in itself might make it distasteful to the worker. In East Maharashtra (Berar) in the 1880s, the attached labourer (termed 'salkaroo'), it was reported,

> always earns more than one who is not a Salkaroo. The demand for Salkaroos is also in excess of the supply. But a Salkaroo must be always at the beck and call of his master, and has to do much harder work. The labouring classes choose to secure their comparative independence, and are content with what their mini-

[18] E. Lawrence, Annual Administration Report, 2 July 1881, para.22, in Bombay Archives, Revenue Dept., vol.17 of 1881.

[19] *Report from a Sub-Committee of the Poona Sarvajanik Sabha* (Poona 1873), pp.14, 16.

[20] L.B. Jagalpure and K.D. Kale, *Sarola Kasar: study of a Deccan village in the famine zone* (Ahmednagar 1938), p.352.

mum capacity of working brings them....[21]

Domestication and enserfment as strategies of labour management also had drawbacks from the employers' point of view—notably their inflexibility—and other forms of labour contract existed alongside them for centuries. As holdings grew smaller through the nineteenth and twentieth centuries, farmers' need for additional hands for short periods would increase. One arrangement by which this could be met was by labour contracted for a share of the harvest. In Nagpur in in the 1820s, many of 'the labours of husbandry are executed by contract; the labourers engaging for a certain share of the produce to perform certain tasks'.[22] This was an old expedient: the *Arthashastra* had recommended that those who contributed only their labour should receive a fourth or a fifth of the output.[23] Observing a similar system in North Karnataka in 1821, Thomas Marshall commented that where a tenant had no agricultural stock, his acknowledged share amounted 'to no more than a bare subsistence...'. But the drawbacks of this arrangement were also clear: 'this compact is not common, being considered as a bad one for both parties, leading to slovenly and imperfect tillage, as well as waste and pilfering of the crop'.[24] The system obviated the need to measure the quantity of work, but required closer monitoring of other aspects of it.

A better-controlled variant was payment by results. Jenkins noted that labour paid by piece-rate 'is always more quickly executed than labour paid by the day. The cultivators are well aware of this circumstance, and when any of the operations of husbandry are required to be speedily executed, they are performed by contract.'[25] Thus, for example, Sykes found harvesters paid five sheaves out of every hundred that they reaped, bound and stacked. The exact share would, of course, vary according to the nature of the task; in harvesting groundnuts, 'from the

[21] L.S. Saunders, Commnr. Hyderabad Assigned Districts, to Secy. for Berar, 4 April 1888, in *Reports of an Enquiry into the Condition of the Lower Classes in India 1888*, pp.3-4 (India Office Collection W3018).

[22] Jenkins, *Nagpur*, p.25.

[23] R.P. Kangle (trans.), *The Kautiliya Arthashastra* (Bombay 1972), part II, p.150.

[24] T. Marshall, *Statistical Reports on the Pergunnahs of Padshapoor, Belgaum, Kalaniddee and Chandgurh, Khanapoor, Bagukot and Badamy, and Hoongoond, in the Southern Maratha Country* (Bombay 1822), pp.16-17.

[25] Jenkins, *Nagpur*, p.26.

labour and tediousness of digging it up, and the cheapness of the produce, the labourers were allowed one-fourth of the whole'. Only when the crop was exceptionally valuable, such as sugarcane, were cash wages paid.[26] This simple system of payment by output existed for cotton-picking, and indeed for a variety of operations where output was easily measurable. In the twentieth century, when large-scale *gul* (crude sugar) production had spread in Western India, the whole complex set of operations from cutting the cane to converting its juice into *gul* could be contracted out to a gang of eight men and twelve bullocks at a certain rate per unit of output.[27] Similar gang-labour systems are widespread even now in many parts of India. Here earnings depend substantially on effort, and there is little need to monitor the time spent or regulate the intensity of effort.

In addition to all these, of course, there were many tasks for which a fixed amount of cash or grain was paid. In 1827 cash wages were paid for weeding—an operation usually performed by women, who were paid four to six pice, and worked from dawn to dusk. Men might earn eight pice a day.[28] Cash wages, of course, had to vary with effort. As one official noted, 'the actual rate varies with the length of the day's work…'. Annual labourers were paid from Rs.70 to Rs.100 in the early twentieth century 'according to the nature of the work required to be done; thus a higher rate would be paid to a labourer working on irrigated land than to a man working on jirait [dry-crop] land'.[29]

In short, a number of labour arrangements and types of payment have existed (and many still exist) in India—the comfortable maintenance of the confidential slave, the skimped measure doled out to the indebted serf, the fixed share of the harvester, the cash daily-wage of the weeder, and others too numerous to mention. This creates severe problems for anyone attempting to measure labour-time. Atchi Reddy has pointed out one aspect of this:

> Figures of average hours of work are not a sufficient index of the level of exertion or volume of work, or of the wage rate per hour, since the quality of labour and the intensity of its application vary from individual to individual

[26] Sykes, 'Special Report', pp.320-5.

[27] P.C. Patil, *The Organisation and Cost of Gul (Crude Sugar) Making in the Deccan Sugarcane Tracts* (Agriculture Dept. Bulletin no.147, 1927), p.1.

[28] Sykes, 'Special Report', pp.322-3.

[29] *Selection from the Records of the Bombay Government*, no.572 (new series), Settlement Report of Yaval Taluka 1915, p.15.

even in the same operation and for...the same individual in different operations.[30]

Furthermore, the level of exertion, or the degree of care exercised, may vary depending on the institutional arrangement under which labour is supplied. The employer of an annual labourer would try to keep him busy, without exact regard to the value of the work. The Banking Enquiry Committee's investigator recorded in 1930:

The servant is not allowed to remain idle. On rainy days he is either given the work of fencing, rope or basket making, cleaning or repairs of cowshed etc., or attend if necessary to the repairs of the house. For leave days the general practice is to take service for thirteen months instead of twelve months.[31]

Smaller farmers had the same attitude towards household labour. P.C. Patil commented, after several years of study:

A farmer, especially a peasant farmer, looking upon farming as a means of employment, quite often spends so much home labour as would not bring sufficient additional yield to pay the wages of that labour at current rates. ...The farmer often cannot find sufficient work to employ his spare labour more gainfully and he, rather than idle away the time, utilises it without looking to the increasing gains due to that labour. When the rains are poor and there are few fields bearing crop, the question arises whether labour should or should not be employed in watching the crop against birds and stray cattle. If the farmer calculates the cost of such labour, and considering it uneconomical does not watch the crop, what little crop there is may be lost. If he watches the crop, the calculated extra cost cannot be repaid. If there is spare home labour, the farmer without looking to its cost makes use of it, and he cannot be blamed for doing so.

Watching is a good example of work at low intensity but long duration. Crops might be watched day and night for two, or even three months. Neglect of watching could have severe consequences: of two *bajri* (millet) plots studied by Patil in 1925/6, one yielded 538 pounds weight per acre, and the other 287 pounds, the difference being mainly accounted for by the 'neglect shown in watching...'.[32]

As the village system broke up over the nineteenth and twentieth centuries, watchmen would watch only their own, or their employers'

[30] M. Atchi Reddy, 'Work and leisure: daily working hours of agricultural labourers in Nellore District, 1860-1989', *Indian Economic and Social History Review* 28 (1991), pp.73-4.

[31] Report in Bombay Provincial Banking Enquiry Files, File no.67/2 (Bombay Archives).

[32] P.C. Patil, *Preliminary Studies of Important Crops in the Bombay Deccan* (Agriculture Dept. Bulletin no.168, 1932), p.12 and section on dry-crop *bajri.*

fields. The labour time expended per acre would vary enormously depending on the size of the field. In the 1880s, the Director of Agriculture estimated that in clear ground unobstructed by trees, a man could watch a ten acre field;[33] by the twentieth century many fields were much smaller than this, with a consequent increase in labour time and decrease in its intensity.

To sum up, labour-time would be a very imperfect measure of work done in Indian agriculture, since the relation of these two hypothetical quantities would vary according to season, topography, operation, equipment and the institutional arrangement under which the work was performed. Simply counting the number of days spent, or reported to have been spent, in different activities can be very misleading; and among those recently misled has been Utsa Patnaik, who based an entire analysis of class relations in Indian agriculture on the assumption that man-days could be treated as homogenous, with woman- and child-days being uniformly equivalent to two-thirds and one-half man-days respectively. A re-analysis of her data showed that labour-days were in fact extremely heterogenous, and carrying out arithmetical operations on such unlike units was like subtracting a lemon from a grapefruit and dividing by an orange.[34]

Is it then necessary completely to abandon any effort to measure labour inputs in Indian agriculture? It is not necessarily so, since one way out of this dilemma was found by one of the pioneers of Indian economics, D.R. Gadgil. His method is best described in his own words:

> The standard labour unit was considered to be equivalent to payment of Rs.150 p.a. in irrigated villages and Rs.120 p.a. in dry villages, which according to our data represent the average payment made for the services of a permanent farm servant for a year. All hired labour was converted into standard labour units by dividing the total cost of the labour (in rupees) by 120 or 150. The money cost of unpaid family labour was first ascertained by noting in each case the rate of wages at which labour similar in character could be hired. In doing this due regard was paid to the permanent or casual character of the labour employed and the consequential difference made in the rate of payment to it. The cost of unpaid family labour was also converted by the method given above into labour

[33] E.C. Ozanne, Director of Agriculture, to Secy. to Govt. of Bombay 15 March 1884 (Bombay Archives, Revenue Dept., vol.344 of 1884).

[34] For a full discussion see S. Guha, 'The labour-exploitation ratio as a measure of peasant differentiation: its fundamental fallacy', *Economic and Political Weekly* (21 August 1993).

units.[35]

This method avoids many of the difficulties discussed earlier, but it also requires us to abandon any attempt to measure labour *time*; instead we get homogenous units of actual or imputed *cost*, which is then turned into standard units with reference to the cost of a certain specific kind of worker—in this instance, the male labourer hired by the year.[36] Social conventions of payment are thus incorporated into the estimation: for example, the widely occurring convention that daily wages for women are around half those for men, or (in some cases) that the wife and children of an attached labourer may be called on to render petty services to his master. Furthermore, this method presupposes an agrarian order in which wage-labour is widespread, if not predominant, so that it is easy to ascertain the wage paid for every kind of farm work. When applied to large samples of farms within a single region at one point in time—the situation for which Gadgil devised it—this method is probably the best way of measuring labour inputs objectively. When applied over long periods of time, its accuracy would be affected by changes in the nature and functioning of households, in the structure of markets, and in the relations between the working and employing classes. But it would certainly be superior in almost all circumstances to the simple counting of days or hours spent in this or that activity, over thousands of years, and under a variety of changing and contested forms of the social organisation of labour.

[35] D. Gadgil, *The Economic Effects of Irrigation* (Poona 1948), pp.182-3.

[36] Interestingly, when a number of Maratha hereditary officials in the former Bhosle territories in the 1820s were asked to estimate the costs and returns of cultivation for the owner of a single plough and pair of oxen, they uniformly estimated labour requirements in terms of so many year-round labourers with each plough, each of whom would receive one or more rupees, plus a fixed quantity of grain for his annual subsistence. Their answers (in Marathi) are to be found in Mss.Mar D.46 of the India Office Collection, London.

Chapter 9

DISCIPLINE AND CONTROL: LABOUR CONTRACTS AND RURAL FEMALE LABOUR

Karin Kapadia

Introduction

Conditions of employment of agricultural labour in eastern Tiruchirappalli (or Tiruchi) District in Tamil Nadu provide some clues to two important questions. Firstly, why are some labour contracts very open to negotiation, while others are far less so? And, secondly, is there 'collusion' between labourers, and, if so, what results does it have on labour contracts? I argue that the steady increase in the use of piece-rate instead of daily wages in the payment of labour signals two major changes in the mode of employment. The first change is in the relationship between labourers and employers: here we find that the traditional 'rights' of workers and the customary obligations of employers are giving way to a more blatant exploitation of the workforce. The second change is in the relationship between labourers and their work: a subtle 'internalisation' of discipline is evident, suggesting that employers' control of workers is now manifested in workers' own self-discipline. This new work-discipline is both a sign of 'modernity' in its impersonalised control of labourers and an indication of the growing vulnerability of labourers. The discussion is focused throughout on the work of landless women labourers of the 'untouchable' Pallar caste in Aruloor village in eastern Tiruchirappalli District.[1]

I focus on women's work groups in this analysis, not only because my anthropological research has concentrated on women labourers, but also because a *feminisation of the labour force* is apparently under way in Aruloor (see Kapadia 1992). Significantly, important recent research on agricultural conditions in Tamil Nadu suggests that this phenomenon

[1] Fieldwork was carried out in Tiruchi District, Tamilnadu, from January 1987 to May 1988 with further visits in December 1988, December 1990 and January 1991. Aruloor is a pseudonym. The author is particularly grateful to Jean Dreze of the London School of Economics for his encouragement and generous advice. She also thanks John Harriss and Gillian Hart for their helpful comments.

is not limited to Aruloor village. Athreya, Djurfeldt and Lindberg (1990: 151-2) observe that women's work participation has apparently increased more sharply than men's. Ramachandran (1990: 106-7, 123-5, 129, 134), Harriss (1982: 97-8) and Athreya (1984: 96) go further, providing statistics that emphatically show that the absolute number of days of employment obtained by women is higher than the number of days obtained by men. However, none of these writers provides an account of the organisation of women's work groups. This is investigated here in order to elucidate the striking paradox that while women workers co-operate in order to raise wages and to resist exploitation by their employers, they compete against each other in their daily struggle to find employment. Their solidarity against employers is discussed in detail elsewhere (Kapadia 1993a), showing that wage rates have gone up through strikes initiated by women labourers in the face of strong resistance from employers. Differences in the patterns of organisation and resistance by female and male labourers suggest that women workers are far more willing to challenge their employers than men are. This intriguing parallel to Gillian Hart's important findings regarding women's work groups in Muda, Malaysia (1991; 1992) raises issues that I discuss in detail elsewhere (Kapadia forthcoming, a and b): in brief, both socio-cultural factors relating to lower-caste perceptions of women as financial providers, and political-economic factors relating to the shrinking labour market for male labourers, are central to the Aruloor situation.

There has been a lack of attention in the literature to the co-existence of different contractual arrangements. There are no theoretical models of labour contracts that incorporate both daily-wage and piece-rate contracts. Mark Rosenzweig's survey of 'Labor markets in low-income countries', in the *Handbook of Development Economics* (Chenery and Srinivasan 1991), for instance, has only two lines on piece-rate contracts (Rosenzweig 1991: 735). Binswanger and Rosenzweig deal only very briefly with piece rates in their article 'Behavioural and material determinants of production relations in agriculture': they state that 'for short-term contracts, piece rates dominate time rates: they are alike in all cost aspects except for the fact that piece rates provide superior incentives compared to time rates' (1986: 520). Aruloor workers agree that piece rates increase their incentive to work. However, the steadily increasing use of piece rates for various agricultural tasks in Aruloor, is also due to the fact that employers state that piece rates *reduce their supervision costs*. Both rates are not alike in all cost aspects.

Stiglitz's early paper on the structure of employment contracts in industrial organisation argues that piece rates are often used to screen out less productive workers (1975: 561). Roumasset and Uy (1980), in their attempt to extend this theoretical discussion to contracts in agricultural labour, state that three factors are central to the increased use of piece-rate contracts: 'In summary, we expect the incidence of piece rates to be higher the easier it is to monitor quality-shirking, the easier it is to set piece rates to equate the implicit wage with the opportunity wage, and the greater the heterogeneity of the work force' (1980: 350). But these are insufficient grounds for explaining the growth in piece rates in Aruloor. It is, rather, the sheer increase in agricultural production, due to the Green Revolution, that has increased the demand for labour—this increase in productivity is also documented by Ramachandran (1990), Athreya *et al.* (1990) and Harriss (1982). But Roumasset and Uy's observation that 'the incidence of piece rates tends to increase with higher wages' (1980: 354) is likely to hold true in Aruloor, because increased demand for labour (particularly female labour) has resulted in higher wages being won by women workers.

It is their discussion of 'piece rates with teams' (1980: 353-4) that is most relevant to my arguments here. Unfortunately no clue is given as to the gender of team members. They state:

> Piece rates with teams motivate decentralization of three types. First, there is an incentive for decentralization of supervision.... [Secondly] Teams also facilitate decentralization of decision-making about internal organization.... [Thirdly] By selecting their own members, each team can take advantage of its own information about a small part of the work force. That is, where information is diffuse, decentralization economizes on information costs (1980: 353).

I, too, argue that 'decentralization of supervision' is a central reason why piece-rate work—which is closely associated with work groups in Aruloor—is increasing. However, I argue that this 'economy' and the so-called 'decentralization of decision-making' also have an insidious effect on workers, who therefore have to 'internalise' discipline and thus 'supervise' themselves. I also argue that, in selecting their own members (the third 'economy' or 'decentralisation'), workers use exclusionary practices (see Hart 1986) that *serve to divide the workforce*. This important issue is not investigated by Roumasset and Uy.

Binswanger and Rosenzweig's edited book on rural labour markets in Asia (1984) has proved influential. However, the various articles collected there describe situations that are very different from those in parts of Tamilnadu. Referring to these articles in their introductory chapter, Binswanger and Rosenzweig state:

All these studies suggest that in developing as in developed countries, rural labor markets exhibit differential patterns of male and female employment and earnings.... *Female unemployment rates are higher*...and in general women have access to contract work paid on a piece-rate basis...*only through male members of their families* (1984: 32-3; emphasis added).

Neither of these generalisations applies to the labour market in Aruloor.

Through the following anthropological analysis of labour contracts in Aruloor, I hope to show that a gendered view that takes account of socio-cultural factors is essential for a better understanding of the issues involved.

Changes in the mode of employment

As in most of rural India, there are three broad categories of casual labour contracts in Aruloor, namely, daily-wage work (or time-rate work), piece-rate work and harvest work. Dreze and Mukherjee point out that 'The coexistence of daily wage, piece rate and harvest share contracts... appears in most microstudies' (1989: 249). The first two categories of work are paid for in cash, while harvest work is paid for in kind. Traditionally, rice was the main crop in this 'wet-soil' area, irrigated by tributaries of the Kaveri river. For the landless labourers of Aruloor, harvest payments of paddy have undoubtedly been their largest earnings, as their value exceeds the low wages they earn through daily-wages labour. Dreze and Mukherjee similarly note that during harvest, workers in western Uttar Pradesh 'earn far more per day than they do during the rest of the year' (1989: 243).

However, the two main kinds of casual labour contract in Aruloor fall into the other two main categories, on daily or hour rates, or on piece-rates. Locally the latter is termed 'contract' work or *pesigitta vele* which literally means 'talked-about work'. The term implies that rates of pay have been discussed or bargained for. Informants often used the English word 'contract' to translate *pesigitta* and also 'Tamilised' it, regularly speaking of *contract-vele*. This compares with the use of the term *kontrek* which Hart defines as 'literally contract labour' (1992: 815). Aruloor workers stated that this was a relatively recent form of labour contract, but that its use was increasing very rapidly.

The term that Pallar women commonly used for daily-wage work was *senjaya vele*. This appears to be their colloquial form of the words *senja vele*, which simply mean 'work done'. They also—more rarely—used the term *kuli vele* for daily wages work: it implies hard manual labour. The term *kuli* means 'daily wages', and so Pallar women would often speak of the *kuli* they had earned that day. Daily-wage work remains the commonest kind of agricultural work. Pallar women also spoke of the *kuli* (daily wages) paid for a piece-rate 'contract' job that was finished in one day. But 'contract' or piece-rate work often takes more than one day to complete and so a different term was used for its payment, namely *panam* (money).

It is a significant characteristic of this labour market that small and marginal tenants *also* work as casual labourers, as their holdings require little labour. Consequently, they understand the concerns of landless labourers well, because they share them to a large degree. They also know the tricks of the trade of landless labourers, particularly regarding daily-wage work.

This, traditionally the commonest form of labour contract, means that the labourer is paid a fixed wage for a fixed number of hours of work. Typically, daily-wage work consists, for female labourers, of jobs like weeding paddy or banana, and carrying manure to the fields. The period of work is generally from 8.00 a.m. to 12.00 noon. This four-hour session is considered a full day's work and the labourers are exhausted when it ends.

As they are paid by the hour and as work is scarce, the women see it as in their interest to drag out a job for as long as possible as this results in more employment. Weeding one acre of paddy normally should take a team of eight women two or three days, but they generally try to draw out the work to four days. Because employers know that labourers are always trying to delay the completion of daily wages work, they usually supervise them. One Pallar caste female employer from a better-off household compared hour-rate work with piece-rate rate work thus:

> Because it's daily wages work you *have* to be by them, otherwise they don't work—they just stand around and talk or work very, very slowly. You have to be there, standing above them, nagging at them. Contract work causes no such problems—they work fast and finish the job—because they know they won't

get any more by dragging it out.

Sometimes supervision is not exercised by the landholder. Instead authority is vested in the Pallar woman who was responsible for recruiting the others in the work group. Given that the normal period necessary to complete a task is well known, if completion is unreasonably delayed the recruiter is held responsible by the employer and is not rehired. Relations of patronage and control between employers, recruiters and labourers will be examined later. Significantly, Pallar women generally considered daily-wage work to be more advantageous to the labourer and less so to the employer. I will examine this perception in my discussion of negotiable contracts.

If hour-rate daily-wage work is considered less profitable to the employer, the reverse is true for 'contract' work. The need for supervision is considered to vanish altogether, for an agreed payment is made for the satisfactory completion of a piece of work by a work-team, no matter how long it takes. Thus here the opposite of hour-rate work is true—it is in the labourers' interest to finish the job as quickly as possible. On the other hand, a minimum standard of work is demanded: though a women's work-group may try to rush a job, they know that the employer can insist on its satisfactory completion before paying them.

One other form of payment for work should be noted, though it has fallen into disuse. Traditionally, among the poorer smallholders, such as Mutharajahs and Pallars, a system of labour exchange existed by which no monetary payments were made at all. It existed primarily between kin and solely within the same caste. When a smallholder needed labour this was provided by other members of the same caste, and the smallholder's family reciprocated by working for them when they needed labour. However, virtually all daily-wage labourers have been paid in cash for more than 30 years now. Interestingly, Ramachandran too notes that in Gokilapuram (Cumbum Valley) exchange labour is in decline and that 'its reduction to insignificant levels...has run parallel to the development of the demand for hired labour' (1990: 115-6).

The major development in recent decades has been the casualisation of work. Because of the unavailability of other forms of employment, marginal landholders have turned to casual labour. Further, there has been a drift of tied labour to the ranks of casual labour in the last few decades. Today there are hardly any *pannaiyal* or tied (or bonded) farm servants left in Aruloor, whereas 30 years ago there were many, with the Pallars represented strongly among their numbers. These *pannaiyal* acted as agents and supervisors for their employers, and they held land from them too. But the *pannaiyal* system began to disintegrate in Aruloor when owners started taking back their holdings from their *pannaiyal* tenants in the 1950s, because of their fear that the Tamil Nadu government's land reform legislation would strengthen the rights of tenants (see Gough 1989: 20-4).

Ironically, both in Aruloor and in other parts of Tamil Nadu, landless labourers today regret the passing of the system of bonded labour, because they regard it as having provided a measure of job security that they now entirely lack. This ironical situation has come about because of the context in which landless labourers find themselves: alternative and better-paid employment is not available to them, and there is an abundance of labour. Hence Ramachandran has rightly noted labourers' readiness to enter into tied-labour status in Cumbum Valley: 'In G.A. Cohen's insightful formulation "the worker here agrees to be a bit of a serf in order to qualify for proletarian status"' (1990: 254). Only one Pallar *pannaiyal* was left in Periyar Street (the Pallar street) in Aruloor in 1987, and he was envied by others for having this 'tenured' job, even though it paid little.

This casualisation of labour is part of a wider trend in which employers in the Aruloor area have been gradually shedding their traditional obligations even to their casual labourers. My Pallar friend and research assistant, Siva, described what was happening: 'About two years ago, everybody—even those with only a quarter acre—would provide lunch at 4.00 p.m. at the end of transplanting for all the women. But today nobody provides this lunch—not for the last two years.' She added that this was due to the continuing drought: landholders were now harvesting only one rice crop, not two. It seems very unlikely, however, that the meal will be provided once again even if crops improve. All traditional prestations and gifts from employers to workers—and from the wealthy to the poor—are declining. Significantly, food is no longer provided for the impoverished lowest castes at the end of the grand household ceremonies of the middle and upper castes.

Even those jobs that replicate the *pannaiyal's* job are now paid by the month and offer little job security. The *tannirpaccararvar* ('water irrigator') or irrigation supervisor performs one of the major jobs of the erstwhile *pannaiyal*: he has to walk around his employer's planted fields every day to see that the water is flowing in the irrigation channels, and he has to build them up or dam them as necessary. We may compare the discussion of *nirpaichi* by Athreya *et al.* (1990: 147). Several landless Pallar men were irrigation supervisors to higher caste employers in 1987—and so were two Pallar women. However, though they can be dismissed at any time, most irrigation supervisors continue to hold their jobs over several years—and the great benefit their position offers is that they also act as recruiters for work on their employer's fields. As we will now see, this offers a major source of power and influence within the Pallar community.

Recruitment and organisation of female waged labour

South of the Panguni river, which divides Aruloor, Pallar women monopolise the task of transplanting paddy. Transplanting is piece-rate work but it is not bargained for; like daily wages it has a standard scale of pay, which, in January 1987, stood at Rs.100 for one *kani* (acre). Transplanting requires specialised skill and is recognised by women of other castes to be the particular skill of Pallar women. Other work, especially daily work such as weeding or carrying manure, is more easily done by women who participate in agricultural work irregularly. This is the case with poor women from the Muthurajah, Pathmasaliar and Christian Paraiyar castes north of the river. However, because they work for employers who live north of the river (especially the Chettiar landlords), there is no conflict of interest between these women and the Pallar women south of the river. I discuss the segmentation of the female labour market and of wage rates between north and south Aruloor in detail elsewhere (Kapadia 1993a).

Because the Pallar women and men of Periyar Street constitute the regular agricultural workforce of those landholders who live south of the Panguni, they see themselves and are seen as controlling the recruitment of labour in this area. Pallar women therefore claimed that if a landlord were to dare to recruit cheaper labour from elsewhere and 'import' this labour into the fields, then they would disrupt such work because the landlord was morally obliged to employ them as they were his regular workforce. Similarly, a landlord would expect that the Pallars would work for him whenever he needed them. However, when labour demand peaks—as it does seasonally for transplanting (October-

November for the *samba* paddy crop) and for the paddy harvest (January-February)—this 'closed-shop' policy is relaxed. This is particularly true at harvest time when both Pallar women and men seek work outside Aruloor as well and tolerate the presence of migrant harvest labourers.

Pallar women perform virtually all their work in work-groups or teams, both for daily-wage work and for piece-rate 'contract' work. They almost never allow women of other castes to join these work-groups. The only occasion when I saw this happen was at the paddy harvest in February 1988 when two poor Odaiyar (caste Hindu) women, a mother and her married daughter, were allowed to join a harvest gang. However, harvest work was exceptional in that the demand for labour normally far exceeded its supply, and consequently it was not a problem to allow a couple of extra hands to join in. It is extremely unlikely, however, that these Odaiyar women would have been allowed by the Pallar women to participate with them in daily-wage work. They were allowed to join because they claimed that they were destitute migrants, and because they had won the favour of the Pallar women by providing a regular sale of cooked snacks in Periyar Street every morning. Muthurajah women were never allowed to join in the Pallar work-groups or at harvest time. At harvest, the poorer Muthurajah women and men would create their own mixed sex work-group. They worked quite separately from the Pallars, in other fields.

Recruitment to daily wages work

Recruitment and organisation differ sharply for daily-wage work on the one hand, and 'contract' work (piece-rate work) on the other. For daily wages work it may or may not be a local Pallar woman who tells the others about the availability of work, for the caste-Hindu landholder or, more usually, his 'irrigation supervisor' may come directly to the Pallar street to recruit labour. As noted above, the irrigation supervisor is often of Pallar caste and is generally used by the landholder as his agent and labour recruiter. More precisely, he recruits male labour while his wife recruits female labour.

The Pallar word meaning 'to recruit to work' is simply 'to call' (*kuppidu*). Thus the women or man who recruits others to work is simply referred to as 'the one who "calls" others'. This Tamil word is an exact parallel to the Hindi term used in Uttar Pradesh, as Dreze and

Mukherjee note (1989: 240): '...the local term for offering employment to labourers is simply *bulaanaa* which literally means "calling" or "fetching"'. They rightly point out one implication of the term: no wage-bargaining ensues where labourers are 'called' to daily wages work. But, as we will see, the same term is used for recruitment to piece-rate work as well, where bargaining is of the essence. Hence, at least in Tamil Nadu, the term does not connote solely a fixed wages context.

Considerable powers of patronage belong to the Pallar woman who acts as recruiter, because in a context where labour means food and survival, the one who provides labour has power and influence. However, the strongly egalitarian tendencies of Pallar society have limited the development of positions of leadership among the recruiters, and the fact that many different women act as recruiters lessens the power of the position. Recruiting for daily-wage work is normally done in the early morning, when news of the day's jobs comes to the street, either from caste-Hindus, Pallar irrigation supervisors or their wives. Any Pallar woman can be a recruiter; all she needs is knowledge of the existence of a particular job for that day.

Theoretically, for daily-wage work, an employer does not mind how many women volunteer for a job by presenting themselves at the field concerned because the more workers there are, the more quickly the job will be done. Actually, however, there are sometimes restrictions on the number who can be employed. The landholder may choose to restrict numbers because he has to give a full daily wage (Rs.5 after mid-October 1987) to all the women who work, no matter how quickly they finish the job.

With daily wages it may appear that *individual* Pallar women seek work, for each woman volunteers for work at the work-site: no team is formed beforehand. However, here too a sense of collective interests is present, because all the women present know each other and live on the same street—and only allow other Pallar women to present themselves for work. No Muthurajah woman would even bother to try for daily-wage work in south Aruloor, because she would know that she would be rejected out of hand by the Pallar women. Further, *all* Pallar women, including those who are old and slow, or young and incompetent, are readily allowed to join the group because their slowness can only drag out the job to the next day, which is entirely beneficial (to the workers) as it provides them with another day's work. Thus both Pallar women's self-interest and their collective interests are served well by daily-wage work.

Recruitment to piece-rate 'contract' work

'Contract' work or piece-rate work is much sought after, because though it requires much harder work at a faster pace, it pays considerably more than time-rate or daily-wage work. Recruitment to 'contract' work is very different from daily-wage work because the work-group is constituted not on the work-site itself, through *ad hoc* membership, but in advance, amongst the Pallar women, before going to the field-site. A far tighter control over recruitment exists, and the influence of the recruiter—the woman who 'calls' the others for work—is correspondingly greater. She is not normally referred to by any special term in Aruloor, but my Pallar informants hesitatingly agreed that the term *kottukari* could perhaps be applied to her. They stated this when I asked them if they had heard of this term, reported by Athreye *et al.*, particularly in its male form of *kothukarar*, for the piece-rate work-group leader in western Tiruchi District (1990: 142). Significantly, Athreye *et al.* point out that the term *kothu* means 'contract' (1990: 318), indicating that piece-rate work is perceived as distinctively 'contractual' work in western Tiruchi District, just as in Aruloor.

Informants claimed that work-groups had no leaders and that though a particular woman happened to call the group for work, she was not considered its 'leader' in any respect. When the job was done 'anyone could collect payment' from the employer, that evening or the next day; the recruiter did not have to do so. What all my female informants emphasised was the democratic ethos of the Pallar women's team: it had no hierarchy, they claimed, and the recruiter gained no special privileges at all for she collected exactly the same wage as the other women.

However, with time, my closer familiarity with the Pallar women made it obvious to me that this emphasis on absolute equality was more ideology than fact. The recruiter did benefit greatly from being the one who constituted the group, because she had a greater say than anyone else on who got in. Because she distributed contract jobs, and because these jobs earned the highest pay, her powers of patronage were considerable. If, like the redoubtable Sarusu, she regularly recruited contract labourers, then her influence was great. This was because a recruiter like Sarusu was perceived to have the ability to offer almost continuous employment, and so the women she recruited did not question her actions. She regularly recruited her own pre-pubescent daughter for contract work, in order to get an extra wage, even though the young girl worked very slowly. But none of her work-group dared to complain, despite the fact that it was their work that 'covered' for the incompetence of Sarasu's daughter. Thus, on the one hand, a strong ideology of symmetrical relations exists within women's work groups, but, on the other, the reality is that one woman obviously functions as leader. Gillian Hart notes that exactly the same tensions exist among the members of poor women's work-groups in Muda (personal communication).

When pre-pubertal girls were recruited, this was normally solely for light daily-wage work, such as carrying manure or weeding. They were very seldom employed in 'contract' work, such as transplanting, because these were tasks where speed and skill were essential and where the inexperience of the young girl would slow down and annoy the entire group. For exactly the same reason, old women who could no longer work fast were not normally included in contract labour work-groups. At best, they were allowed to participate in daily-wage work. However, as in the case of Sarusu's daughter, young girls and old women were sometimes allowed to join a 'contract' work-group when it was a kinswoman or kinsman who organised it.

For instance, Alamelu, an elderly widow who was piteously thin, told me: 'I went to Lalgudi today to pull out seedlings—but there was no work. They had enough people.' This was at the end of November 1987, when some transplanting work was still going on: having pulled the paddy seedlings out of their nursery, the women transplanted them in the fields the next day. Siva explained:

> These days the women leave very early for the fields, even by 5.00 am, while it's still dark. By the time they get to the field it's a little light and they start work at once—pulling out the seedlings and tying them in bunches. This way they corner the job—and those who come later are not allowed to join in. So Alamelu probably turned up a bit late—and these days that's fatal because the other women don't allow you to join in. They do this because this way you can earn more—the fewer workers there are, the more you get because you're paid a fixed rate—Rs.110 for one acre. So if there's twenty of you, you get only Rs.5.50 each—but if there's just ten of you, you get Rs.11 each! That's a lot—

so the women try to keep others out.

Athreya *et al.* have suggested, following Hart (1986), that the exclusion of incompetent, slow, disabled and inferior workers from work-groups identifies this process as what Hart termed an 'exclusionary labour arrangement' (1990: 145). This is because the manner in which work-groups are organised 'excludes a number of potential wage workers from sharing in a given quantity of employment opportunities' (Athreya *et al.*, 1990: 145). It is quite true that the constitution of 'contract' work-groups excludes many categories of 'less-capable' workers (as will be seen below). However, in adopting Hart's characterisation of employment trends in rural Java without qualification, Athreya *et al.* appear to have overlooked a major difference between Hart's Javanese context and their own. They describe two types of work-groups in western Tiruchi District: (1) 'gangs' organised by labour contractors, and (2) 'gangs' organised 'fraternally' (1990: 142-3). In Aruloor only the latter type of work-team exists, that is, a so-called 'sororal' group organised by an ordinary Pallar who shares equally in the work. In Java it is only the former type of work-group that is discussed by Hart, who points out how *employers* recruit labourers in an exclusionary manner, favouring some workers and excluding others (1986). In the Pallar women's work-groups (and in the 'fraternal' work-groups described by Athreya *et al.*) 'exclusionary practices' are practised by workers against other ('less competent') workers. This is very significant, as I will show.

Sarusu was the recruiter par excellence for female contract work because she—unusually—held the 'male' job of irrigation supervisor for Sundaresan, a Muthurajah who hired labourers not only for his own fields but for other landholders too. Being Sundaresan's agent, she could almost always provide work and so her work-group exhibited the most constant membership of any work-group. Most other work-groups had temporary, fluctuating memberships, for when a job was done a woman often joined another group. Only those groups where the recruiter was able to provide regular work, through her excellent contacts with landholders or through her husband's good contacts, had a fairly stable membership. None had as constant and loyal a membership as Sarusu's team of women, precisely because she literally fought—as we will see—to get work.

In contract work collective interests—narrowed to the interests of the work-group—are taken much further than in daily labour. Here the team is regularly formed beforehand, in the Pallar street, usually in secret, because the women do not wish to antagonise those women who are left out. The team limits its numbers because the women wish to maximise their earnings: each woman earns an equal share of the sum stipulated in the verbal contract. Because more workers mean less pay, the women deliberately choose to remain a smaller number than the optimum—this will mean harder work and longer hours for them, but the higher share of pay makes this desirable. But because work is scarce—and this crucially determines production relations—each work-team is implicitly in competition with every other work-team within the Pallar street.

The constitution of contract work-groups

Though the membership of work-teams fluctuated there were certain broad patterns discernible in membership, particularly the fact that women who lived very close to each other tended to coalesce in a group. Rajalakshmi, who was unusual in being high-school educated and from Tiruchi city, explained what the criteria for membership were. I had asked if women preferred to recruit their own kin. She replied,

> No, we don't. What's important is that the woman should be a good worker—she must be fast, because if she's slow she'll delay us all. So we don't take anyone who's old—because the old women have become slow—and we don't allow anyone very young because they don't know how to do it. And yes—if she's called me for work, then that's a good reason to call her too.

I asked if kinship wasn't important. 'Only a little', she replied. 'If she's fulfilled our criteria we take her—if she's also kin that will help her a little, but not much.' This, however, is not quite true, for kinship is clearly of great importance to the Pallars (see Kapadia 1990, 1993b).

Another factor, proximate residence, is of great importance too. Rajalakshmi explained why:

> You've got to rush off for work early in the morning and can't spend time going to houses further away. So you call the other women for work very discreetly. If you wake the others they'll want to come too and when you refuse them they'll get angry and abuse you. So we take close neighbours; we wake them quickly and go off, with no fuss.

The discretion required in calling others for work is important and for this reason the recruiter usually keeps the job secret, telling only those women she wants on her team. Such discretion is essential not only to avoid ill feelings, but also (informants said) to avoid the danger of *dirushti*, the evil eye, which is caused by envy.

Some women claimed that, in their work-groups, decisions about new members were taken jointly by the group and not unilaterally by the original recruiter. They stressed that women moved from one work-group to another merely as a matter of convenience. Thus A moved from B's group to C's because C had a better-paying contract for the next day's transplanting, or because A could not go for work with B's group for one day. On the other hand, absence for even one day could mean that a woman was not allowed to rejoin her group, even if her absence was not due to 'deserting' to another group but due to illness. Young Kannagi, smiling, but upset, made a sad joke about it: 'I wasn't able to go with them for two days because I was sick. Now they won't let me rejoin. They've rejected me just like a monkey mother who rejects her baby because it fell to the ground!'

This suggests that for Pallar women work is strictly business with no considerations of kinship or friendship intervening, as Rajalakshmi contended. But this is not true because kinship actually plays an important role in the selection of workers by a recruiter. The first women whom she employs are usually her own close kin: this is clear from the actual constitution of the teams, because sisters, co-wives, mothers-in-law and daughters-in-law are always in the same groups. Kin tend to live next door to each other and this might explain why my informants stressed proximity of residence rather than kinship.

It is significant that in theory the Pallar women gave salience to a meritocratic ideology, rather than stressing kinship. This ideology seems to serve two purposes: on the one hand it makes it seem just and reasonable that 'incompetent' workers are not allowed to join work-teams. On the other hand, this stress on competence disguises the fact that some of the team-members are there solely because of kinship considerations.

The struggle for work

Sarusu clearly emerged as the most important leader of the work-groups—a tirelessly hard-working woman who, during the months of transplanting, was gone for work from 6.00 a.m. to 6.00 p.m. every day.

However, several Pallars, both women and men, were antagonised by her—they tended to portray her as domineering, unwomanly, and 'like a man', because of her unrelenting pursuit of higher wages for herself and her team, even if this meant lower wages for others. But her doggedness and aggressive pursuit of employment ensured that her work-group almost always had work, even when no others did. Hart notes a very similar phenomenon when quoting Boedhisantoso's (1976) description of how privileged workers were regarded: 'The holder [of an exclusive right to harvest] may suffer from social isolation...by being condemned as greedy or anti-social by his fellow villagers, who are themselves fighting for their day-to-day livelihood from limited opportunities to work' (Boedhisantoso 1976: 24, quoted in Hart 1986: 690). This reverberates strongly with the way Sarasu was perceived and with the two conflicts I shall now describe.

Significantly, both were associated with and blamed on Sarusu. The period between January 1987 and June 1988, was a time of immense hardship for the Pallars because in 1987 the monsoon failed for the third year and the widespread drought was acute. Agricultural work was very scarce and the competition for work, between Pallars, was exacerbated. In both the conflicts to be considered, the prevailing opinion was that Sarusu and her work-group had broken the unwritten laws which dictated, first, that no restrictive practices should be allowed in the recruitment of harvest labour, and secondly, that no one should seek an unfair advantage in getting a job.

Sarusu's group broke the latter norm by offering a bribe of Rs.50 to Mani, the Muthurajah irrigation supervisor of the biggest Brahman landlord. In return Mani was to give to their group all the transplanting work on five acres of paddy land owned by Venkat Aiyar. The other women in Periyar street were outraged when the story of the bribe became known, and angrily exclaimed that they knew that bribe-giving pervaded all other areas of life, but had never before heard of a bribe given for an agricultural job. Old Ramaiyee had actually given the bribe, probably under orders from Sarusu. Recrimination poured down on her head: 'Ramaiyee and her group should be ashamed! Do they want to start a new fashion, where we all starve ourselves further, in order to bribe the irrigation supervisor?'

The second conflict, which occurred during the *kar* harvest in September 1987, was far more serious. Because of the acute shortage of water only the wealthiest landholders, who owned diesel pump sets, had been able to plant the *kar* crop. There was great anxiety among the Pallars to get what little harvest work was available. Sarusu's refusal, at this time of acute need, to allow other Pallar men and women to join in harvesting a field that 'her' women and men were reaping, consequently caused an enormous furore because it went against custom, which allowed anyone to join in harvesting who was strong enough to carry the immensely heavy head-loads of paddy. Interestingly, Hart notes that in Java, 'The open harvest, in which all who wish to participate are paid a share of the paddy they reap, is commonly regarded as the archetypical "poverty-sharing" institution' (1986: 686). This was certainly how the harvest was regarded in Aruloor and why Sarusu's behaviour provoked such anger.

Two fields belonging to a rich Chettiar were to be harvested. When other Pallars entered the field in which her group was already working, Sarusu shouted that they should get out of 'her' field and go to work in the next. Her argument was that her team had been reaping for half-an-hour before the other harvesters arrived and so had done more work than the others. She demanded that the paddy reaped in 'her' field should be bundled, threshed and paid for separately by the Chettiar. As Sarusu's group was small, it was obvious that it would receive far more paddy per head than the later, much larger group. But when her team put their case to the Chettiar's manager—also a Chettiar—who stood, tall and fat, at the threshing-floor, he refused point-blank and said he would allow no division: all the paddy would be bundled, threshed and paid for as one harvest. If they did not agree, they could leave right away, he said, and he would get other harvesters. As he could have done so easily, Sarusu was forced meekly to accept his edict.

But tempers were still at boiling point when the harvesters returned to Periyar Street, and Kalyani, known for her forthright speech, denounced Sarusu (in absentia) in the middle of the street. 'Who does that woman think she is?' yelled Kalyani. 'Is she queen of this country? The Chettiar's wife? Does she own that field? Let that whore watch what she says! Is only she to eat—and the rest of us to starve? She wants *more* work and *fewer* workers so that she can get the highest wages! And the rest of us can just starve!' Kalyani clearly voiced the sentiments of many others who also felt that Sarusu often broke the cardinal rule of at least a modicum of sharing. Everyone recognised that work-groups competed with each other, but not to this degree. But Kennedy's mother, Lingamma, in her humorous way commented both generously and perceptively, 'That Sarusu! She's not afraid of anyone, is she? She's a *man*, really, in her courage!' And this grudging admiration must have tempered the envy that many others felt, because through Sarusu's initiative her team always had work.

Workers' awareness of collective interests

Though no one explicitly commented on the sad fact that the Pallars were fighting each other, instead of fighting their exploitative employers, it is clear that an implicit consensus existed to that effect. In both the quarrels described above, the common theme was the plea: 'Are we going to compete so ruthlessly that we all starve ourselves further?' Everyone recognised the fact that the Pallars had to stick together or they would simply be even more grossly exploited. This recognition was given concrete shape in the absolute unity of the Pallar women when demanding higher wage rates (see Kapadia 1993a). Rudra has argued that agricultural workers have little awareness of their own best interests. He says:

> Even labourers belonging to the same village do not reveal any capability of acting in a collective fashion. Even in the villages where there were strikes they were precipitate actions without any collective bargaining; that is to say, they were in the nature of wild-cat strikes typical of undeveloped working-class consciousness (1984: 260).

This comment does not apply to the Pallar women whose strikes were always backed up by collective action. That is why, even though the women's work-teams did compete with each other for work, they can still be seen as constituting at least 'a *proto-union* form of organisation', as Athreya *et al.* have put it (1990: 145).

The data from Aruloor present us with a striking paradox: Pallar women workers are quite conscious of their common economic interests, and are therefore willing to see their internal conflicts as secondary and their struggles with their employers as primary—when occasion demands it. Yet their loyalties do not traverse caste-street and kinship boundaries—because production-relations themselves are organised along the axes of kinship and caste. Further, their exclusionary practices clearly serve to divide the labour force and to weaken its solidarity. In short, 'micro' competition for work significantly weakens 'macro' class solidarity even though it does not entirely undermine it.

Finally, a brief note on terminology. If the Pallar women unite collectively, as they clearly do, is this rightly termed 'collusion'? This term, commonly used in the literature (see Dreze and Mukherjee 1989: 246, 251), is actually defined by the *Shorter Oxford English Dictionary* as meaning 'secret agreement or understanding for purposes of trickery or fraud;...deceit, fraud, trickery' (1973: 369). Surely the very negative connotations of this term make it entirely inappropriate when used to describe the solidarity and mutuality of workers.

The move to negotiable contracts

The research of Athreya, Djurfeldt and Lindberg (1990) provides important evidence for the rapid increase in the use of piece-rate 'contract' labour and work-teams in Tiruchi District. They state:

> Labour relations are changing in the wet area. *Kothu* or contract gang labour seems to be spreading and gradually replacing *atta coolie* (daily wages work) in all major agricultural operations involving much labour.... Contemporary studies have noted that gang labour has become common in modern Indian agriculture. Desai, for example, says, 'Time wages have been replaced by piece wages for all major operations without major resistance' (1990: 139).

These observations accord exactly with the changing conditions in Aruloor (which lies east of the area studied by these three researchers) where, too, daily-wage tasks are becoming piece-rate tasks—and where virtually all piece-rate work is done by female or male work-teams. But the spread of 'contract' labour has a further, very significant implication—it signals the spread of what Dreze and Mukherjee have termed a 'non-negotiable' contract.

As the Pallar women pointed out, daily-wage work benefits workers because they can be lazy and yet earn their hour-rate wage. Dreze and Mukherjee point out that daily wages are remarkably rigid:

> The rigidity of daily (money) wages...remains a substantial issue in spite of the availability of alternative contracts. We have also argued that this rigidity is usually found to be more pronounced *downwards* than upwards. Moreover, the resistance to wage cuts clearly comes from *labourers*, not employers (as in the efficiency theory of wages) (1989: 259).

These observations fit Aruloor exactly: I discuss elsewhere the manner in which the solidarity of Pallar women workers maintains the 'stickiness downwards' of daily wages, despite the constant desire of employers to pay less (Kapadia 1993a). But the crucial point about daily wages is that they are '*non-negotiable*'.

Dreze and Mukherjee explain this term as follows:

An examination of the nature of different contracts in Palanpur reveals the existence of two distinct contract 'types' which could be called 'negotiable' and 'non-negotiable'.... With non-negotiable contracts, standard contract terms apply to all partners *and* these tend to exhibit a strong rigidity over time; examples are daily wage employment.... For negotiable contracts, on the other hand, terms are flexible *both* between persons and over time; examples are piece-rate labour contracts... (1989: 260).

It is because daily-wages contracts are 'non-negotiable' that subsistence theories of wages are particularly suitable in explaining them. As Dreze and Mukherjee point out, such theories 'consider real wages to be determined by norms or forces largely extraneous to labour market conditions.... Examples of such norms are...common moral perceptions of what constitutes a "fair wage"...traditional "rights"...' (1989: 253). But these supposedly 'common', and 'non-negotiable', moral values of 'fairness' and labourers' 'rights' are disappearing today—though it must be asked to what degree 'common' and consensual norms *ever* existed in the Indian village. Rather, it was *employers*' views of labourers' rights (views which are unlikely ever to have coincided with the labourers' own perceptions) that traditionally prevailed and that were represented as consensual—and it is these employers' views that are changing today. There is not space here for a detailed analysis of why these changes are occurring: what is obvious, however, is that the relationship between labourers and employers is consequently becoming more *nakedly* exploitative, for the traditional 'rights' of workers disappear as soon as they lose their usefulness to employers. Thus Dreze and Mukherjee aptly point out that 'Traditional patron-client relationships have disappeared... [and] a clear trend exists towards the "casualisation" of labour transactions' (1989: 244). In short, a move to piece-rate, 'negotiable' contracts is correlated with the casualisation of wage labour—and both signal the increasing vulnerability of the labour force.

Discipline and control

It is against this background that two sets of correlated changes emerge in Aruloor. On the one hand there is a steady decline in non-negotiable labour contracts, typically in the form of daily-wage labour, together with a decline in employers' obligations to workers. On the other hand, there is a steady increase in piece-rate labour contracts, whose crucial characteristic is that they are negotiable. But, due to the large surplus of labour, the labour market of Aruloor ensures that this negotiability is *primarily to the benefit of employers*.

Piece-rate work, as already noted, is viewed by Aruloor employers as a boon, because they claim that it removes the need for supervision. But because workers recognise that they are fortunate to have work at all, and because of the quality control exercised by employers, workers are forced to institute a regimen of supervision where they *supervise themselves*. This 'internalisation' of discipline, subtle though it is, signals a powerful extension of employers' control of the agricultural workforce. Thus not only do ordinary Pallar women recruit work-teams, but they discipline them as well. This new process of self-discipline, enacted by the worker on herself, as though always observed by the now invisible employer, is a sign of the 'modernising' of agrarian production relations: the 'movements, gestures, attitudes, rapidity..., the economy, *the efficiency of movements*, their internal organization' all these constitute 'the modality: it implies an uninterrupted, constant coercion' (Foucault 1987: 137, emphasis added). In the old regime, labourers routinely were brutally punished by their employers for infringing work rules (see Gough 1981, 1989). Today a less spectacular but more insidious system of social control exists. Because the chronic unemployment of landless labourers makes them insecure and fearful, the 'privilege' of belonging to work-teams makes them value these jobs greatly, thus ensuring 'not only an adequate labour force, but also a hard-working and docile one' (Hart 1986: 689). All the signs indicate that landless labourers are at risk. Given that Pallar women tend to be major breadwinners for their families, it is entire communities who are being steadily impoverished while they, ironically, are forced to participate in disciplining and controlling themselves.

References

Athreya, V., 1984. *Vadamalaipuram: a resurvey* (Madras).

Athreya, V., G. Djurfeldt and S. Lindberg, 1990. *Barriers Broken: production relations and agrarian change in Tamil Nadu* (New Delhi and London).

Binswanger H. and M. Rosenzweig (eds.), 1984. *Contractual Arrangements, Employment and Wages in Rural Labour Markets in Asia* (New Haven).

Binswanger, H. and M. Rosenzweig, 1986. 'Behavioural and material determinants of production relations in agriculture', *Journal of Development Studies* 22, 3, pp.503-39.

Chenery, G. and T.N. Srinivasan (eds.), 1988, 1991. *Handbook of Development Economics*, vol.I (Amsterdam).

Dreze, J. and A. Mukherjee, 1989. 'Labour contracts in rural India: theories and evidence', in S. Chakravarty (ed.), *The Balance between Industry and Agriculture in Economic Development*, vol.3 (London).

Foucault, M. 1977, 1987. *Discipline and Punish: the birth of the prison* (Harmondsworth).

Gough K. 1981. *Rural Society in Southeast India* (Cambridge).

—, 1989. *Rural Change in Southeast India* (Delhi).

Harriss, J. 1982. *Capitalism and Peasant Farming: agrarian structure and ideology in northern Tamil Nadu* (Bombay).

Hart, G., 1986. 'Exclusionary labour arrangements: interpreting evidence on employment trends in rural Java', *Journal of Development Studies* 22, 4, pp.681-96.

—, 1991. 'Engendering everyday resistance: gender, patronage and production politics in rural Malaysia', *Journal of Peasant Studies* 19, 1, pp.93-121.

—, 1992. 'Household production reconsidered: gender, labor conflict and technological change in Malaysia's Muda region', *World Development* 20, 6, pp.809-23.

Kapadia, K., 1990. 'Gender, caste and class in rural South India', unpublished PhD thesis, University of London.

—, 1992. 'Every blade of green: landless women labourers, production and reproduction', *Indian Journal of Labour Economics* 35, 3, pp.266-76.

—, 1993a. 'Mutuality and competition: female landless labour and wage rates in Tamil Nadu', *Journal of Peasant Studies* 20, 2, pp.296-316.

—, 1993b. 'Marrying money: changing preference and practice in Tamil marriage', *Contributions to Indian Sociology* (NS) 27, 1, pp.25-51.

—, forthcoming (a). *Siva and her Sisters: gender, caste and class in rural South India* (Boulder, Colorado).

—, forthcoming (b). 'Women earn, men eat!: gender and household production in rural Tamilnadu'.

Ramachandran, V.K., 1990. *Wage Labour and Unfreedom in Agriculture: an Indian case study* (Oxford).

Rosenzweig, M., 1991. 'Labor markets in low-income countries', in H. Chenery and T.N. Srinivasan (eds.), *Handbook of Development Economics*, vol.1 (Amsterdam).

Roumasset, J. and M. Uy, 1980. 'Piece rates, time rates and teams: explaining patterns in the employment relation', *Journal of Economic Behaviour and Organization* 1, pp.343-60.

Rudra, A., 1984. 'Local power and farm level decision-making', in M. Desai *et al.* (eds.), *Agrarian Power and Agricultural Productivity in South Asia* (Delhi).

Shorter Oxford English Dictionary 1973. C.T. Onions (ed.) (Oxford).

Stiglitz, J.E., 1975. 'Incentives, risk and information: notes towards a theory of hierarchy', *Bell Journal of Economics* 7, pp.552-79.

Chapter 10

PEASANT SUBSISTENCE AND FOOD SECURITY IN THE CONTEXT OF THE INTERNATIONAL COMMODITISATION OF PRODUCTION: THE PRESENT AND HISTORY

Utsa Patnaik

Introduction

Over much of the ex-socialist European world today, there appears to be an unquestioning acceptance of the proposition that great benefits are to be derived from the pursuit of 'market-friendly' policies, of promoting unfettered private enterprise and market exchange. The collapse of socialism and the effective political counter-revolution, has involved a withdrawal of the state in most of these countries, to a greater or lesser degree from production and distribution, and even in some cases from that degree of regulation of the economy which characterises many advanced capitalist countries. The results so far have been rather distressing though predictable in terms of hyperinflation, downward-spiralling production, rising unemployment, and in parts civil war. The restive populations of these countries are being told, somewhat in the manner of patients in India undergoing homeopathic treatment, that the measure of the medicine's efficacy lies in the symptoms getting a great deal worse before they get better. Unfortunately the analogy ends there for, while homeopathy never killed anybody, the unfettered market has historically proved itself an efficient mechanism for dispensing death. When the vital statistics are eventually put together, a significant rise in morbidity and mortality rates, compared to previous trends, may be confidently expected in the European ex-socialist world from 1990 onwards; for no country in history has ever experienced substantial G.D.P. declines and rising inequalities without an unfavourable shift in vital statistics and welfare indicators.

On the other hand, the resistance to the *simpliste* ideas of the dispensability of state intervention and systematic regulation of the economy, is much stronger in a latent way if not yet overtly, in Third World economies in Asia, like India, which have never claimed to follow socialist policies in any serious sense but have had a good deal of state regulation of the economy in the post-Second World War period. Perhaps there is a more realistic recognition of the effects that the unfettered operation of the market has produced in the past in promoting various types of inequality, and an implicit recognition that the stabilising of the existing high level of inequality—a level which is of immense benefit to those at the top—is necessary for the system to continue to function at all. The margin of safety in experimenting with 'market-friendly' policies is very narrow in a country like India where such a large part of the population lives at subsistence level and a substantial fraction is already below it. Market-friendly policies are seldom people-friendly; as we will try to establish in this essay, policies which are over market-friendly have destroyed Third World peasants' food security and created the conditions for famine in the past, and continue to have this effect in the contemporary world.

This essay is in three parts. In the first part we explore the underlying structural reasons for the fact that the market orientation of agricultural production, in Third World countries like India, is associated almost invariably with a decline in domestic consumption of basic staples, with reference to the world division of labour in this sphere. The second part looks at the mechanism of commercialisation in the colonial period and the undermining of the concept of peasant subsistence. The third part draws a theoretical parallel between the colonial and the contemporary experience of increasing agricultural market orientation in indebted Third World countries.

Agricultural commoditisation, the international division of labour and of production

The primary focus of discussion in this part is the factors underlying the strong historical trend towards commoditisation of agricultural production in today's ex-colonised, developing countries, which are located almost without exception in the tropical and sub-tropical zones of the earth. Such a preliminary discussion is necessary in order to understand the reasons behind the present renewed trends of rapid commercialisation and export-orientation of Third World agriculture.

It would appear to be a fairly obvious proposition that agricultural production differs from manufacturing in a fundamental respect, namely that crops are climate- and soil-specific while manufactures are not. The economic implications of this self-evident proposition, are not, however, kept in mind to the extent that they should be. Shoes, cement, steel or textiles can be produced anywhere in the world, on the basis of imported raw materials if they are not locally available; the fact that Britain could grow no cotton owing to its cold climate did not prevent cotton textiles from emerging as the leading manufacturing sector of the first industrial revolution. The fact remained, however, that British manufacturers' access to a semi-tropical crop like cotton had to be ensured. The fact that Japan could grow very little sugarcane did not prevent the emergence of an efficient and large sugar-refining industry in Japan; but access of the Japanese *zaibatsu* to ample sugarcane supplies had to be ensured. In both cases state-induced production and exports from colonies provided the bulk of supplies of the relevant agricultural raw material.

Without prohibitively expensive hot-house cultivation, Hampshire cannot grow bananas, the Dordogne cannot grow sugarcane, nor can Idaho grow coffee. India, as indeed other parts of the bio-diverse tropical world, can grow not only bananas, sugarcane and coffee, but potentially if not actually, every crop that Hampshire, the Dordogne and Idaho are capable of growing. The European summer crops are easily grown in winter in tropical regions, and varied topography—that is, land at varying elevations—allows practically every crop from strawberries, asparagus, coffee and tea to rubber, mango, melon and avocado to be grown in India as in many other tropical countries. This basic fact has determined the world division of labour in agricultural production over the past three centuries, a division of labour arising not from Ricardo's 'comparative advantage' (which is perhaps the cleverest *ex-post*-theoretical rationalisation of national self-interest ever put forward by an economist) but from the use of active intervention including force, to secure for European populations (including those of mainly European origins as in the U.S.A.) settled in temperate regions, access at the cheapest possible rates to the richness of tropical bio-diversity. The consumption basket of the average citizen of the countries of the North has been transformed over the last two centuries, and continues to undergo diversification; part of this has arisen from improvements in their own agrarian production, but the bulk derives from securing access to cheap tropical sources of supply.

Reliance on the products their own agriculture was capable of supplying had meant a very restricted consumption basket for Europeans, adequate in terms of nutrition but not balanced, and monotonous to a degree. Analysing the data on the diets of the nobility and serfs alike on feudal estates in Germany and in Sweden in the sixteenth and seventeenth centuries, Slicher van Bath finds that the average intake of about 3,500 calories per head was made up of large quantities of bread, rancid salted butter and a little cheese, and meat of which nine-tenths was preserved and heavily salted. Food not roasted was cooked in animal fat. Fresh fruit was similarly limited. The highly-salted food generated an 'oceanic thirst' and beer consumption per head in Germany and Sweden was 48 times the level today (Slicher van Bath 1963). Vitamin-deficiency diseases (rickets, night blindness and so on) must have been common. This contrasts sharply with the availability of a large range of products in the contemporary tropical world. Abu'l Fazl's *Ain-i-Akbari*, written in the late sixteenth century during Emperor Akbar's reign in North India, lists the market prices of 35 varieties of fruit and 26 types of vegetables, and says the list is not exhaustive (see *ain* 27 and *ain* 28, *Akbarnama*).

The agricultural improvements in Europe from the eighteenth century onwards leading to convertible husbandry, and the introduction of the glass cold-frame, improved matters a little, while Latin Europeans had the saving grace of the vine, the olive and some citrus fruit. The real transformation in the diets and clothing of the Northern populations, however, awaited their seizure of tropical colonies and the widespread dissemination of the commercially-valuable tropical crops to distant areas of similar climate for the purpose of plantation production for European consumption. The dissemination of Brazilian *hevea* to Malaya and Indian sugarcane to Java and Brazil, amongst many others, hardly needs recapitulation. Europeans had access now to cheap and comfortable cotton clothing besides the traditional leather and wool, and the gentry could sport tropical hardwood furniture, wood-panelled houses, and ornamental plants and fish. The consumption basket of the working class too underwent substantial modification with the use of sugar, tobacco, rum and cotton cloth. Britain, as the earliest and most successful coloniser, was already dependent on agricultural imports for final use (that is, excluding raw materials) to the extent of 40 per cent of its own corresponding primary sector production by 1830; and by 1850 this proportion had increased to a staggering 60 per cent (calculation from basic time series data in Mitchell and Deane 1966).

Any developing country with even half this degree of import dependence today would at once plunge into a deep balance of payments crisis. No payments problems arose for the European colonisers because their import surplus of tropical goods was financed by taxing the colonised population, and thus represented the commodity equivalent of a part of the tax revenues which was transferred. In essence, these import surpluses were thus costless; the production costs at the point of origin were not recompensed to colonised producers in equivalent goods, and the cost of transport, where transport was always statutorily monopolised by metropolitan shipping, was borne by final consumers in Europe, helping to sustain the home country's shipping industry with its attendant backward linkages. In their writings, most economic historians in Northern countries, apparently quite innocent of the rudiments of open-economy macroeconomics, continue to treat colonial transfers as normal trade.

Today, the high standard of life of Northern populations, and especially of its better-off classes, relies even more heavily on the physical availability of tropical goods. If trade in these goods was for some reason to cease overnight, these living standards would collapse to near-medieval levels. Certainly they would collapse to at least to the levels of the Eastern European countries, which historically had no access to transfers from tropical areas, and continue to have more primitive consumption patterns, with local goods predominant, than do Western Europeans and North Americans. The only things that temperate region agriculture can produce, and in abundance owing to past productivity-raising investment, are cereals, potatoes and other root crops, a small range of summer vegetables and fruit, and dairy products.

The Northern consumer gave up herrings and ale for breakfast long ago. The imported tropical and semi-tropical goods consumed daily in the typical Northern household include a range of beverages like tea, coffee, cocoa, citrus and other fruit juices; sugar and honey in several forms; fresh, canned and frozen vegetables and fruit; fish, prawns and other sea food; lean meat and processed meat products; sugar-based alcoholic beverages, cocoa and sugar-based chocolate, confectionery and ice-cream; flavourings, spices, nuts and vegetable oils. The Northern summer varieties of vegetables and fruit are available in winter owing to imports. A range of imported tropical primary products additionally enter, as part of the raw materials, into the Northern food-processing industry and into manufacturing. A look at the by-no-means exhaustive list of final consumption goods given above will indicate how severe an impact would be felt on the high import-dependent consumption standards to which Northerners have become habituated, if availability of these goods were to drop to zero.

It is sometimes argued that tropical raw materials are by now a very small fraction of value-added in manufacturing in advanced countries, and that tropical final consumption imports are similarly a small fraction of average Northern incomes. These are spurious indices of the importance to the North of access to tropical primary products: one may as well say that the cost of the air we breathe bears a zero fraction to our income and hence air is unimportant. The reason tropical products, despite their pervasive physical presence in Northern life-styles, make up a small fraction of average Northern incomes, is that they are obtained on very favourable price terms. Another way of saying that real incomes in the North are high, is to say that prices are very low for a large range of by-now essential imported consumption goods. If these imported goods' prices reflected the true cost of production to the societies which supply them, they would be several times higher, and average Northern real incomes would be correspondingly lower.

What are the costs to the Third World societies of supplying the exportables which are an essential basis for present high Northern consumption levels? At best increasing the supply of exportables has involved historically a reduction in domestic absorption—that is, a fall in the availability of these goods for the domestic population in order to make them available to richer foreigners. At worst it has involved a reduction in basic food-staple production for the local Third World populations in order to divert land and resources to the production of exportable goods, and it has created the preconditions for local famines. The international prices of the exportables certainly did not in the past, and do not today, reflect even in the smallest measure the immense cost to local Third World populations suffering a decline in consumption—including, in limiting cases of sudden economic shocks coming on top of the trend, quite catastrophic falls leading to a rise in mortality rates. Debt has been the main economic mechanism through which Third World producers are induced to grow and sell exportables at the expense of their own consumption. We will discuss the working of this mechanism in colonial India in the next section, and its working in the contemporary world in the last section.

It will be noted that we are putting forward in a strong form what might be termed a zero-sum game hypothesis of the end-use of agricultural production: expanding exportables to meet the market demand of the North is at the expense, necessarily, of lowering domestic absorption in poor tropical developing countries. This has been historically the case under old-style direct colonial systems, and as we will see this is being replicated today in highly export-oriented developing countries.

It might be argued that this is far too pessimistic a view: that there is no reason why, *if the rate of growth of agricultural production is high enough*, increasing exports should not be compatible with maintaining domestic availability. It is here that the specificity of agricultural as opposed to industrial production again has to be borne in mind. Tropical land supporting crop-diversity is as scarce as tropical forest supporting bio-diversity in general. While fossil fuels are obviously a non-renewable resource, it has to be recognised that the supply of specific types of land supporting plant-diversity is almost equally limited: no doubt there can be some land-augmenting technical progress in tropical agriculture, but the limits to this are reached far sooner than is the case with technical progress in industry.

Almost all the possible tropical land which can be cultivated is already under cultivation, and large additions to area are not feasible, as this can only be at the expense of further slashing of forests, with serious consequences for long-term rainfall trends and the survival of species. If cultivated area cannot be laterally expanded to any very great extent, then the other possible avenues are to raise the productivity of land so that an acre does the work of two or three through multiple cropping in a given year—in short, to raise the ratio of gross sown area to net sown area; and to raise, somehow, output per unit of gross sown area as well. Under tropical conditions, raising output significantly by raising the ratio of gross sown to net sown area, requires substantial irrigation investment. The maximum possible growth rate achieved in the past (1950 to 1975) in Indian agriculture as a whole, for example, when considerable state investment in irrigation and infrastructure was taking place and the multiple cropping index was rising, has been about three per cent per annum. Private investment has long been found to be positively associated with state investment, and when the latter slows down so does the former.

Now it is the case that the same set of policy measures (so-called 'structural adjustment' programmes) implemented via indebted Third World governments today by world lending-agencies, which stress increase in exports including agri-exports from Third World countries, also require them to go in for deflationary cuts in budgetary expenditure, and thus are conducive to a fall in public productive investment including in the rural sector. The regime of export-growth from agriculture in indebted Third World countries, in short, has been associated invariably with an overall decline in investment in agriculture, even after taking into account growing investment in the export sector. The real objective behind these deflationary policies is not so much the declared one of correcting payments imbalance (since removal of all restrictions on imports—complete trade 'liberalisation'—is advised at the same time), but rather the undeclared one of restraining the growth of domestic incomes, and hence market demand and the absorption of their own resources and products by Third World populations, in order to release these products for exports for the benefit of the North. This has led almost invariably to falls in domestic absorption of necessities per capita in order to increase exports. Where the initial levels were reasonably high, some societies have been able to absorb consumption decline without overt manifestations of distress; where initial levels were low to begin with, however, consumption decline has had and are having serious consequences.

This argument is coterminous with the argument that sustainable development of the world economy requires a decline in excessive First-World consumption. The main difference is that while energy consumption, especially of non-renewable fossil fuels and the over-exploitation of tropical forests, have been the focus of the 'sustainable development-cum-conservation' viewpoint, this argument is much more general, comprising within it the whole of tropical agricultural production.

We will return in the last section to the connections between external debt, the obligation to follow structural adjustment programmes, and the drive to increase exports at the expense of domestic absorption, illustrating the argument with data from Africa and Latin America. In the following section we go back to the experience of forced commercialisation in India in the colonial period and its impact on peasant food security.

Commoditisation in colonial India and food security

Colonised Indian agriculture experienced increased agri-exports under the obligation to pay cash taxes, the resulting diversion of resources away from domestically-consumed products, and hence a falling food availability per head of the domestic population and increasing vulnerability to famine. The experience was not unique to India by any means. All colonised countries were historically subjected to the same or similar mechanisms of transfer of a portion of their taxation or other revenues via an unrequited export surplus of mainly agricultural and mineral products required by the metropolitan countries. The economic essence of being a colony was that there was complete and obligatory openness to the world economy, through explicit prohibition of tariff autonomy; and the regulation of the systems of production, taxation and monetary policies in the interests of the colonising power. Thus during the relatively brief period that it colonised Formosa and Korea (1896 to 1945, and 1910 to 1945), Japan developed these countries as a source of rice, sugarcane and minerals, products in which it was itself deficient. Within two decades one-fifth of Japan's rice supplies were from colonial imports, and three-fifth of the imports were from Korea alone (Schumpeter 1940; Grabowski 1985). Despite a reasonably high rate of growth of Korean production, under the relentless pressure to export owing to heavy tax burdens and government policy, the per capita cereals available for Koreans fell substantially, its composition worsened with sorghum substituting for rice, while the poorest rural Koreans were reduced to eating wild grasses and tubers for a part of the year (Hayami and Ruttan 1970; Cummings 1981).

India, which was colonised for a much longer period by Britain, saw a similar process of largely forced commoditisation of agricultural production under the pressure of cash taxation and resulting debt, combined with government policies to expand production of exportables rather than the domestically-consumed crops. It may be noted here that cash taxation of the colonised population plays a dual role. First, it puts pressure directly on the producers to sell output to pay tax (or rent in the case of the tenant of the tax-paying owner), and thus involves them in indebtedness and offers an entry-point to traders for ensuring that exportables are grown against cash advanced. Second, it deflates domestic incomes to limit absorption and thus releases products for export.

The second effect is the same as operating a surplus budget, to the extent that a part of the tax revenues is earmarked in the budget as 'expenditures incurred abroad', and is actually used to pay in rupees the producers of the (mainly agricultural) export surplus. Thus the expenditures out of revenues in India were not only the normal ones that any sovereign country would also have, but contained a part specific to colonial relations, namely paying for an export surplus out of domestic taxation, which export surplus therefore was nothing but the commodity equivalent of the transfer of that part of revenues. No claims to the foreign exchange they had earned accrued to the Indian producers of the export surplus; this foreign exchange was paid to the Secretary of State for India in London where it was held, and its equivalent was remitted in the form of 'council bills' encashable only in rupees, paid to Indian exporters by the Treasury out of the Indian budgetary revenues set aside for that purpose (see Patnaik 1985, for a simplified exposition of the real-cum-financial mechanism). The Indian nationalist writers such as Dadabhai Naoroji and R.C. Dutt grasped and explicated the essence of the mechanism of transfer; the reason we do not use the traditional terminology, 'drain of wealth', is that it is slightly confusing. Wealth is an asset or a *stock* concept; while we are really talking about annual *flow* variables—income, tax revenue, export surplus embodying transfer of a part of revenue, and so on. Of course, a steady annual transfer abroad of a part of economic surplus necessarily lowers domestic annual asset formation, and over the decades its effects are cumulative in terms of a negative investment-expenditure multiplier. This is probably why it was referred to as a drain of 'wealth', understood as potential wealth. Our argument is that, quite apart from taking away much of the annual potential for capital formation, the deflationary impact of the annual transfer was so large, and the distribution of the burden such, that it also affected domestic consumption per head adversely.

The ratio of the revenues transferred through export surplus to the total revenues of British India, varied from 16 per cent in the 1880s to as much as 25 per cent in the late 1920s (see basic data quoted by Dharma Kumar 1985; however, she does not interpret these data analytically). This is at the same time the measure of budgetary surplus, exercising a very strong deflationary impact on the economy and ensuring that employment and incomes were depressed sufficiently to limit domestic market demand and release a large enough part of domestic production for export (see Sen 1992, for a discussion of the deflationary effects). For the period for which national income estimates are available, from 1900 to 1949, we find a constancy in real income per head in three out of four estimates and a rise only in one.

At the level of individual peasant producer, the pressure to pay tax (or rent where the producer was a tenant of the tax-paying owner) was felt directly in the imperative need to grow crops which could not be directly consumed but necessarily had to be sold, even when price was low and barely covered costs. For example, the growth of cotton production in Madras for export, was not undertaken by the majority of growers because it was 'profitable' but because cash land revenue or rent had to be paid. A. Saradu Raju in her *Economic Conditions in the Madras Presidency*, quotes the Report of the Superintendent of Government Farms, 1876:

> When asked why they continued growing it [cotton] at all under such admittedly unfavourable circumstances, the agriculturists of Coimbatore gave the curious [*sic*] explanation that it was done in order to pay the land revenue, and if they cultivated nothing but grain 'they would have nothing wherewith to pay the Government rent as they and their families could and would consume the whole produce of the land, they having to content themselves at present with only two meals instead of three' (Raju 1941: 91).

This is of course a clear and direct example of altering the crop-mix and restricting consumption under the pressure of the obligation to pay the government demand, by choosing a non-foodgrain exportable crop. The mechanism was far more general, however, and included the commercialisation of the foodgrain crop itself. It operated through the pervasive system of advances by merchants against contracts to grow and supply exportables including cereals.

Opium production in the early decades and indigo production up to the middle of the nineteenth century, were carried out in large measure under sheer physical duress, through both the threat and the actual use of force by government and by private planters (Chaudhuri 1965). In the case of indigo, oppression by the planters, some of whom were former slave-drivers from the Caribbean, provoked the famous indigo revolt of 1859. As the second half of the century progressed, however, the more indirect economic mechanism, of debt-induced contracts with traders and landlords, appears to have replaced the use of overt force. Elsewhere we have explored at greater length the economic logic of the processes of 'forced commercialisation', noting that the thin stratum of the better-off or rich peasants may well have voluntarily produced more for the market, especially of exportables, in order to make profits, but that, for the majority of peasants subject to debt, the involvement in exportables' production was under economic duress and on very unfavourable terms (Patnaik 1975).

To summarise the mechanism briefly, the introduction of a regime of high cash-revenue demand by the government (when it first undertook a land settlement in a region annexed to British India), combined with rigidity in the *timing* of the revenue demand, meant that all peasants, other than those few who had accumulated savings from previous production cycles, necessarily had to borrow. Non-cultivating landowners passed on the demand to their tenants as rigidly-timed rental demand. The timing of the demand coincided with the harvest, and did not allow sufficient margin for crops to be marketed and cash realised, while whatever could be sold realised lower than average prices; the timing imposed financial stringency, made borrowing unavoidable except for the small well-to-do section, and thus provided an entry point to traders and landlords who could influence the grower's cropping pattern because they were prepared to give immediate cash advances tied to contracts for growing the next season's crop, which was a specified exportable. Taking advances from trader or landlord creditors appeared less burdensome to cultivators than taking loans from pure moneylenders, since the former carried either no explicit interest rate or a lower-than-market rate (the implicit interest element, in a contract price which was lower than market price, was not so clear to them). These advances were taken and partly used by peasants for settling immediate revenue payment obligations, and they then used the balance, plus the proceeds of their existing harvests, to meet the input requirements of the next season's contracted crop. Where the latter was an exportable, these input requirements were often higher since exportables were generally higher-cost and higher value than the traditional staples.

While credit certainly smoothed out the inter-temporal imbalance in the demand for and availability of cash, it did so only by exacting the heavy price of tying producers to a debt cycle from which very few could escape. Making use of the cultivator's enforced financial stringency arising from rigidly-timed rental and revenue demands, the trader-creditors, when giving the advance, pitched the terms of contract as much in their own favour as possible: they specified that the crop had to be sold only to them, and sold either at the low post-harvest price, or an even lower price than this (the implicit interest element entered here), leaving in peasant hands nothing of the increased surplus arising from the shift to higher-value exportables. Not uncommonly, planters, traders or creditors acquired superior tenure rights over the peasants in order to bring the pressure of rent directly to bear.

Diverse sources, discussing cash-crop production in different parts of British India, have similar things to say which bear out the mechanism we have outlined. On the general question of financial stringency imposed by rent and revenue timings, and the advantage for the creditor, the Madras Provincial Banking Enquiry Committee informs us that:

> The main paddy harvest is in December and January and into February, and there is generally a slump in the market at that time. Also it is at that time that in such areas the Government's kists and the landlord's rents fall due. This is naturally the case as the produce is the security for the rent or kist which is the first charge on the land....

Writing of the market methods followed by the non-indebted cultivators, it goes on to say:

> Such methods can be improved upon but in these cases the ryot is at least free to sell or not as he sees fit. The ryot who has taken advances is in a much worse position. It is a common practice to take a loan on condition of repayment at the time of harvest. This, in its simplest form, compels the ryot to sell at once though the market may be glutted and the prices offered preposterously low; but it is common when the loan is taken from the village moneylender who is also the shopkeeper and general merchant, or from the bigger merchant, to stipulate that the crop should be sold at a rate sometimes prearranged but more frequently agreed on as the market rate at the time, or in many cases at a fixed amount, about 5 per cent, ...less than the prevailing market rate.

C.W.B. Zacharias comments:

> Merchants and traders indulge in moneylending, partly in furtherance of their trade and partly to seek employment or investment for their surplus funds. Often the understanding is that the crop, on the security of which loans are given, will be sold through them. The terms offered by them are found to be easier, as they hope to make up for the loss in the lower price they invariably pay for the crop sold to them (1950: 177).

Within this overall agrarian economic environment, as the demand for exportables from metropolitan centres increased, the cropping-pattern gradually shifted in favour of the exportables (cotton, sugar-cane, indigo, jute and wheat) and away from the locally-consumed staple crops; this change was brought about through the agency of traders' advances tied to growing particular crops. In short, the existing system provided an easy entry point for those wishing to expand the supply of exportables, at the expense of domestically-consumed products. These cash crops, although they were of high value, appear by all accounts to have given hardly any additional income to the growers compared to what they might have been growing earlier; except for a small minority of well-to-do farmers who were free to choose whether to grow them or not, the vast majority of raiyats grew these crops under unprofitable conditions and grew them nevertheless because of the pressure of debt.

The growth of indigo and opium production in Bengal towards the middle and later decades of the last century has been discussed by B.B. Chaudhuri (1965). The government constituted itself as the monopoly purchaser of opium, advancing cash to peasants, obliging them under threat of reprisal to sow plots to opium, and buying the crop at a very low price, while reselling at a substantial profit to the export merchants engaged in the China trade. Indigo expansion was carried out by private so-called 'planter' manufacturers, who seldom planted directly using hired labour; they purchased the plant from peasants to whom they advanced cash for the specific purpose of growing indigo, at a very low contract price barely covering costs to the grower. Chaudhuri says:

> The price given to the peasants for their produce was generally low and unremunerative; such a system which tied the peasants to a low price, nearly completely stifled the peasants' initiative. The problem was to keep the reluctant peasants bound to the system. The solution was relatively easy where the pressure of rent made the peasants eager to utilise the ready advances offered by Government or the planter.

Clearly, the solution was easier still if the 'pressure of rent' could be brought to bear directly by the trader-manufacturer himself, and this is precisely what in many cases he tried to do by acquiring some form of superior rent-collecting right over the raiyat—whether as occupancy raiyat, through other sub-tenures, or zamindar. In such situations, the grower ended up paying his entire surplus, and sometimes a bit more by cutting his own consumption, to a single agent with no very clear distinction between rent, interest and profit.

Indigo and sugarcane in the United Provinces similarly appear to have expanded under the familiar mechanism of revenue-cum-rent payment inducing acceptance of advances tied to unfavourable delivery contracts. E. Whitcombe tells us that, before 1840, government required the revenue, which was divided into eight *kists* (instalments), to be paid regardless of whether crops and been harvested or not. Indeed: 'Collectors and...*tehsildars* were empowered to appoint their own watchmen to prevent the removal of crops before the government demand was paid...' (1972: 155). After 1840, the number of *kists* was halved, and timed to allow, in theory, for the harvesting and sale of crops, two *kists* being set for the *kharif* season and two for the *rabi* season. In practice, however, the timing allowed only a fraction of the output value to be realised through sale. All growers other than a handful comprising the rural rich were obliged to take loans for paying their landlords or the government; since interest rates were at a seasonal high at these times of financial stress, and loans had to be repaid within three months at most, namely after realising the value of the crop, the growers turned to those who appeared to offer relatively easier interest terms and whose advances did not have to be repaid until the end of the next crop cycle, namely the traders dealing in exportables. The fact that these advances entailed growing a different crop or altering the crop-mix proportions, and were tied to contracts whose terms were ultimately unfavourable, would be outweighed by the imperative immediate need for cash.

Commenting on the lack of profit in indigo for the growers, Whitcombe says:

> Given these minute profits of cultivation, it may well be asked why farmers contracted to grow indigo. The answer lies in the time at which the indigo advances were paid.... With these cash advances paid in January, cultivators could clear the charges due to the zamindars, and the latter in turn could meet their revenue obligations—the cultivators' payments falling due before crops could in most cases be marketed.

Nor was sugarcane production, which grew fairly fast in some districts of U.P., particularly profitable for the growers. This crop needed much higher cash outlays than the food crops, and was grown on advances from the traders, zamindars or both. Being tied to a contract price even lower than the open-market post-harvest price, the grower could barely cover costs. But as with indigo advances, the cash advances given for growing sugarcane 'proved irresistible for the regular remittance of rent charges' (Whitcombe 1972: 123).

We do not have reliable and comprehensive time-series of the area under the different crops before 1880, to help us to establish conclusively that the exportables grew at the relative expense of, and not in addition to, the domestically-consumed crops. But the associated trends in the macro-economy are consistent with this view. First, the overall budgetary policies were deflationary, as earlier noted; the large proportion of expenditure earmarked in the budget as expenditure abroad required a merchandise export surplus to be sent out of the country every year as the real counterpart of revenue transfer, and hence all policies were geared to promoting exportables, mainly agricultural at this time, and expansion of the colonised domestic economy took second place. India's role as a market for Lancashire goods was satisfied by displacing its domestic textile manufactures—that is, not by expanding Indians' purchasing power, but by the converse mechanism of deflationary, income-reducing policies forcing a shift to cheaper imported machine-made substitutes, in the context of a completely open economy without tariff protection, thus inducing domestic manufacturing unemployment. As Keynes put it, for a country to export goods which compete with the receiving countries' products is tantamount to its 'exporting unemployment' abroad (Keynes 1970). In India, these decades were marked by a rising proportion of landless dependent on the market for food.

Secondly, the incidence of serious domestic food consumption shortfalls, culminating in the worst years in starvation for large numbers of people, appears to have risen at the same time that exportables were rising, and peasants were being drawn 'into the orbit of world prices'. Famines of varying severity were experienced in different parts of the country in the 1860s and the 1880s, and quite widespread and devastating famines took place in the 1890s. Insightful contemporary analysts such as W.W. Hunter and R.C. Dutt noted both the increase in the relative numbers of the famine-vulnerable, and directly related the increasing incidence of serious famines to the enforced commoditisation of agriculture under the pressure of high cash revenue payments, particularly in the raiyatwari areas. W.W. Hunter and R.C. Dutt's macroeconomics was a great deal more sound than that of the contributors to the *Cambridge Economic History of India* (Kumar 1985), who have nothing to say either on the colonial transfer in regard to famines.

Hunter writes:

> After a minute comparison of rural India at present with the facts disclosed in the manuscript records, I am compelled to the conclusion that throughout large tracts the struggle for life is harder than it was when the country passed into our hands.... Among a people whose sole [*sic*] means of subsistence was agriculture, a landless proletariat [is] springing up, while millions more [are] clinging to their half-acre of earth apiece, under a burden of rack-rent or usury... (1903: 98, 147).

Hunter had earlier compiled and published *Famine Districts of Bengal* (1874), which estimated the relative numbers of the famine-vulnerable in each district, identified as the landless labourers, poorer artisans, fishermen and similar groups who were net food purchasers.

For the period after 1880 and up to independence, the time-series on crop areas, output and population collected systematically for British India, have been extensively analysed, the most notable estimations of trends being by Sivasubramanian and Blyn (and Heston in the *Cambridge Economic History*, with, however, such unsupported assumptions that this estimate is seldom quoted). The results are similar: Blyn finds that foodgrains output value at constant prices was virtually stagnant, growing at an annual trend rate of only 0.11 per cent, whereas the non-foodgrains grew over ten times faster at 1.31 per cent per annum. The latter rate is not very impressive by modern standards (the present-day indebted poor countries are expanding agri-exports at up to 8 per cent per annum for the benefit of Northern consumers, as we will see in the last section), but the niggardliness of colonial government expenditure on crop research, irrigation and other investment, has to be borne in mind.

The distinction between food crops and non-food crops is not exactly coterminous with the division between crops for domestic consumption and exportables, because the superior irrigated foodgrains like wheat and fine rice varieties were also exported. The non-food to food differential of ten to one would be exceeded if we considered the growth rate of exportables versus domestic consumables. The important point is that the availability of foodgrains per head of population (where 'availability' is domestic production plus net imports) declined by about a quarter for British India as a whole over the period, but the regional declines varied, ranging from a 38 per cent decline in Bengal to an 18 per cent decline in Punjab. There was an absolute fall in the area sown to rice in Bengal and a rise in the production of exportables, at the same time as the domestic requirement for the staple was rising with faster population growth after the first world war. However, the banal neo-Malthusian argument that accelerating population growth was 'responsible' for falling per capita food-availability simply will not do. For if rising population growth rates were the main factor, then all crop output should have declined equally on a per capita basis. Using a simple neo-Malthusian argument it is impossible to explain why per capita exportables' output was rising quickly at the same time as the per capita production was declining in the basic food staples which the growing population needed so badly; in short, it is impossible to explain the structural shifts. It may be noted that three out of four estimates of national income find that over the relevant period per capita income in real terms was either constant or slightly declining, while only one estimate finds a negligible rise in real income. Given this, there is no scope for arguing that an Engel-effect was operative, such that people were becoming better-off and substituting basic staples with higher-value food. On the contrary, the growth of domestic market demand was restricted by the deflationary and income-restraining policies referred to earlier, thus 'releasing' scarce agricultural land and resources for growing exportables, and this was intensified with the inter-war depression.

We have argued elsewhere (apropos Sen's thesis, 1981) that it is wrong to ignore the question of long-run food availability even while it is right to focus on changes in purchasing power (see Patnaik 1992). Imperialism has historically entailed strongly operative long-run changes in crop-mix and food availability in the tropical countries it dominated. In India this long-run structural shift in the output composition in agriculture towards exportables at the expense of the basic staples, so typical of the colonial process, and the resulting well-documented trend decline in food availability per head combined in the interwar period with sharply deteriorating international terms of trade for farmers, so as greatly to enlarge the relative numbers of those vulnerable to famine. This is what we termed a 'pre-famine conjuncture', meaning thereby a situation of increased vulnerability with larger numbers of rural classes in poverty, who could be pushed over the edge into starvation by any economic shock (a series of harvest failures, sudden inflation in food prices, sudden deflation in prices of products exchanged by petty producers against food). We suggested that the large toll in the Bengal famine of 1943/4—up to 3.5 million dead—arose from the fact that a sudden shock to the economic system (deficit-financed war expenditures imposed on Indians, quadrupling rice prices in 18 months in Bengal where most of the expenditures took place) came on top of the longer-run trend of declining foodgrains availability and worsening terms of exchange for farmers, a decline which had already enlarged the number vulnerable to famine (Patnaik 1992).

The colonial structural patterns discussed so far are being replicated in the contemporary developing world to a remarkable degree, albeit in new forms given the absence of overt political control by the ex-metropolitan powers. The main control mechanism appears once again to be debt, this time national external indebtedness of developing countries which obliges them to implement specific economic policies as a condition for receiving loans. The nature of these loan-conditional policies is remarkably reminiscent of colonial conditions: complete openness of the developing economy to trade and investment flows, restriction of the freedom to impose protective tariffs, removal of domestic subsidies protecting producers and consumers, domestic demand-restricting budgetary policies, and encouragement of exports with especial emphasis on agri-exports, Verily, '*plus ça change, plus c'est la même chose*'. There are additional features not present earlier as well, such as adherence to a patents regime applicable to research with tropical genetic diversity.

The last section of this paper explores briefly the reasons from the demand side for the new pressure on developing countries to increase agri-exports to the North; discusses some salient results of implementing 'structural adjustment programmes' so far; and argues that all indications are of falling basic foodgrains' production, associated with increasing exports. In regions where initial food security systems were fragile, it is possible to observe a rapidly increasing vulnerability of local populations to famine.

'Structural adjustment programmes', trade liberalisation and the loss of food security

Since the import dependence of the North on tropical agricultural products was already high at the time of post-war decolonisation, it might be thought that little scope existed for a further relative increase in the demands made on scarce tropical lands by temperate-region populations. In the course of the last two decades, however, a substantial shift in the consumption patterns of rich Northern populations has been gathering momentum, one which involves fresh demands on tropical bio-diversity. This shift is taking place largely owing to growing knowledge of what constitutes a healthy diet and life-style, which from being the preserve of a few practitioners earlier considered eccentric, is now spreading rapidly to the general populations of North America and Europe.

Prone to obesity, heart diseases and digestive tract disorders, under medical advice, the well-to-do consumers in Northern countries are cutting down on the high-fat dairy and meat products that their own agriculture produces, introducing more fish and white meat as well as more fibrous green vegetables and fruit into their diets, and replacing fat as a cooking medium with vegetable oils low in polysaturates. Only a small fraction of the new products demanded by the health conscious can be supplied by Northern agriculture, and this consists mainly of reducing the refining of existing grain output in order to retain more germ and fibre. Furthermore, as the cumulative adverse effects of using synthetic chemical pesticides, fertilisers, preservatives, flavours and food dyes become evident, there is now a premium on biological pest control, use of organic fertiliser, natural vegetable sources of preservatives, flavours and dyes. All this involves a new phase of an increased demand by the North on tropical agriculture, and a sustained attempt to acquire control, through revised patent laws and other means, over research with the basic genetic material underlying tropical bio-diversity. Thus there is today a veritable new onslaught on the limited productive capacity of Third World agriculture, combined with attempts to transfer tropical genetic materials to Northern laboratories.

On the other hand, domestic agriculture in the advanced countries has seen such high past levels of investment and productivity increases, that there have in most years been large gluts of the wheat, meat and dairy products that the temperate regions produce. This situation is being further exacerbated as, given higher incomes, and diversification of diet away from locally-growable products, closely related in turn to the increased health consciousness mentioned above, the demand per head for domestic grain is now declining, as indeed also is the demand per head for items like full-fat milk, butter and fatty red meat. Lean meat on the hoof can by definition be produced only in warm climates since fat is nature's protection to large non-hibernating animals against cold. Hence a relocation of lean-meat production to warmer countries is also taking place, with very adverse effects by way of displacement of human foodgrains by animal feedgrains. Animal fats like lard and butter are no longer widely used for cooking as they were only two decades ago, and are being replaced by vegetable oils including oils in hydrogenated form.

Faced with an unprofitable product line, a capitalist in manufacturing would simply switch to another product. Owing to the climate-soil specificity of crops, however, this is not possible in agriculture except to a very limited degree. If potato-growing in Idaho under free-market conditions does not pay, then farmers in that region cease to survive as farmers, unless they are helped by the state to survive. Hence it is that massive annual subsidies have been doled out to the farmers in North America and Europe, paying them to restrict acreage. (Under the April 1994 GATT agreement, subsidies in the North will not go, as is mistakenly believed by many; it is merely the form that has been changed already, in order to evade the specific form of subsidy prohibited by GATT, namely area-restricting subsidy.)

Clearly it is in the direct interest of the temperate regions' advanced countries to export their surpluses of wheat and dairy products to the rest of the world. The main players in the world wheat export market are the U.S.A., Canada and the European Union. Other than the former Soviet Union, which entered as a large purchaser in recent years, the largest potential buyers of foodgrains are the populous Asian, Latin-American and African economies, whose per capita incomes may be low, but who spend a high fraction of that income on basic foodgrains and constitute a very large potential market in absolute terms. However, for potential buyers to become actual buyers the developing countries concerned must lose their self-sufficiency in food production, and become net food importers. An important part of the impact of 'structural adjustment programmes' (S.A.P.) which are implemented under the guidance of the international lending agencies, is in fact precisely this, that the policies mandatorily followed under S.A.P.—of opening up the indebted countries' economies to unrestricted trade, removal of subsidies and promotion of agri-exports—lead to a decline in the output per head of basic foodgrains for local populations. Furthermore, worsening international terms of trade mean that the capacity to import food out of agri-export earnings lags behind the increasing requirements for food imports, as land and resources get displaced from domestic food production. If the decline goes far enough it increases the vulnerability of already poor populations to famine, which can then be precipitated by any moderately severe economic shock. The reasons for this and some of the emerging evidence are discussed next.

The basic reason for the loss of food self-sufficiency by the indebted Third World country which opens up its economy fully to international trade and removes all protective devices (implements trade 'liberalisation') is quite simple: the country thereby subjects its agricultural cropping and output pattern to the pull of international purchasing power and international demand. Three-quarters of the world's purchasing power is concentrated in a handful of rich capitalist countries in Western Europe and North America which, as we have seen, have based their high consumption standards historically on their privileged access to tropical goods, and whose demand for these goods has entered a new phase. The average consumer in the United States has a per capita income of $18,000 as compared to $300 in Asian countries, including India. The income of less than one per cent of the Indian population would approximate to the average level in the U.S.A. The richest 'elites' in Third World countries do tend to model their consumption patterns on those of metropolitan centres; and so we may conceptualise the 'metropolitan' demand as comprising two segments of unequal weight: the domestic higher-income urban segment in the Third World country concerned, and the segment representing demand emanating from the foreign metropolitan countries. The latter far outweighs the former in its quantitative impact once full trade 'liberalisation' is introduced.

In the case of Latin America, de Janvry (1981) found a systematic tendency, from the period of green revolution from the late 1950s onwards, for a decline in the growth rates of what he calls the 'wage-goods' portion of agricultural products, of which foodgrains are the most important, relative to the growth rates of the commercial crops. This was traceable in part to rising income inequalities and adverse demand shifts within the countries concerned, but mostly to the increase in export orientation. This has translated in many countries into an actual decline in the production per head of the basic food staples and a rise in production per head of exports to the metropolitan centres, at both the national and world levels. As a result, a marked dualism is observed, with a dynamic capitalist enclave producing in one way or another for export, and a peasant subsistence sector with stagnant yields and falling output per head.

Mexico is an instructive case in this regard. We all know that Mexico pioneered the early high-yielding wheat varieties, and had a successful green revolution in the 1960s; the rapidly growing external indebtedness of the 1970s, however, entailed the implementation of trade liberalisation and S.A.P., which in turn led to a massive invasion of the Mexican agricultural sector by the multi-national companies interested in supplying the U.S. market in particular. The speed of the resulting shifts in cropping pattern and in the final use of grain crops had been quite remarkable. Fodder crops for cattle and other livestock, which took up only 5 per cent of total crop area in the 1960s, have steadily displaced human foodcrops, and by 1980 accounted for 23 per cent of the crop area. While sorghum is a food crop in India, it is exclusively a feedcrop in Mexico, and its output has been growing at 13 per cent per annum. The proportion of all grain output which is fed to livestock had risen from 5 to 32 per cent by 1980, while the proportion going to meet human needs has correspondingly declined from 95 to 68 per cent. Over this period Mexico has become a large net importer of foodgrains (Rodriguez 1980, Sanderson 1986).

Mexican cattle are fattened under the supervision of the U.S. companies dealing in lean-meat supply for the North American fast-food industry and for direct consumption, and are exported on the hoof across the border to their abattoirs. Beef, poultry and other livestock products certainly supply the increasing demand from the Mexican well-to-do as well, but of far greater importance is the pull on output-mix exercised by the immensely superior purchasing power of the American consumers, which has brought about a major structural alteration in a matter of a mere two decades. On the other hand, the basic staple crops of the Mexican peasants—maize and beans—have not had sustained research devoted to increase yield or price incentives so as to maintain output; they have shown a very low growth rate, and in per capita terms their output has been falling, involving the small peasant producers in a spiral of economic deterioration. Many are obliged to leave farming altogether and migrate in search of wage-paid work. In a bizarre re-enactment of the sixteenth-century events in England which had prompted Thomas More's famous remark about the hitherto meek English sheep eating up men, the standard of life of the Mexican peasants has been falling in about the same proportion that the consumption levels of Mexican cattle have been rising, with scientific feeding programmes, over the duration of their admittedly short lives.

Apart from livestock feedgrains, Mexican agriculture has also seen a substantial rise in the area under vegetables and fruit, ranging from asparagus, beans and peas, to strawberries and pineapples, under the American ægis of food multinational giants like Birdseye, Ralston-Purina and Gerber. These products are canned and frozen for sale in U.S. supermarkets. Every time a well-to-do American or Western European consumer reaches out to the supermarket shelves for a packet of frozen beans or strawberries, a bunch of bananas or a tin of pineapple, she or he is helping to bid away not only Mexican land but also the land of other, much poorer countries from production of the basic staples for local populations.

If all these structural shifts arising from increasing agri-exports led to a sufficient rise in foreign exchange earnings comfortably to finance rising food imports, then the case against export orientation would be weakened. The remarkable fact is that despite its greatly increased export orientation, Mexican agriculture is today a net loser of foreign exchange; what it earns from agri-exports is more than offset by what it has to pay out for food imports (Sanderson 1986). This is not unique to Mexico, but common to most developing countries going in for agri-exports, owing to the common feature of *a sharp deterioration in their international terms of trade.* The prices of tropical exportable commodities and minerals have fallen by 40 to 50 per cent during the 1980s, while the price of wheat has risen as also have the prices of manufactures imported from the advanced world. This decline in the international terms of trade for developing countries has continued into the 1990s, and there is no indication of a reversal. This should not surprise us, since in the 1980s governments of 78 developing countries implemented identical trade liberalisation and S.A.P. policies under I.M.F.-World Bank supervision, thereby competing with each other to increase their agri-exports, to the great benefit of the developed world which has been obtaining its tropical goods cheaper than before. In truth, the

I.M.F. dispensation has proved to be a more powerful international homogenising force in the contemporary world than international communism allegedly ever was. From mid-1991 such a large Asian country as India has started following the same programmes, and will contribute further to international price declines through its rapidly growing agri-exports.

The decline of per head food production to a degree severe enough to produce pre-famine conjunctures, as a direct corollary of increased exports against a background of declining external terms of exchange, is observed clearly in sub-Saharan Africa. The six most populous countries—Nigeria, Kenya, Ethiopia, Sudan, Tanzania and Zaire—out of the 46 countries in the region, account for over half of the total production. Each of these countries has been implementing trade liberalisation, export promotion and adjustment programmes throughout the 1980s, ostensibly designed to correct external payments imbalance, policies which have in fact worsened their indebtedness and severely undermined their food security. Taking 1980 as the base year, and comparing with the average of the triennium 1987-89 (which is the latest period for which the data are available), in these six countries per capita cereal production declined by 33 per cent while tubers plus plantain output rose by only 6.8 per cent. The overall decline in foodgrains production for all 46 countries has been 16 per cent per head over this nine-year period, and the decline is continuing into the 1990s. This is, in fact, a faster rate of decline of food production than was experienced in colonial India in the interwar period. As in colonised India, in present-day sub-Saharan Africa the fall in basic food output per head is the result of an overall stagnation in real capital formation rates (owing to the expenditure-reducing, contractionary budgetary policies followed under S.A.P. in this case) within which scarce land and capital resources are being diverted to the exportables under the prevalent regime of completely open economy.

Thus we find that, precisely during the recent period when food output has been stagnating or declining in a number of countries, their agri-exportables in real terms have been growing fast. In Nigeria the non-food output index has been rising at a rate of as much as 11.3 per cent annually from 1986 to 1991, while in Kenya it rose at 6 per cent per annum during 1975-85 and at 3.5 per cent thereafter. In Ethiopia, the food crops grew at only 0.9 per cent annually from 1986, compared to a four times higher growth-rate of non-food crops at 3.7 per cent annually. In Sudan food crops actually decreased absolutely with a negative rate of 2.3 per cent, while the non-food crops grew at a phenomenally high rate of 13.9 per cent during 1980-85 (these data are from *African Development Indicators*, Table 8.7). The downturn in food production per head dates without exception from the mid-1980s onwards, as most of these countries liberalised trade, devalued, and implemented S.A.P.

Since the terms on which these exportables are sold to the Northern countries have been declining (as we have seen, the decline for particular countries in terms of trade has ranged from 40 to 60 per cent over the 1980s, depending on the composition of their exports), it is a case of these countries' having to run faster and faster to stay in the same place. A doubling of export volume will result in no gain in exchange earnings if the unit price of exports has halved. It will give a loss of earnings if the unit price of importables has risen at the same time. The result is that we find that increasing physical volumes of agri-exports are being forced out of these countries merely to maintain stagnant earnings, while this physical or volume increase itself entails long-run cropping-pattern shifts, and shifts in the end-use of products, such that import dependence for food is increased. Since the capacity to finance food imports is reduced, however, one or other of two things happen: external indebtedness rises, or food availability falls—or there is a combination of the two, with some rise in indebtedness and some fall in availability.

The situation in sub-Saharan Africa today is painfully reminiscent of colonised India in the inter-war period when, as we have seen, exactly the same pattern of decline in per head food crops and rise in commercial crops prevailed, when international terms of trade were moving steadily against agricultural producers, forcing larger-volume exports against stagnant earnings, and the number of famine-vulnerable people was being enlarged. We also have to bear in mind that the initial levels of food production with which the sub-Saharan Africans started in 1980 were lower than the Asian levels, so that a given decline has had more serious consequences. If we use the prevalent convention of taking 5 kilograms of tubers as equivalent to one of grain, and aggregate on that basis, the combined cereals, tubers and plantains output per head in 1980 was only 137.6 kilos per capita per annum during the triennium centred on 1982, and by 1987-89 this had declined to *only 122.5 kilos per head*. This may be compared to India which had a gross output per head per annum of 190 kilos in the mid-1980s, corresponding to about 166 kilos, net of seed. China had a gross output of 285 kilos, or, on a net basis, about 260 kilos. (Even if we add up to 20 per cent to the African figure, it is still lower than the Indian level at the initial date before decline starts; further, we have assumed that all African plantain output is domestically consumed, whereas some part of it is in fact exported.)

It is not surprising, given these absolutely low initial levels followed by decline owing to export growth, that much of sub-Saharan Africa should be in a pre-famine conjuncture where larger sections of local populations are more vulnerable than before, and incapable of absorbing shocks to the system such as a series of bad harvest years or a high rate of inflation. In 1992/3, large tracts of Southern Africa (Zimbabwe, Western Transvaal and Zambia) were in the grip of imminent famine owing to a past process of displacement of staple millets by horticultural exportables followed by a series of bad rainfall years. Actual famine was averted only by large infusions of food 'aid' from the North, increasing the region's indebtedness. A similar emphasis on agri-exports in India from 1992 combined with the free entry of Northern multinational companies has already started affecting the cropping pattern in many areas, with paddy land being converted to prawn fisheries, sunflower for oil replacing millet, and the growing of fruit and vegetables contracted for by the multinationals replacing foodgrains. The impact will start showing up clearly in the data in a few years.

In lieu of conclusion, we may recall Paul Baran's observation (1957) to the effect that what is cooked in the kitchen is not decided in the kitchen. What peasants in the developing countries will cook in their kitchens, or whether they will have anything to cook at all, is being decided increasingly by what is cooked in the kitchens of the advanced countries.

References

African Development Indicators, 1992.(U.N.D.P., Washington D.C.).

Abu'l Fazl, 1872-7. *Ain-i-Akbari* (trans. Blochmann; Calcutta).

Baran, P.A., 1957. *The Political Economy of Growth* (New York).

Blyn, G., 1966. *Agricultural Trends in India 1891-1949: output, area and productivity* (Philadelphia).

Chowdhury [Chaudhuri], B.B., 1965. *The Growth of Commercial Agriculture in Bengal* (Calcutta).

Cummings, B., 1981. *The Origins of the Korean War* (Princeton).

De Janvry, A., 1981. *The Agrarian Question and Reformism in Latin America* (Baltimore).

Dutt, R.C., 1901, 1903. *Economic History of India* (2 vols.) (London).

Grabowski, R., 1985. 'A historical reassessment of early Japanese development', *Development and Change* 16, pp.235-50.

Hayami, Y. and V. Ruttan, 1970. 'Korean rice, Taiwan rice and Japanese agricultural stagnation: an economic consequence of colonialism', *Quarterly Journal of Economics* (November).

Hunter, W.W., 1903. *The India of the Queen and other essays* (London).

—, 1874. *Famine Districts of Bengal* (London).

Kumar, D. (ed.), 1985. 'The fiscal system', *et al.* in *Cambridge Economic History of India*, vol. 2 (with M. Desai; Delhi).

Madras Provincial Banking Enquiry Committee 1929-30. *Report*, vol.1.

Mitchell, B.R. and P.V. Deane, 1966. *Abstract of British Economic Statistics* (Cambridge).

Naoroji, Dadabhai, 1901. *Poverty and Un-British Rule in India* (London).

Patnaik, U., 1975. 'The process of commercialisation in agriculture in the colonial period' (working paper, C.E.S.P., J.N.U.).

—, 1992. 'Food availability and famine: a longer view', *Journal of Peasant Studies*, 19, 1.

Raju, A. Sharada, 1941. *Economic Conditions in Madras Presidency 1800-1850* (Madras).

Rodriguez, G., 1980. 'Tendencies de la production agropecuaria en las dos ultimas decades', *Economia Mexicanan.*

Sanderson, S.E., 1986. *The Transformation of Mexican Agriculture* (Princeton).

Schumpeter, E.B., 1940. *The Industrial Development of Japan and Manchukuo* (New York).

Sen, A.K., 1981. *Poverty and Famines: an essay in entitlement and deprivation* (Oxford).

Sen, Sunanda, 1992. *Colonies and the Empire* (Calcutta).

Slicher van Bath, B.H, 1963. *Agrarian History of Western Europe AD 1600-1850* (London).

Whitcombe, E., 1972. *Agrarian Conditions in Northern India* (Berkeley).

Zacharias, C.W.B., 1950. *Madras Agriculture* (Madras).

Chapter 11

AGRICULTURE AND ECONOMIC LIBERALISATION: THE INDIAN OUTLOOK

Abhijit Sen

Introduction

A curious aspect of the debate on the liberalisation policies being implemented in India, with verve, and amidst much debate, is that there is almost no discussion about the likely impact on agriculture, barring the short-run concern about possible monsoon failures and their impact on inflation. It is true that, with industrial production declining, exports sluggish in the face of a continued rise in external indebtedness, and inflation rates well above the two-digit level, the economic situation in India may turn dramatically worse if there is a poor monsoon, on top of the already grim conditions following the stabilisation and structural adjustment policies adopted in July 1991. At the beginning of 1992, there was a widespread feeling that it would be miraculous if there were a fifth good monsoon in a row. In fact, with the monsoon progressing fairly normally after a somewhat delayed start, there was no real cause for panic.

However, there is cause for concern about the food situation in India even when the monsoons turn out to be good, because the government's foodgrain stocks, especially of wheat, are uncomfortably low, because of low procurement and high off-take in 1991 despite a record crop, and the fact that they were not replenished adequately in 1992. That year's wheat procurement needed to be almost 10 million tonnes, 2 million more than the previous year's actual, to maintain adequate supplies with which to run the Public Distribution System (P.D.S.). But at the time of writing, with procurement practically over, this had been only around 6.3 million tonnes. Moreover, for one reason or the other, plans to import wheat seemed to have been delayed, and only subsequently was the decision taken to import one million tonnes, probably not enough to maintain P.D.S. supplies. Foodgrain prices increased by almost 25 per cent in the course of twelve months. And how the government dealt with the short-run problem of managing the food economy was bound to be crucial in what happened to the economy later in 1992, and to have an important bearing on the outcome of the government's liberalisation policies. But more important is the converse question of what these policies mean for the agricultural sector. This is the focus of the present paper.

Here, the most remarkable aspect of the economic reform-strategy of the government was that, while it purported to be a virtual overhaul of the entire economy which would revolutionise its workings, there was (as said) effectively no serious consideration of agriculture. This lack of official comment and the absence of any systematic approach to the agrarian question and the possibilities of raising productivity in agriculture in future, are noteworthy because they marked a departure from the past and changed government perceptions about what really matters in the economy and how growth patterns can be altered. In the liberalising world-view, most economic problems can be resolved by a greater recourse to markets and by allowing the price mechanism 'free' play. Presumably a similar position governs the attitude to agricultural growth, in that it is supposed that relative price movements and profitability ratios will be sufficient to ensure that supply responsiveness in agriculture will lead to higher rates of growth. Not only is the price factor seen as dominant, but it is assumed that deregulation and withdrawal of subsidies, exchange-rate flexibility and reduction of government intervention in the agricultural economy, are the best ways to achieve the desired price movements. In India, this 'hands-off' policy towards the agricultural sector will of course always be tempered by political reality, in the shape of the strength of the rich-farmer lobby and other factors, but it seemed the chief mode of thinking in government in the early 1990s.

Thus, the situation is that very little was being done in terms of any definite policy-initiative within the liberalisation framework. But, elsewhere, bodies like the World Bank have articulated their view of 'desirable' agriculture-sector policies. This paper reviews this literature to anticipate the agricultural strategy likely to emerge in India under the tutelage of the international financial institutions, and assess its probable impact given recent trends in the performance of the agricultural sector. At the time of writing it was apparent that Indian agriculture and India's food security would also be affected by agreement on the Dunkel Draft for conclusion of the Uruguay round of GATT, and an attempt is made to indicate the expected impact of this. By way of a conclusion, the final section is an effort at a more general prognosis.

II

The World Bank view on agriculture, both in general and as it related to structural adjustment in a country like India, can be found in three recent documents. *The World Development Report 1986*, dealing with the general issue of trade and pricing in world agriculture, addressed both the issue of appropriate policies in developing countries and the international trade issues which are the focus of GATT negotiations. *Restructuring Economies in Distress* (edited by Thomas *et al.*, 1991) is a World Bank publication based on its own evaluation of its structural lending programmes, and contains a section on agriculture. The second volume of the Bank's 1991 *Country Economic Memorandum* for India, subtitled *Agriculture—Challenges and Opportunities*, is the latest Bank document specifically dealing with Indian agriculture. Taken together, these documents provide a fairly detailed outline of the sort of policy-thinking that the Bank and other multilateral agencies are likely to urge upon India's agriculture policy-makers—probably, on the evidence of 1991/2, with considerable success. For this reason, it is useful to begin with a brief review of these documents.

The general tenor of the 1986 *World Development Report* (W.D.R.) was that, although world agricultural output grew fairly rapidly during 1970-85, and growth in food production was higher in the developing countries than in the developed ones, this was a result mainly of the availability of new—particularly 'green revolution'—technology. On the other hand, according to the report, the general economic policies followed by developing countries limited the growth of agricultural production and hampered efforts to reduce rural poverty. In particular, the *Report* identified two critical areas: first, the restrictions on international trade, and the quota and tax-subsidy regimes which result in a 'bias against agriculture' caused by a divergence of domestic prices from international parities; and, second, government interventions at all stages of the production, consumption and marketing of agricultural products and inputs, that have frequently resulted in greater inefficiencies and lower output and incomes.

The Bank's critique is transcendental in that, according to it, the discrimination against agriculture 'is very much an integral part of development strategies that promote domestic industries behind high trade barriers'. In other words, the Bank disapproves, as being inimical to agricultural growth, of any industrialisation process which involves an attempt to accelerate industrial growth by turning the internal terms of trade between agriculture and industry so that agriculture is worse off than it would be if domestic prices were aligned with world prices. And it argues further that 'government programs for price stabilization, consumer subsidies, and producer input subsidies...are far less effective than they are thought to be in promoting either a more efficient allocation of resources or a more even distribution of income'. Moreover, so runs the argument, the discrimination against agriculture is exacerbated by ways in which countries attempt to cope with changing economic circumstances, failing to adjust exchange rates sufficiently in periods of inflation, and relying instead on excessive foreign borrowing and on *ad hoc* exchange and trade controls.

In the main, the W.D.R. asserts its 'get prices right' position rather than argues its case, embellishing the assertions through examples, mainly from Africa, where ill-designed government interventions did disrupt agriculture. It is silent on the role of government in providing agricultural infrastructure or initiating reform in tenancy or ownership structures, and ignores the glaring fact that during the colonial period prices were in fact 'right', but agricultural performance abysmal. It also ignores two obvious points. First, if protective industrial policies do lead to rapid growth of non-agricultural employment, as in Japan earlier this century, per capita agricultural incomes can rise much faster than would be the case without protection. And second, since what matters for output is the ratio between output and input prices, subsidising the latter is an effective corrective to any lowering of domestic agricultural prices relative to world prices, consequent to the trade policy followed. On the contrary, the W.D.R. picks up for particular criticism the prevalence of fertiliser and credit subsidies, arguing rather awkwardly that if output prices are 'right', such subsidies can only be justified temporarily, and that they are usually bad because, besides being costly to the exchequer, they may involve inefficient distribution by parastatal agencies.

In addition to these observations regarding agriculture in developing countries, the W.D.R. also criticises the developed countries for subsidising and protecting their agriculture excessively and for providing cheap food aid which keeps agricultural incentives low in the developing countries. Thus, while developing countries are accused of discriminating against agriculture, the developed countries are accused of discriminating in favour of their agriculture. And the whole nexus is seen as one where world agricultural trade suffers, along with the ability to gain from comparative advantage. The perspective is similar to that espoused in the Uruguay round of GATT talks, of reducing agricultural subsidies. Buttressing the case is the now quite familiar argument that the biggest gainers from withdrawal of farm subsidies in the developed world will be the developed countries themselves. Awkwardly, however, the simulations presented to substantiate this case show that, while liberalisation of agricultural trade by the developed countries would have brought gains of $49 billion (in 1980 prices) for the developed countries in 1985, this would actually have meant a *loss* of $11 billion for the developing countries, mainly because the latter are net importers of developed-country produce whose production would fall and whose international prices would rise consequent upon liberalisation. Although this is glossed over by claiming that both developed and developing countries would gain if everyone liberalised agricultural trade, the fact remains that it is not at all obvious that land-short developing countries, such as India, have comparative advantage in agriculture over many land-rich developed countries. If the latter remain net exporters even after liberalisation, it is also not clear whether or not a part of the farm subsidies in developed countries would be a net transfer to consumers in developing countries. Yet the W.D.R. concludes by claiming that getting prices right, cutting subsidies and liberalising world agricultural trade would make everyone better off, and towards this end recommends its own Structural Adjustment Loans (SALs) and sector adjustment lending (SECALs) which it claims 'have proved to be important instruments for supporting reform'.

However, the Bank's own analysis of SALs and its agricultural SECALs, as reviewed by Knudsen and Nash in Thomas *et al.* (eds.), *Restructuring Economies in Distress*, is less than comforting: 'Since prices are the main determinant of incentives for agriculture, pricing has received special attention in adjustment loans.' Almost 60 per cent of SALs and all but one of the agricultural SECALs have included agricultural pricing conditions, with requirements to increase output prices about as common as those to increase input prices. In addition, 60 per cent of SALs and 40 per cent of agricultural SECALs required some liberalisation of agricultural trade, and 71 per cent of SALs and SECALs contained conditions related to privatisation or deregulation of product or input distribution. In most cases, these conditions were successfully implemented in the sense that prices were raised, trade liberalised, and distribution privatised or deregulated. But the outcome in terms of output growth has been mixed at best. Half the countries showed lower agricultural growth after the loans than before. In defence, the authors suggest cautiously that 'adjustment is a long-term process', but they are also constrained to note that 'a certain fatigue is evident', with very few agricultural SECALs likely to be continued in future in the countries where they have already been tried. They note that the experience shows that 'correction of relative prices alone will not lead to sustained increases in agricultural growth', and that 'the experience with price and non-price factors and agricultural growth also demonstrates the importance of maintaining public and private investment'. Generally, the review, although couched in optimistic terms by the authors, who clearly do subscribe to the overall perspective, is an expression more of failure than of success. The comments on the paper by Nurul Islam (in the same volume) go further, questioning the very basis of some of the conditionalities imposed; for example, 'that the reduction of subsidies is often accompanied by a reduction rather than an increase in public investment in agriculture, with adverse effects on growth and employment in agriculture'.

Yet the latest World Bank publication on Indian agriculture (vol.II of the 1991 Economic Memorandum for India, *Agriculture—Challenges and Opportunities*), concludes as follows:

> In the short-term, priority actions include: (a) reducing agriculture subsidies; (b) regaining control of public expenditure; (c) improving the safety net offered by food programs while restraining costs; (d) initiating credit reforms to prevent the collapse of the agriculture credit system; and (e) ending coercive marketing and trade restrictions. While these urgent actions will occupy the attention of policy-makers, it is just as important to begin medium-term reform programs, particularly in the areas of agriculture pricing and trade policy, input production and marketing (including restructuring of the fertilizer industry), research and irrigation. These adjustments are essential for laying the foundation for future sector growth.

It is noticeable that short-term priorities are the stuff that most agriculture SECALs have been made of, amounting as they do to reducing input and credit subsidies, along with reduced but more targeted coverage of the P.D.S. and the rural banking network, but with no real increase, and probably a cut, in public investment. The medium-term reform programme echoes the 1986 W.D.R. in putting emphasis on liberalising trade and privatising input production and marketing, the passing reference to research and irrigation being somewhat deceptive. In both cases the thrust in the main report is also towards privatisation, with an accent on improving intellectual property rights and imports of technology in the case of research, and on government divestment of implementation and investment in the case of irrigation.

So far India has not taken any agricultural SECAL, and progress on the priorities identified by the Bank has also been slow. All that has happened so far is some cuts in the food and fertiliser subsidies. However, further cuts in these subsidies were implied in the latest budget (at the time of writing), and a Committee on the Financial System (the Narasimham Committee) had recommended major reductions in the quantum of credit directed towards the agricultural sector by the public sector banks. Thankfully, however, the Government of India, unlike the World Bank, did not tout these measures as 'agriculture reform', recognising that they are at most efforts aimed at reducing the fiscal deficit and at improving the profitability of the financial sector. And thus, although there was some talk of encouraging agro-business and thus boosting exports of processed agricultural products, there was then precious little which could be termed as an agricultural policy proper.

III

This agricultural non-strategy of the government merits serious concern because the basic structural and other problems in this sector have by no means been resolved. While the trend rate of growth of agricultural output over the 1980s was more than satisfactory, at over 3 per cent per annum, this was mainly due to the sharp increase in 1988 when an excellent monsoon broke a period of relative stagnation during 1983-87. After 1988, output seemed to have reached a higher plateau during the subsequent four years of above-normal monsoon, but there appeared to be little in terms of a sustained growth impetus, with the output in 1991/92 less than that in 1989/90 despite a similar monsoon. And within this not very remarkable aggregate output growth, there were a number of disturbing tendencies with implications for current output and future growth potential.

The first was the increasing regional concentration of growth in output and marketed surplus, in a context in which there were already severe regional inequalities at the beginning of the decade. The south and west had been performing badly in terms of both foodgrain and cash-crop output, especially compared with past performance. The northern region had become even more dominant in the production and marketing of foodgrains. Other than this, only some parts of the eastern region, in particular West Bengal, showed any signs of dynamism in agriculture. Indeed, West Bengal, with a growth rate of over seven per cent in agricultural value-added—more than two-and-a-half times the national average—can be described as the agricultural success story of the 1980s. This has important lessons for policy, because the major planks of this success have been tenancy reforms, decentralised and village-level planning of infrastructural investment, and conscious attempts to improve minor irrigation sources. Despite this performance, the basic all-India picture has been one of significantly greater regional inequalities within agriculture. This reflects a major failure of the government in encouraging a horizontal spread of growth, which was the explicit aim at the beginning of the decade.

The second disquieting feature is that such growth in output as has occurred has not come from any extension of cultivated area through more intensive cropping. Rather, land-use intensities have remained stagnant or have even fallen, putting all the onus of the impetus to growth on yield increases. This was not entirely a positive development reflecting improved yield-increasing practices, because at least to some extent it reflected also a long-term problem of degradation of land due to over-cropping and other ecological misuse. This is a feature which is likely to become even more marked in future, and points to the crucial need for much more investment in wasteland development and developing the laboratory-to-farm outreach that is now almost non-existent in most dry-land areas. Additionally, minor irrigation works are threatened as groundwater tables are falling in a number of regions, requiring more effective local control over aquifers and the development of better watershed management. All these measures will require time, and will succeed only if local institutions are allowed to develop and given adequate resources.

Thirdly, while output growth was satisfactory in the 1980s, agricultural investment showed a downward trend, having been reduced in real terms by almost 20 per cent since the peak reached in 1978. This occurred despite a term-of-trade improvement for the agricultural sector, and was mainly attributable to a decline in public investment in agriculture. There are strong linkage effects and it is evident that private investment increased precisely in line with public investment, far from being a substitute for it. Thus the decline in public investment in agriculture and in rural infrastructure had both direct and indirect effects through the concomitant decline in private investment. This has disturbing implications for the future pattern of growth, especially given the longer-term processes of ecological and soil degradation which are likely to affect agricultural productivity in future.

Fourthly, as a result of reduced public investment and the consequent reduction in the horizontal spread of infrastructure, Indian agriculture also became progressively more dependent on intensive use of current inputs, such as fertilisers and pesticides, in the already developed regions. So strong was this effect, especially for fertilisers, that today the principal determinant of short-run yield changes are the use and pricing of this input. In the 1980s the data show an almost perfect negative correlation between the price of fertiliser relative to foodgrains and the foodgrain yield per hectare, with the latter having an elasticity of over 0.5 with respect to the former. This extremely large sensitivity of output to the real fertiliser price reflected not only the increased commercialisation of input use among farmers, but also the fact that other sources of growth were much more muted during the 1980s than earlier. An important corollary of this is that unless the other growth impulses are restored through technological improvements and stepped-up public investment in necessary areas, any rise in fertiliser prices is likely to have a serious adverse impact on yield and output.

It is clear that the agricultural sector is by no means in a sufficiently healthy state to allow the government to be complacent about its prospects. A systematic strategy is required, without which agriculture is likely to face continuing problems of rising costs and falling input productivity. The government seems to be aware of this at least with regard to the need to develop and improve oilseeds cultivation (probably because of the high costs of import), and in this sphere some degree of success has been achieved in recent years. But for almost all other crops, and for the sector as a whole, there appears to be sublime unconcern about the incipient problems and the need to formulate strategies to overcome them. And, even more significantly, the current economic policies of the government are such as to aggravate several of these problems, as well as to reduce the standard of living of large sections of the agricultural community.

Consider one of the basic elements of the stabilisation-cum-adjustment package: the reduction or elimination of subsidies. This means particularly food and fertiliser subsidies, which at the time of writing had already been cut significantly in the preceding two Union budgets. It is true that the fertiliser subsidy is partly a transfer to an inefficient domestic fertiliser industry. But, since the strategy seems to be to raise domestic fertiliser prices to cover the higher cost of domestic fertiliser production, it remains likely to have a negative impact on fertiliser use by farmers, with corresponding effects on yield. Simply allowing procurement prices to rise along with the price of this particular input is no real solution because fertilisers are now widely used for the cultivation of crops which are not covered by government procurement, as also by groups of farmers whose output is marketed only to a limited extent. In any case, what matters from the point of view of output is the ratio of fertiliser price to the product price, and maintaining this in the face of higher fertiliser price would require larger increases in output price than are being talked about as compensation for the resulting loss of farmers' incomes. The simple fact which is being lost sight of is that maintaining relative prices unchanged in the face of a one rupee cut in the government's fertiliser subsidy will require almost a ten rupee transfer to farmers from either the budget or from consumers. Since food subsidies are also being cut, the burden will have to fall mainly on consumers, particularly the poor. Clearly, the attitude towards fertiliser pricing has to be a much more careful one which takes into account all the implications for agricultural production and for food security. The oft-repeated mantra that such subsidies are necessarily bad, ignores historical and present examples of most agriculturally successful nations.

However, reduction of fertiliser subsidy would not be a real cause for concern if cuts in input subsidies were balanced by an increase in public investment in agriculture and rural infrastructure, and if greater attention was being paid to the old but still relevant issue of agrarian reforms. Public investment, particularly in irrigation and so on, has been the main motive force behind output expansion in the past, and, although irrigation and fertiliser use are complementary, there is a case to be made for a trade-off between cutting input subsidies and increasing public investment, because the latter, by spreading horizontally the use of even a given amount of fertiliser, can enhance the efficiency with which it is used. Moreover, the West Bengal example, where value-added has grown faster than gross output, contrary to trends elsewhere, suggests that greater efficiency in input use is possible through reform and devolution.

The unfortunate reality is that (at the time of writing) both fertiliser subsidies *and* public investment were being cut simultaneously, with almost no concern for either devolution or agrarian reform. Capital spending by the Union government was being slashed with a major squeeze on its agriculture-related investments. Furthermore, the effective decline in financial transfers to the states would affect the states' ability to increase or maintain investments in agriculture, a state subject. Of course, resources are scarce and need to be increased. But there is little sign of any co-ordinated effort by the centre and the states either to tap rural surpluses through a landholding tax or to raise electricity tariffs and irrigation charges, all of which (being more in the nature of fixed costs) would affect agricultural output less than increased fertiliser price or reduced public investment.

If any further disincentive to private agricultural investment were required, it would be provided by the implementation of the recommendation of the Narasimham Committee that loans to priority sectors (including agriculture) be cut to only 10 per cent of total bank lending from 40 per cent today. This move is likely to be particularly disruptive. It is ill-conceived in that, instead of reducing the subsidy element in the interest rates charged, it seeks to cut the volume of agricultural lending and reduce the reach and coverage of the banking system in rural areas. Essentially all this means that investment in agriculture, which has already been declining over the past decade, is likely to fall further, with adverse consequences for output growth.

Can the government afford to ignore this possibility? Or, to put it another way, how important is the performance of this sector for the overall success of the government's economic strategy? There is, of course, the obvious point that agriculture continues to be the largest employer of the workforce, and that productivity increases here are of major weight in the economy. But it has also been noted that over the

1980s the rest of the Indian economy became progressively less dependent on the behaviour of agricultural output. This is evident from the fact that the period of relative stagnation in agricultural output from 1983 to 1987 was nevertheless marked by high growth rates in non-agriculture, and the earlier dependence of aggregate economic growth on the behaviour of the monsoon seems to have diminished. There are two major reasons for this. The first is the decline in the employment elasticity of the output of non-agricultural sectors, which means that increased output in industry and services today involves a much lower concomitant increase in the demand for wage goods, particularly foodgrains. The second is the falling share and stagnation of the traditional agro-based industries, implying that agricultural raw materials now play a less significant role as industrial inputs than earlier. The boom sectors of the 1980s—chemicals, consumer durables and 'high-tech' services—have very little linkage to agriculture. Despite this, however, the importance of agriculture cannot be wished away easily, and a fall in agricultural output can still have severe negative implications for the economy, in terms both of output and inflation.

Indeed, what is striking about the experience of the past decade is that, despite the declining dependence of non-agricultural sectors on the performance of agriculture, the prices of agricultural goods have risen faster than the general price level. This has meant a reversal of the earlier movement in the terms of trade against agriculture, and has also gone against the current international trend. This increase in agricultural prices did not have an unbearable inflationary implication, partly because of the tendency described above of weakened inter-sectoral linkages, and partly because of the way the food procurement and public distribution systems functioned. Government procurement of foodgrains was more than adequate and government stocks generally sufficient. The procurement prices were typically close to the market prices, and domestic prices were also not too far from world prices, so that there were fewer speculative pressures. The P.D.S., along with the government's free market operations, worked to some extent to keep the prices of essential foodgrains under control. In fact, an important trend during this period was that, while agricultural prices as a whole increased faster than the general price level, foodgrain prices increased more slowly, so that it was possible for real wages in agriculture to rise without increasing the product wage. This was an important contributory factor behind the decline in rural poverty which occurred during the period.

Recent policy changes have altered these particular equations as well. Devaluation and the partial convertibility of the rupee (wherein foodgrains exports and imports are to be at the market rate) have made the international prices of food much higher in rupee terms (by almost 50 per cent since June 1991), fueling expectations of rising domestic food prices. Food prices had already increased by almost 25 per cent in the twelve months before the time of writing, and expectations of further rises are such that there were then strong pressures on the government to raise procurement prices immediately, so that farmers shared the gains and these were not reaped entirely by traders. This was already clear from the extremely low wheat procurement in 1992, and from farm-lobby demands, supported by militants in Punjab, that Indian farmers should be paid at least as much per quintal as the government paid for any import of wheat in terms of rupees. At the same time, the procurement system and P.D.S. were being affected by fiscal pressures which have meant real cuts. The government was then caught in a bind: an increase in the procurement price would be a signal to farmers to hoard in anticipation of further increases; a failure to increase the price would mean that higher procurement would require zoning restrictions and coercive measures which the government was unlikely to be able to sustain in the face of big-farmer opposition; management through imports of foodgrains had its own problems, not only because of the embarrassment that it would cause *vis-à-vis* the farm lobby, but also because it would be a considerable charge on scarce foreign exchange resources.

Of course, the World Bank-type argument in this connection would be that the very rise in domestic agricultural prices would provide an inventive to farmers who would be stimulated to greater production and investment. And, in a related but separate context, the Government of India's Commerce Ministry touted the Dunkel Draft on the Uruguay round of GATT talks as being beneficial to India because world foodgrain prices would rise as a result, and India would emerge as a significant exporter, presumably because domestic prices would be allowed to rise and this would increase production and restrain consumption. One difficulty with this line of reasoning is that if manipulation of relative prices is the only necessary corrective, it would be less disruptive and more efficient to influence input rather than output prices, especially when the latter feeds into other prices in the economy. And, more significantly, relative prices alone are not sufficient to ensure agricultural growth. The experience in India as well as in a number of other developing countries has been that other non-price factors are far more significant in affecting production conditions and output—especially technology (which incidentally the GATT proposals will make more expensive to acquire), and public infrastructural investment and institutional changes. Relying solely on the output price incentive may lead to a situation in which prices keep rising but growth impulses in agriculture remain low.

Indeed, in this context, the Dunkel provisions on agriculture in the Uruguay round of GATT, far from being favourable to India, should be a matter of considerable concern. The TRIPS provisions require providing *sui generis* protection to plant varieties, and this may mean that Indian farmers have to pay some form of royalty on new varieties in the future. The minimum access provision in Dunkel's draft would necessitate large imports of edible oils, undoing some of the progress so far made on this front, with adverse effects concentrated in the poor, dry regions where these crops are spreading most. The requirement that government purchases and sales be at market prices would mean that the government would have to follow the market and not lead it, thus restricting its ability to intervene effectively in the market for maintaining food security. Dunkel provisions also can be interpreted as requiring that the P.D.S. would have to be limited only to those falling below some specified nutrition level, and thus be considerably diluted. But the most important adverse consequence of the Dunkel provisions may follow precisely from the cut in agricultural subsidies and the consequent decline in crop production in Europe, combined with free world trade in agriculture, both of which the Indian government sees as an opportunity. It is not at all obvious that the withdrawal of land from agriculture in Europe and U.S.A., following from the implementation of Dunkel's proposals on cutting output-related subsidies, would be positive for the Third World, which is a net food importer, and particularly for India which suffers from a low land:man ratio, and where any long-run comparative advantage is unlikely to be in those temperate crops whose world production would thereby be reduced. In any case, the net short-run result would be higher prices of temperate crops, mainly grains and edible oils. Of these, India is deficit in the latter and barely self-sufficient in the former, so that the net balance of payments and food security implications are likely to be negative unless very large breakthroughs in output quickly take place. Moreover, rising world prices will put pressure on domestic prices, and the resulting inflation and worsening in the balance of payments are likely to lead to further cuts in public investment. Together, these will intensify the syndrome of high prices and low output growth towards which the government's current non-policy is already headed, even without Dunkel. It is therefore rather odd that the Government of India supported the U.S.A.'s position on this issue rather than the European one, especially since in the GATT context this is the only area where there is strong developed-country opposition—from European farmers—to the interests of the multinational companies with which India has major differences over other GATT provisions, mainly in TRIPS and TRIMS.

All this may or may not embarrass the government, whose keenness to ingratiate itself with the world powers-that-be is now matched only by its seeming inability to do any independent economic analysis; but the consequences for the poor in the country are undoubtedly serious. As almost all studies of poverty in India show, a combination of rising food prices and decelerating output growth is a sure prescription for increased poverty. And recent studies (for example by the International Institute for Applied Systems Analysis) show that the global liberalisation of agricultural trade will improve incomes of rich farmers and may increase exports, but will do so not so much by increasing production, as by increasing prices and reducing consumption. The consequence will be increased poverty, since those most affected adversely will be the poorest groups in the country—landless agricultural labourers, marginal farmers and urban unorganised workers—whose main item of consumption is food. The real victims, whether of the domestic non-strategy of neglecting agricultural investment while setting prices free, or of global liberalisation of agricultural trade *à la* Dunkel, will thus be those who are already living at the very margins of survival.

IV

What does all this imply for the future of India's agrarian economy and her polity? Ultimately, the issues which arise concern the fundamentals of the development process. In traditional development economics, the essential development problem has been viewed as being how to increase the per capita incomes of the agricultural population while simultaneously reducing the proportion of the total population dependent on agriculture. Except in the period of the first three five-year plans, from 1950 to 1985, when both organised sector employment and agricultural G.D.P. grew much faster than population, India's record in solving this essential problem has been extremely poor. In the 1980s, although the all-India average agricultural income per capita did grow, its unequal regional distribution meant that it did not do so in most regions. And, despite the rapid growth of organised-sector incomes, organised-sector employment remained almost stagnant.

Although in principle liberalisation and a more export-oriented strategy can create more non-agricultural employment if they encourage the growth of labour-intensive manufactures, the current world scenario and the exact mix of strategies being followed in India do not look too promising on this account. The world economy seems to have entered a period of prolonged relative stagnation, and this, along with the resulting protectionist pressures in the developed countries, make export prospects less than rosy.

In the meantime, the stabilisation policies being followed domestically are impinging hardest, through low domestic demand and high interest rates, on the growth of labour-intensive small-scale industries. And, as far as large-scale industries are concerned, the accent of present policy seems to be on reducing rather than on increasing employment. Moreover, the strategy of reducing industrial protection, particularly for capital goods, and allowing food prices to rise, will have the effect of encouraging the substitution of labour by machinery. The scope for withdrawing labour from agriculture is thus now even narrower than earlier, and is likely to become even worse as protection to industry is withdrawn in further rounds of liberalisation. If, on top of this, agricultural growth does slow down for the reasons given above, not only will the growth of industrial employment fall further but per capita agricultural incomes may also decline, contributing to a vicious circle of increased underdevelopment.

As far as growth is concerned, some of these adverse features may be mitigated somewhat if there is indeed a massive inflow of foreign investment which creates employment and builds linkages with agriculture or develops export markets for horticultural and aquacultural produce. But the prospects of this happening on the required scale are negligible. It is certainly possible that particular regions may benefit from such development but, since it would lead only to lower production and higher consumption of essential food, it is unlikely to do much for the rural poor.

The prognosis is no better when it comes to the more recent developmental concern regarding the effects on ecological degradation. Currently, because of geographically-concentrated growth, pockets of affluence have higher pollution levels than would have occurred with more dispersed growth. Simultaneously, these pockets make demands on the forest and other reserves of less affluent regions which are difficult to resist because often the latter have few alternative sources of growth. And, in any case, the lack of sufficient non-agricultural employment and alternative infrastructure puts unbearable strains on grazing, forest and other common property resources, such as water and wood for fuel. The emerging trends, outlined earlier, are such as to intensify all of these problems, and indeed to add a few more negative features. The trend towards agro-business, for example prawn-culture on paddy land, has the potential of severely degrading prime land, as observed in Thailand recently. And the general climate of privatisation, and of giving priority to global environmental issues over local ones, through the device of watch-and-ward protection of the Forestry Department type, is likely to make the problem of common property resources even worse.

It is clear that the real solution lies in some combination of less rapid population growth and a more rapid, but sustainable rate of agricultural growth. The crying need is for greater public investment in rural infrastructure, both physical and human; and for institutional reforms which would allow better combinations of surplus labour with scarce non-labour resources, both within and across regions. This has two requirements. First, funds should be found, by increasing taxes on the rich, to expand infrastructure (roads, irrigation and electricity, and also health, education and family welfare) horizontally at a rapid pace, but with due concern for economic efficiency and ecological consequences. And, second, a fresh agenda should be set out for the reform of ownership and control over property, while giving greater regional and local autonomy in matters of intra-region allocation of resources. Unfortunately, the liberalisers' priorities are rather far removed from these.

However, whatever the ideal solution or the priorities of those currently making policy, the eventual shape of things to come will depend on the impact that the policies being followed now have on the nature of the political balance in the medium-term future. This is obviously the great unknown about which economists' guesses are unlikely to be more informed than those of others. But certain pointers can be offered. First, it is fairly clear that, although the rural poor will lose, large sections of the rural rich will reap windfall gains because of higher agricultural prices. Second, such gains for rich farmers will, however, be regionally quite differentiated. The gains will be larger in regions growing crops like rice, whose current domestic price is much lower than the world price. They will be lower or non-existent in regions growing crops, like oilseeds and sugarcane, whose parity is the other way round. Third, and most importantly, these windfall gains will be accompanied by a cut-back in the flow of resources from the government to rural areas, and, along with this, a diminution in the flow of resources that the rural rich currently get from the government.

In other words, rural India is likely to see increased income disparities both within and across regions while, simultaneously, the sources of patronage by which the rural elites are currently linked to the national political system are likely to get weaker. The implication of this is that the ability of the central government to carry out distributive transfers will decline even as the need for such redistribution increases. More generally, there will be a further decline in the influence over rural elites by the existing *modernising* state, whether liberalising or dirigiste. New, more local mechanisms of coercion and patronage will almost certainly become more important. Moreover, since the rural elite is unlikely to be able to develop these on its own, and since the need for mechanisms for conflict resolution in the countryside will become more acute, the role of the *mofussil* elite and middle classes is also likely to become more important.

Short of a reassertion of the centre through some form of dictatorship, which is unlikely, only two outcomes seem possible. One of these is the possibility of heightened sub-nationalism leading to more Punjabs, but another is a natural move away from the type of top-down political culture of the Congress towards the more diffused style of the non-B.J.P., non-Communist opposition today. These groupings have hitherto played the politics of distributional coalitions, with little by way of a programme for mobilising resources or development. An increase in their strength, and the fact that they are almost all currently opposed to or ambivalent about the liberalisation moves being carried out, are likely to create obstacles for the liberalisers. Whatever their protestations to the contrary, they are even more dependent than their opponents on the centralised state that they seek to dismantle. Equally, the opponents of liberalisation are likely in future to face far more complex challenges on their agenda for agrarian reforms and public investment. The outcome of the tussle between these two variants of *modernising* influence will depend ultimately on which of them can cope better with the complexities that liberalisation will introduce into a polity where the established urban interests have so far only just about managed to hold back the political challenge of the emerging middle castes.

Chapter 12

THE MEANING OF SOCIAL ORDER IN THE TAMIL COUNTRY

S.S. and Chitra Sivakumar

This chapter presents evidence which reflects on a theoretical discussion to be published elsewhere.[1] It considers the organisation of agrarian society in Tondaimandalam, in relation to economic imperatives. The purpose is to show that many of the agrarian institutions of the region served important economic functions. The discussion has to be historical, since the institutions that exist today are an admixture of pre-colonial and colonial and post-colonial elements. However, our concern is with a social order that was transformed over time.

Theoreticians have mostly been confounded by the agrarian structure of India. In Tondaimandalam, however, we find a web of interconnected relations of production which can be explained in organisational terms. The variables include information and the politico-legal environment, the 'transaction regime'. They set the parameters of the producers' possible responses, which are established according to the minimum return to secure a particular input (the marginal production condition) and the limit at which proceeds cover costs (the marginal cost condition). The two calculations include both direct and indirect (social or transaction) costs at a given distribution of wealth, entitlements and resources. Hence arise the differences within and between societies and over time. The variables produce different production regimes; they are the key to understanding the economic process, at the macro as well as the micro level. In agriculture, however, they cannot be precisely ascertained or predicted; this indeterminancy results from unknowable labour productivity, especially within teams of workers, from externalities (inputs which are not paid for), and from uncertainties of output. These factors enhance the risk of all inputs and the need for insurance, affecting agricultural credit in particular. Along with asymetry in access to information and other resources, these factors

[1] S.S. and Chitra Sivakumar, 'Organisation in agriculture, society and ideology: perspectives from the Tamil country, India'. Aspects of its argument are summarised below in the second paragraph, written by Peter Robb.

make for hierarchical societies.

On this basis we will consider the evolution of agrarian society in Tondaimandalam from 1780 to 1980. The time-frame of the discussion needs to be explained. The colonial era began in 1780 when the British East India Company (known in Tamil as *kumbini*) was beginning to take stock of the organisation of agrarian society and of agriculture in the region. This was the time during which serious official debate on land revenue systems in the southern part of the British colonial territories began, and available official records of the period offer rich insights into the way pre-colonial agrarian society was organised in the region. The end-point of our discussion is 1980. By then the agrarian society of the 1780s had been subjected to the powerful influence of colonial rule and the even more far-reaching influence of the post-colonial administration of independent India. What follows will discuss first some broad features of the region and its systems of production, and then pre-colonial agrarian society in Tondaimandalam, the effect of colonial rule, and the pressures and compulsions of the post-colonial administration on agrarian society.

Ecology, irrigation and paddy cultivation in Tondaimandalam[2]

Ancient Tamil literature had two types of classification of regions. One was on the basis of *mandalam*, a broad political subdivision according to which there were five regions.[3] The other classification was of eco-cultural types known as *tinai*, consisting of *neidal* (coastal plains), *marudam* (irrigated plains), *mullai* (forests), *kurinji* (hills and mountains), and *palai* (arid and semi-arid regions). We do not have information on the classification of *mandalam* into *tinai*. We surmise, from data available for the eighteenth century, that Tondaimandalam might have consisted mainly of *marudam tinai* and partly of *neidal* and *palai tinai*, for a large part of the area consisted of tracts irrigated by rain-fed

2 This section is reproduced from Sivakumar, 'Organisation in agriculture'.

3 They were Tondaimandalam (consisting of the present-day districts of North Arcot, South Arcot and Chengalpattu in the state of Tamilnadu, and parts of Nellore and Chittoor districts of the state of Andra Pradesh), Cholamandalam (Tiruchirappalli, Pudukkottai and Thanjavur districts of Tamilnadu), Pandyamandalam (the southern districts of Tamilnadu—Madurai, Tirunelveli and Kanyakumari), Kongumandalam (Coimbatore, Salem, Dharmapuri, Periyar and Nilagiri districts of Tamilnadu and parts of Chamarajnagar and Mysore districts of Karnataka), and Cheramandalam (present-day Kerala state). Most of the discussion in this part derives from Sivakumar and Sivakumar (1979, 1993).

tanks catering to at least a single crop of paddy, while the rest was comprised of unirrigated and coastal tracts. These tanks collected the water from the rains which occurred (most predictably) in October-November and (less predictably) in June-August.

According to official statistics, there are around 32,000 rain-fed tanks in Tamilnadu of which around a fifth are in Tondaimandalam. This region consists of plains with gentle undulations, interspersed with small hills that extend down the Eastern Ghats. These gentle undulations have been utilised to create a complex network of man-made storage tanks (ranging in size from 0.2 to 9 sq.km.). Usually the water flows from a hilly or rocky catchment area into a reservoir at the lower end of which is a man-made dyke (*bund* or *karai*) varying from 0.5 km. to 5 km. in length. These dykes are interspersed with one or more 'gates' or sluices made out of large slabs of granite buried vertically along the dyke. A hole in the 'gate' is stopped with a large wooden peg rolled up in cloth or sacking material. When the fields are to be irrigated, the sluice attendant dives into the water and removes the peg. From each sluice a channel distributes water to the fields.

There are three organisational implications underlying this irrigation system. First, not all fields abutt the channels. In most cases irrigation involves a complex method of temporarily damming the channel with earth (to raise the level of water), and making a temporary channel to one's own field.across the field abutting the main channel. Second, seepage leads to considerable loss of water along the channels (present-day agricultural engineering estimates put seepage and evaporation losses at 20 per cent), thereby increasing the water-use per unit area of fields as their distance from the sluices increases. Moreover the most distant fields have a less assured water supply than the ones closer to the sluices. Finally, the heterogeneity in the quality and size of plots of land constituting the fields imposes varying degrees of economies of scale (or diseconomies as the case may be) on the farmer's irrigation activities.

Irrigation management of this type is very complex and involves calculation of water-in-storage (which varies according to rainfall and the duration of storage), length of channels, quality of soil, distance of fields from channels, and the crops. Since there is a great deal of variation in these parameters from village to village, a regionally centralised management regime is neither feasible nor conceivable. Thus irrigation created the need for organisations at the village level that would plan and manage water allocation as per local requirements.

In a system as complex as this, disputes are inevitably a possibility. Taking water from the channels is a matter of necessity for an agriculturist. From his point of view, his entitlement over water acquires primacy over those of others. Violation of norms of water-supply constitutes a constant temptation, and hence a potential source of friction and dispute. Further, taking water from the main channel to fields at some distance from it involves the implicit permission of the owners of the intervening fields. Disputes between the respective owners, pertaining to other matters, could be manifested as overt or covert denial of this permission. Organisations relating to water management are therefore necessarily involved in adjudication and arbitration in disputes regarding water entitlements and the enforcement of rules of entitlement. It is natural that these organisations should be called upon to adjudicate and arbitrate in other matters as well. The extent to which they are called upon to do so depends upon the relative importance of water management activity. By and large, the greater the water resource to be managed, either in terms of months of possible irrigation or in terms of quantum of water-use per unit of land, the greater the role of the organisations in matters outside water management. Conversely, in sparsely and infrequently irrigated villages, the water management organisations may find it difficult even to enforce water management norms.

Given the uniformly warm-to-hot climate of the region and the existence of suitable soil conditions, the availability of water makes possible the cultivation of paddy. The stringent requirements of irrigation management ensure uniform cultivation of paddy when water is available; the diversity of wet-land crops grown in the command areas of tanks enhances the already-complex problem of water management, and increases transaction costs.

The major agricultural season coincides with the cultivation of the paddy crop that lasts from July to January. This season, or *bhogam*, is known today as the *samba* season, named after a variety of paddy that came into use in the 1930s. Eighteenth-century records do not have any reference to *samba*, but the cultivation year (known as *fasli*, from *fasl*, the Persian word for crop) began in the months of June and July. In the second traditional season, known as *navarai* or *madippu* in most parts of the region, cropping was much more varied, ranging from paddy in areas where the tanks contained enough water to irrigate shorter duration varieties, such as *kar* and *kullakar*, to groundnuts or millet, sorghum or pulses. The third crop, very rare in the area, was almost always a dry-land crop and was known as *soranaiwari*.

The *samba*, being the major season, dictated the labour-use pattern in the area. Most of the labour demand occurred in three operational peaks: during ploughing and transplantation of the seedlings; during weeding, and during harvesting and threshing. Since everyone's crop-calendar followed the dictates of irrigation management during *samba*, seasonal fluctuations in labour demand also followed a uniform pattern. The peak demand in the village as a whole occurred during the Tamil months of *Masi* (15 January to 14 February); *Aadi* (15 July to 14 August), and *Aipasi* (15 October to 14 November). For the rest of the year paddy required very little labour. If demand and supply alone had determined wages, those for the peak-season would have risen beyond the reach of most farmers.[4] In addition, wage determination in agriculture in general is a problem. Not surprisingly, therefore, rewards for labour had to be decided by other means. While doing so the following considerations were important: (a) to stabilise the peak-season rise in wages; (b) to ensure subsistence during the off-season; (c) to ensure a built-in bargaining mechanism (since the first two considerations represented employers and labourers respectively); and finally (d) to provide for the spread of risk among employers and labourers.

Further, the arrangement for fixation of rewards for labour had to emphasise the symbiotic nature of the relationship between employers and labour. Since family labour alone was insufficient to cultivate paddy on more than 0.25 acres,[5] the employers could not afford to force wages so low as to discourage labour participation; at the same time, they could not afford to be too liberal because this would reduce their share of the output. From the point of view of labour, the symbiosis consisted of its dependence on employers for off-season subsistence and its superior bargaining power during peak seasons. In other words,

[4] Economic theory identifies a unique set of problems relating to the non-intersection of demand and supply functions. For instance, if the lowest price acceptable to suppliers is higher than the highest price offered by buyers, price becomes indeterminate (Menard, 1989). Peak-season wages are of this type: if labourers are aware of the acute demand for labour, they can demand wages which only a few employers can afford.

[5] See Sivakumar (1980) for data pertaining to the region under consideration.

organisation of the labour market had to reflect two contrary compulsions of organic solidarity and conflict of interests. Agriculture would cease if employers and labourers did not co-operate, while at the same time the interest of each was essentially at the expense of the other.

The arrangements eventually arrived at, would depend upon the degree of labour scarcity, the cropping pattern, irrigation and the local bargaining practices. For our present purpose it will suffice to conclude that factors that determined one aspect of organisation of agriculture, such as irrigation, also affected the others. For instance, while the cropping pattern was influenced by the availability of water, it influenced the labour market on the demand side; at the same time, the quantity and quality of labour influenced the efficiency of utilisation of water and of cultivation.

The most important of these inter-linkages was between land and water. The individuals who felt that they were entitled to water were obviously those who had prior entitlements to land. The greater their land entitlements, the stronger their claims over water entitlements. Thus, not only did water use influence land use, but land entitlements also influenced water entitlement. Finally, since a given structure of land entitlements constituted the demand side of the labour market, we may infer that land, labour and water were linked through webs of use as well as webs of entitlements. In short, they constituted the basis of the hierarchy of interests, the economic basis of the transaction regime at the ground level. Wealth, as a source of credit, was ubiquitously present as an over-arching aspect of hierarchy.

Kaniyatchi *as a system of entitlements*

Official records of the British East India Company refer to the system of land entitlements in eighteenth-century Tondaimandalam as *kaniyatchi*. Under this system there were two types of entitlements over land (*kani* in Tamil). The first of these was the superior one comprising not merely arable land, but all lands in the village including building sites, the dykes (*karai*) and irrigation tanks, roads and public gardens. The holders of these entitlements also had a number of other privileges. These individuals, known as *kaniyatchikkarar*, were said to be the descendants of the original settlers and held their entitlements in the form of a share (*pangu*) in the village. By the end of the eighteenth century not all were such descendants; since *kaniyatchi* could be bought and sold, we might surmise that a number of the eighteenth-century *pangu* had been purchased rather than inherited from original settlers.[6]

The inferior entitlement was the *user's* entitlement. Not all *kaniyatchikkarar* cultivated their land. Nor did all of them live in the village where they had a share. In such cases, someone else could acquire the tilling entitlement known as *payirkarar*. The *payirkarar* (tiller) paid a fee to the *kaniyatchikkarar* known as *swamybhogam* or *tunduwaram*. There was considerable variety in the types of person who held the inferior entitlements. There were those who held superior and inferior entitlements over different plots of land in the same village. Others held superior entitlements in one village while holding inferior entitlements in another village. The third category of people held only inferior entitlements. Among these were erstwhile labourers, who were made *payirkarar* after some years of service.

Inferior entitlement-holders were further classified as *ulkudi* (resident) and *purakudi* (non-resident). The line dividing the *ulkudi* from the *kaniyatchikkarar* was also broad. Often a *ulkudi* of long standing gained a status close to that of a *kaniyatchikkar*; his *swamybhogam* declined in quantity as his tenure lengthened, and might be merely a token fistful of paddy. In some places such *ulkudis* were described as *kanikudi*, thereby acknowledging their near-*kaniyatchikkarar* status. Thus while the formal system of land entitlements drew a clear line between superior and inferior status, the actual practice of the system permitted considerable scope for contextual interpretation of the relationship between the two.

This tendency was reinforced by the custom regarding the sharing of net output between the *melwaram* (the upper or the ruler's share) and the *kudiwaram* (the subject or resident's share). It provided that the fraction denoting the *melwaram* was lower on lands designated as being under the cultivation of a *payirkarar*. Thus practices akin to those of present-day tax-evaders were adopted by some *kaniyatchikkar* who declared themselves *payirkarar* over parts of some other *kaniyatchik-*

[6] Many villages of eighteenth-century Tondaimandalam were as ancient as the Pallava dynasty (c. seventh to tenth centuries AD) which ruled the region. In a system of land entitlements as old as this, 'original settler' had little meaning. Yet, as official records show, *kaniyatchi* entitlement could be legitimately claimed by heirs or descendants up to 100 years after the death of the original holder of these entitlements. As will be seen later, there were many social preconditions attached to these claims.

karar's land. This enabled both of the *kaniyatchikkarar* to understate some of their *melwaram* dues.

However, whatever the fluidity in the actual status of *payirkarar vis-à-vis kaniyatchikkarar*, the absolute primacy of the latter in all matters relating to the village was undisputed. In fact, one of the synonyms of the word *kaniyatchi* was lordship over the domain. It included *kaniyatchi manyam*—denoting a certain area of land in the village over which no ruler had any entitlement, even by way of *melwaram*. The extent of land under *manyam* varied. There were villages whose land consisted entirely of *manyam*. They were known as *swastyam* or *shrotrium* villages. Others had around half the village land under *manyam* and were known as *ardha manyam*. In most villages, however, *manyam* land seems to have averaged around 20 per cent of the total cultivable area.

Official data on *manyam* in late eighteenth-century Tondaimandalam distinguished between *sannad manyam* (grants by Muslim rulers of the eighteenth century), and *tarappadi manyam* (ancient grants). *Manyam* were granted to individuals, institutions and groups. *Kaniyatchi manyam* consisted of lands granted to *kaniyatchikkar* as a group, according to the *tarappadi manyakanakku* (village accounts relating to ancient grants). *Kaniyatchi manyam* was a collective grant made to the body of co-sharing (*pangali*) *kaniyatchikkarar* in the village. This grant emphasised not only the economic and political autonomy of *kaniyatchikkar vis-à-vis* the ruler, but also their collectivity. Tondaimandalam villages were characterised by the absence of village headmen and the existence of collective governing bodies called *ur* (village) or *sabha* (assembly).

The principle of collectivity was explicit in the meaning of the word *pangu* (share of a *kaniyatchikkarar* in a village). As already noted, this was a fractional share of the joint entitlements of *kaniyatchikkarar* over all lands (arable, non-arable and residential). The most emphatic form of collectivity was the widely prevalent version of the eighteenth century, the *pasungarai*, wherein all the fields were periodically redistributed under different *pangu* (shares). *Pasungarai* (also known as *samudayam* or community) ensured that every *kaniyatchikkarar* tilled the best as well as the worst land in the village. From our discussion in the previous section, it is easy to see the organisational rationale of *pasungarai*. First, it integrated the organisation of land management with that of water management, thereby maximising managerial economies of scale, or minimising transaction costs relating to land and water management. Secondly, it integrated labour management in such organisation, since the *kaniyatchikkarar* also constituted the collectivity of employers. This is a crucial point if we recall that labour inputs in paddy occurred in teams, thereby resulting in indeterminacy of wages. In addition, the possibilities arose of intra-family labour-sharing in the form of *maattu-aal* (labour exchange) and *maattu-yer* (plough-sharing) due to the fact that in most cases the *pangali* (co-sharers) were kinsmen and fellow-castemen; to this day, long after the formal demise of *kaniyatchi*, the roles of different kinsmen are clearly spelt out through custom and ritual in all foreseeable contingencies, such as marriage and other rites of passage.[7] In other situations, such as agricultural or financial misfortune in a family, the kin-group came to the aid of the affected family through temporary or long-term assistance.[8] The immense influence of the caste-fellows, kinsmen and members of the joint family over individual decision-making, pre-empted a large part of decision-related risk for individual *kaniyatchikkarar*. Thirdly, their membership of kinship and caste group enabled *kaniyatchikkarar* to gain access to the main source of dissemination of all types of information, namely the caste-kinship network. When people travelled, they almost always stayed in the homes of distant kinsmen or in *sattiram* (hospices) maintained by their caste organisation. These travels and meetings with kinsmen and caste-fellows were the source of information on markets, goods, and political and economic events across the region. In addition, marriages and other rites of passage were times when caste-fellows and

[7] For an excellent biographical illustration of these practices in the nineteenth century, see Saminada Iyer (1950). For a graphic narrative in a fictional form, see Janakiraman (1981)

[8] Intra-kin assistance is extensive, even today. To cite a few examples: (a) children might stay in a kinsman's house because of better access to education, or because of the poverty—whether transitory or long-lasting—of the parents; (b) during lean seasons, well-to-do individuals might send some paddy to their poorer kinsmen to enable the latter to last out the season; (c) as in the case of certain parts of Tamil country, a poor person in need of money might host a dinner for well-to-do kinsmen, during which these kinsmen would leave gifts of money underneath the plantain leaf from which they had had their meal; (d) well-to-do individuals might allow free use of their bullocks, *picottahs* and farm implements to needy kinsmen; (e) individuals might send their family members or labourers to help a kinsmen in agricultural operations, and (f) well-to-do individuals might 'bless' a ritual occasion in a kinsman's home, thereby implicitly promising to share the expenses.

kinsmen gathered and shared information. Finally, *kaniyatchi* as a system of entitlements formed an organisational hierarchy in the agrarian economy. The *kaniyatchikkarar* were at the top of the ladder with the agricultural labourers at the bottom. At the intervening levels were *payirkarar*, artisans and village functionaries.

The complex web of entitlements under *kaniyatchi* consisted of a combination of institutionalised aspects which were explicit, and other aspects that were amenable to contextual interpretation. In the case of water management, a system known as *vattamurai* (the custom of rotation) dictated the order of priorities in many villages. In such cases water entitlements and land entitlements were closely linked. However, *vattamurai* only indicated who would get the water after whom and before whom. The allocation of the amount of water depended upon the amount of storage in the tanks, the distance of the fields from the channels and—most importantly—the assessment of the stage of growth of crops in different fields. In other words, it depended upon the immediate context of the cropping season.[9] The problems that might have arisen from these dual norms were minimised by a certain level of homogeneity that prevailed in land use during the main season, both in terms of crops and the timing of the sowing. The annual variations in allocation of water to different fields, and of crops raised raised in different fields, were recorded in village accounts known respectively as *vaikal-kattu-karaikkattu*, a chart of channels and fields (*karais*), and *sagupadi-kanakku*, harvest accounts.

Likewise, the issue of fertilising the soil (which consisted of the application of leaves, backyard compost, cow and sheep dung, duck droppings and tank silt) followed the dual norms of institutionalised guidelines and contextual interpretations. Leaves that were to be ploughed into the soil came from trees that grew on areas held in common, as well as from trees on individual holdings. Priority of access to the former was usually in terms of need—that is, individuals with few trees in their outer backyards (*kollai*) had greater access to trees in areas held in common. Tank silt was accessible to all cultivators; yet custom dictated that each cultivator availed himself of this access once in a certain numbers of years—three, four or six. The practice of permitting free grazing in the fields after harvests ensured that the cattle wandered

[9] For an interesting case study see S.S. Sivakumar (1978).

across everyone's fields and fertilised the soil. At the same time the dung collected at home and composted in the backyards varied according to the number of cattle that belonged to the household; there was no common access to these individual compost heaps.

Duck-rearing was a nomadic occupation. Ducks fed on the snails, insects and frogs that lived in the fields and ponds. Wandering duck-keepers visited the village with their flocks, usually when the paddy was growing and the fields had standing water, and the ducks fertilised the soil. Villagers fed the duck-keepers during this period, and paid them a customary cash reward called *vathu mamool* (duck custom) when they left. Since each flock moved along a set route during the year, the eggs were sold to traders along the way who had advanced credit to the keepers. Wandering goatherds visited the village when the crops had been harvested. In addition to the length of time that these herds customarily spent in everyone's fields, they could also be enclosed in fences known as *patti* for a night or two in particular fields. In the latter case, the owner of the fields fed the herder and paid him a *mamool* (reward).

Even entitlement to land had its own variations. We have already seen that under the *pasungarai* arrangement all land was held in common, and that the use of arable land was rotated periodically. The length of time after which rotation was effected varied from four to twelve years across the villages of the region. Another arrangement was *arudikkarai*, under which arable land shares were not rotated, but fixed for ever. Between these two types lay various combinations of the two, whereby some land was rotated and some distributed once and for all.

The managerial efficiency of such an organisation working under heterogeneous conditions necessitated a high level of accountability among members, transparent functioning, and a concern for cohesiveness. These were ensured through several institutions. First, village accounts extensively recorded a range of economic subjects, from accounts of *kaniyatchi* shares to irrigation details and a list of trees and orchards in the village. These accounts were so meticulously maintained that every tree, every rupee of transactions and every alteration in the smallest channel went on record.[10] Accounts were to be made

[10] Many villages had the following accounts: *pangumalai*, an account of *kaniyatchi* shares; *tarappadi kannakku,* an account of fertility grades of fields; *warappattu kanakku*, an account of shares of different individuals in output ac-

available for members of the *ur* or *sabha* (village assembly) whenever any disputes arose. Secondly, accounts were maintained by *karnam* or *kanakkupillai* (accountants) who had hereditary entitlements to the job. Over generations, detailed information was handed down on village land as well as on the history of the families who had entitlements over it. In most villages of the region, a certain portion of land was earmarked for the sustenance of the *karnam* and his descendants so that they would be free from influence. These lands, known as *kanakkan manyam*, were even exempt from revenue obligation to the rulers.[11] Thirdly, the *ur* was an important judicial authority, with a right to impose punishment and penalties. It was also a court of appeal in cases of disputes over the treatment of servants by employers. In addition, in at least some parts of the region, the *ur* could undertake credit transactions on behalf of its members with the *ur* in other villages. A certain measure of banking and investment activity by the *ur* is thus indicated.[12] The collective solidarity of the *ur* is nowhere better demonstrated than in the sizeable volume of records relating to their concern over the admission of 'strangers' as cultivators. This was particularly the case when the question of granting *payirkarar* status came up. Even if a *kaniyatchikkarar* wished to allow some of his land to be tilled by a *payirkarar*, the *ur* had to vet the candidate and grant its approval. On the issue of sale of *kaniyatchi* shares, the *ur* had the major say on who could sell to whom. In fact, theoretically, the sale of *kaniyatchi* was prohibited even in the eighteenth century. Sale deeds were almost

---

cording to *melwaram* (government revenue), *kudiwaram* (villager's share of output after payment of revenue) and *swamybhogam* (agreements); *tarappadi manyakkanakku*, record of ancient grants of revenue-free land; *sannad manyakkanakku*, record of revenue-free land grants by eighteenth-century Muslim rulers; *merei sudantira padam*, a chart of customary payments to different individuals from the harvest; *yer aal attavanai kanakku*, a schedule of ploughs and labour employed during the season; *sagupadi kanakku*, an account of crops raised; *tukai jabida*, a list of cash flows; *tottakkal kanakku*, a record of orchards and their yields; and *vaikkal kattu karaikkattu*, a record of positions of channels and their variations. *Madras Collectorate Records*, vol.1021, pp.334-51.

[11] For an extensive list of different types of *manyam*, revenue-exempt grants prevalent in the region, see Sivakumar and Sivakumar (1993).

[12] See Iyengar (1921) and Thirumalai (1981) for details and illustration of functions of the *ur* during the colonial period and the ninth to twelfth centuries respectively. Sastri (1955a) and Sathyanathaier (1987) discuss the Pallava and Chola 'mahasabhas' extensively.

always couched in the language of *danam* (gift). The village therefore stood as the mainstay of a highly decentralised transaction regime.

Kaniyatchi *as a system of income distribution*

The village output was distributed according to a more or less uniform principle across the region. There were three types of payment which labour received: a subsistence payment, usually known as *padi*, from the employer; a customarily-determined portion of the village output at the time of harvest known as *swatantram* (customary entitlement); and periodic payments in kind from the employer.

The first category was paid at regular intervals (a week or a month), and consisted of a certain quantity of paddy. The second category also consisted of paddy, but was received immediately after the harvest. This was part of a fraction of the gross output customarily deducted and known as *swatantra dittam* (scheme of customary entitlements); it averaged around thirteen per cent of the gross output of the village. It consisted mainly of *swatantram* to the labourers, a smaller *swatantram* to the artisans, and—sometimes—a very small *swatantram* to the *kaniyatchikkarar*.

The third payment was highly varied and was context-oriented. One component, for instance, was *purakalam* or *kalavasam*, which consisted of gleanings from the threshing floor. Evidently this was a bonus given at the time of harvest. The employers allowed a certain amount of paddy, usually around eight per cent, to remain on the threshing floor (*kalattumedu*) after the harvest, so that the labourers could gather it while cleaning the threshing floor. Another component was the clothing provided by employers during the harvest festival (*bogi*). Yet another consisted of expenses for marriages, funerals and other rites of passage which were borne entirely by the employers. Put together, the three modes of payment provided three major requirements of the labour market in an agrarian economy with an indeterminate relationship between labour productivity and rewards: the requirement of subsistence for labour, the requirement of incentives to induce a commitment to work, and the requirement of insurance against risk by spreading it across the community. The *padi* met the first requirement, and was subject to bargaining and negotiations each year after the harvest festival. The second category of payments, such as *kalavasam*, provided incentives, as did other 'gifts' from employers which depended on good will. *Swatantram* represented an institutionalised guarantee of rewards based on the risk-sharing principle; since it was a fraction of the gross output, it varied according to the state of the crop.

Bargaining emphasised not only that the workers and employers were two rival interest groups, but also that the two were indispensable to each other. Official records refer to the elaborate ritual that heralded the period of bargaining every year. Not only were workers and employers required by custom to act as two antagonistic groups in this ritual, but truancy or abstention by individual employers or labourers was strongly censured. The principle emphasised here was that non-participation in this bargaining ritual was detrimental to the village as a whole. The ritual of bargaining consisted of a symbolic mass desertion of the labourers who left the village on a certain day and stationed themselves a short distance away. They then proceeded to abuse the employers. The employers had ritually to pacify seemingly abusive and recalcitrant labourers with suitable gifts, so that bargaining might begin. Workers were required to adopt an aggressive posture whether they were individually well-treated or not. Likewise, employers had to be pacificatory whether they had already been good employers or not. Symbolically this ritual emphasised the dependence of employers and labourers upon each other; it also emphasised the necessity of bargaining to the process of wage-determination.

Carpenters, smiths, potters and other service groups presented an interesting picture. Their service constituted the 'overheads' of the agricultural populace. Their rewards therefore consisted of two parts: a customary fraction of the gross output of the cultivation, and a certain portion of land earmarked for their benefit as *manyam*, free of obligations to the ruler. The *manyam* stressed their importance to the village as a whole, while the harvest share emphasised the recurring obligations of the cultivators to the artisans. Part of the latter's customary payments came under *swatantra dittam*, and part came under another deduction known as *murai* (tradition). The difference between *murai* and *swatantram* lay in the fact that the former involved payment of grain with straw, while the latter involved only grain. Both were institutionalised guarantees of entitlement.[13] *Murai* averaged around twelve per cent of

[13] *Murai* (spelt in colonial literature as *merai*) also consisted of a number of payments other than those made to the artisans. Among other things they almost always contained *yeri murai* (customary provision for maintenance of ir-

the gross output of the village in the region.

Murai and *swatantra dittam* provided for a part of the cost of cultivation. This was the institutionalised collective provision for labour of all sorts. One of the important accounts maintained by the *karnam* (accountant) was the *murai-sudantira padam*, the chart of *swatantram* and *murai* payments. The grant of *manyam* land to artisans and servicing households was another collective provision for the cost of cultivation. However, the fact that yield from land was subject to fluctuations, meant that these individuals were intended to share the risk and uncertainty along with cultivators. At the same time, the collective provision for *manyam* represented both the collective need among the villages for their services, and the superiority of their status in comparison with that of the labourers. The extent of these *manyam* varied from village to village.

This perception of rewards for agricultural inputs as a combination of institutionalised guarantees and individual variations, appears to have been appropriate to the context of the basic indeterminacy of factor rewards in agriculture. When we recall that entitlements over land, water and other goods were subject to the same dual arrangements, we see that economic organisation as a whole was oriented towards three goals: guarantees of entitlement, flexibility in negotiating individual incentives, and a concern for spreading risk across the populace.

The social fabric of economic organisation

The legend of the origin of *kaniyatchi* is as follows: around the tenth century A.D., during the reign of Surya Chola (or Aditya Chola), a

rigation tanks), which averaged around 2.5 per cent of the gross output, and *vershachalam* or *varunaapam* (customary allocation for prayers for rain). Prayers for rain were evidently an important event in the life of the populace; so important, indeed, that the Collector of Chengalpattu district provided a substantial sum of 2,000 pagodas for *varunajapam* when the rains failed in 1799. From the point of view of the economics of organisations, it is evident that the method of payment of labourers and artisans was an efficient way of handling problems of motivation and co-ordination (Simon, 1991). In addition, what Simon identifies as the advantage of the 'docility mechanism' enabled identification of labour with the employers. As we attempt to show in this paper, there was a lot more than 'docility' too in the way the system ensured principles of solidarity, in its rituals, in its normative order and in its emphasis on Simon's 'we'. More generally, it will be seen shortly that the social and ideological organisation of society enabled the possibility of commitment, or what Williamson (1979) calls the 'rules of governance'.

group of individuals from the Vellala caste forced their way from Chola *mandalam* into Tondaimandalam under the leadership of Atandai.[14] The Vellalas are said to have fought against and defeated the pastoral people called Kurumbar who ruled the area. In acknowledgement of this victory, Surya Chola is said to have granted legitimacy in Tondaimandalam to the unique social arrangement of Vellalas called *kaniyatchi*. The major element of this arrangement was the grant of *kaniyatchi manyam* or revenue-free land to *kaniyatchikkarar*, thereby symbolising a considerable degree of autonomy for them.

There is another legend which argues that the Kurumbar themselves defeated earlier Vellala settlers in order to colonise the area.[15] It appears from the two legends that, by the eighteenth century, the region had been under Vellala hegemony for over a thousand years, with an intervening interregnum of Kurumbar colonisation. This impression is strengthened by the fact that there are three major sub-castes among Tondaimandalam Vellalas: the *tuluva*, the *kondai-katti*, and the *chozhiya*. The names suggest that the former came from the Tulu-speaking

[14] Mackenzie Manuscript No.D3287, Government Oriental Manuscripts Library, Madras. According to the legend, the number of Vellalas was 48,000 (*narpatti ennayiravar*, the 48,000 people). This must obviously not be taken literally. It appears to have been the practice of groups of people to add the numerical suffix *ayiram* (thousand) when referring to themselves. Note, for instance, other Vellala groups who referred to themselves as *ayiravar* (the thousand people), such as *nanadesi ennayiravar* (the 8,000 from four *nadus*, Nagaswamy, 1973); or *ashta sahasram* (from Tamil and Sanskrit, *sahasram*, thousand, and *ashta*, eight; Saminatha Iyer, 1950). It is more probable that they represented large kin-groups consisting of a few hundred members each. Thus *narpatti ennayiravar* might have meant 48 large kin-groups. See the ensuing discussion on *nadu*.

[15] Mackenzie Manuscript No.D3285. According to Sathyanathaier (1987), the Kurumbar could be the same as the Pulindas who were a subject tribal people of Ashoka in the southern extremity (Tondaimandalam) of his empire. Sathyanathaier's speculation, while interesting, is not supported by any of the Mackenzie manuscripts available now, nor by any known epigraphic evidence. That the Kurumbar could be the ancestors of the caste of Kurubas from Kannada-speaking Karnataka, is a speculation that Sathyanathaier supports. He finds evidence from the *North Arcot District Manual* which states that in the nineteenth century there were Kurumbar in that area who spoke a mixture of Kannada and Tamil. There is no way of verifying this today, for there are hardly any Kurumbar left in the area. To the best of our knowledge, there are only a few families who call themselves Kurumbar today, and they live in the Mudumalai Reserve Forest in Western Tamilnadu adjoining the Bandipur Reserve Forest in Karnataka.

country on the west coast of the peninsula. The *tuluva* claim superiority of status over the *kondai-katti* Vellalas, as the original Tondaimandalam Vellalas. According to the Vellala *puranas*, the *kondai-katti* Vellalas had been earlier residents of Tondaimandalam than the *choziya* Vellalas who had accompanied Atandai from the Chola country. The claim is lent credibility by the fact that complex organisational structures like the *ur* and *sabha* are epigraphically recorded prior to the reign of Surya Chola.[16]

The two important inferences from these three legends are that neither Atandai nor any other ruler was given special privileges for the conquest of Tondaimandalam, and that the legends do not mention any successors to Atandai in the region. In addition, the only dispute is that regarding which of the Vellalas were founders of *kaniyatchi*. In short, *kaniyatchi* is tantamount to Vellala hegemony. The complex economic organisation of the region is attributed to a single caste.

Tondaimandalam Vellalas have been known as Mudalis. Across the Tamil country Vellalas have been known by various region-specific honorifics such as Mudali, Pillai and Gounder. The generic bond claimed by them is that they are all Vellalas. One of the commonly accepted derivations of the word Vellala is from the root *vel-anmai*, meaning 'command over agriculture'. The implication is that the Vellalas ascribe to themselves the status of the original agriculturists as well as that of a people with ancient entitlements over land in the Tamil country.[17] Following from this is the suggestion that the other peasant castes were later entrants into the field. In short, the evolution of agrarian settlements is seen to coincide with the evolution of caste society in Tamil country.

The above inference gains importance in light of the fact that by the eighteenth century, non-Vellala castes who were holders of *kaniyatchi* shares in Tondaimandalam gradually assumed regional honorifics which were hitherto used only by Vellalas. The two major illustrations

[16] Consider, for instance, the discussion of *Mahasabha* inscriptions in Sathyanathaier (1987).

[17] The term *vel* also refers to chieftains and kings in ancient Tamil literature of the Aham branch (second to sixth centuries A.D.). The same literature also refers to the popular warrior-god *Murugan* as *Muruga vel*. The conjunction of meanings of *vel* with a warrior-god, with agriculture and with kingship has important ideological implications. There is further discussion of this below. See also Chitra Sivakumar (1989) for the myth of *Muragan* also known as *Kandan*.

are that of the Ahamudaiyar, or Agambadiyar, of the Mukkulattor group of castes who assumed the honorific Mudali, and the people of the Vanniyar group who assumed the name Gounder, the Vellala caste honorific in the neighbouring Kongu country. A saying which originated in the region around this period was, 'Kallan Maravan and the sturdy Ahamudayan slowly became a Vellala'.

The assumption of Vellala honorifics was not restricted to peasant groups alone. In Tondaimandalam the weavers of the Sengundar caste began to refer to themselves as Segunda-Mudali; the pastoral Yadava caste members identified themselves Pillai, and the well-to-do among the Paravar caste (who were coastal fishermen) assumed the honorific Chetty (belonging to the trading castes of the eighteenth century, and a subdivision of Vellalas—*Velanchetty*—some centuries earlier). Evidently the process of social mobility, often referred to as sanskritisation, was at work. More importantly, the process acknowledged the widespread belief regarding the superior landed legacy and status of the Vellalas. To be a Vellala in Tamil agrarian society meant an ancient entitlement to superior social status. Caste, therefore, legitimised economic entitlement. It was not enough to have entitlements over resources; it was also necessary, in the Tamil agrarian social order, to have these entitlements legitimised through Vellala status, or at least a Vellala honorific.

Whether a society has written codes of law or not, basic social norms reflect such laws as markers of conduct in day-to-day life. As the society ages, ancient rights and claims acquire a legitimising power of their own. It is in this sense that caste, through centuries of practice, legitimised the entitlement structure of *kaniyatchi*. It is irrelevent for us to go into the question of which pre-dated which, because one was the symbiotic point of reference for the other. Whether caste and economic entitlements were specifically defined in the ancient texts of Manu and *Yagna Valkya* was of secondary importance in a social order which interpreted these contextually over a millennium.[18]

[18] It is quite possible that these *smrti* (or texts) were periodic revisions of codes of entitlement according to changing conditions. In other words, rather than being codified injunctions they were possibly periodic (contextually relevant) restatements of more ancient injunctions. For instance, we are informed that economic entitlements of individuals and groups were quite different in the *smrti* of *Yagna Valkya* from the earlier *smrti* attributed to *Manu* (we are grateful to S. Neelakantan for pointing this out to us). More generally, we find

One of the main features of caste has been endogamy. In the case of Vellalas, rules of endogamy were highly restrictive. Usually marriages took place within the *nadu*, a locality usually consisting of 12 to 25 villages. Even within the *nadu*, the mother's brother (*tai-maman*) and the father's sister's son (*attan*) had the claim of *muraippayyan* (customary claimant) over a girl. Thus, if caste was the legitimising source of *kaniyatchi*, kinship was the source of the reproduction of caste. Of the two lines of kinsmen, namely agnatic and affinal, the former took primacy. The fact that male agnatic kinsmen were referred to as *pangali* substantiates the argument that kinship was the social foundation of *kaniyatchi*. Reference to a group of kinsmen from 'such-and-such' *gumbal* (group or crowd) who came from 'such-and-such' *vazhi* (line of descent), constituted the social definition of individuals.[19] One's place in a line of descent was clearly demarcated as being *pangali onnam-kal* (once removed), *redam-kal* (twice removed), and so on.

Further, the distinction between agnatic and affinal kinsfolk tended to get blurred over generations, due to the custom of marrying a girl to her mother's brother or father's sister's son, within the *nadu*. Numerous disputes arise even today over what should be done on ritual occasions concerning kinsmen who are both affinally and agnatically related. The ritual obligations of agnatic kinsmen (*pangali*) are very different from those of affinal kinsmen (*tayadi*). The bond of kinship thus not only reproduced caste as the legitimising principle of a given structure of entitlements, but also helped to sustain and preserve it over time. More generally, an individual's economic prospects and fortunes were almost entirely contingent upon the social space into which he was born.

Ascriptive aspects of the development of an individual's knowledge, skills, aptitudes and personality were of crucial significance. Further, the fact that almost all learning (be it of agriculture, or medicine, or metal-working) involved the endosomatic approach of apprenticeship, tended to create a self-perpetuating link between ascriptive and achieved aspects of learning and socialisation. Thus social background

---

(Sathyanathaier, 1987) that the '*ur*' or '*sabha*' or '*mahasabha*' did not figure in Kautilya's *Arthashastra*, often said to be the most important text on Hindu public administration.

[19] Even today, the stigma attached to an individual without a known *vazhi* or *gumbal* or caste is powerful. The worst insult or term of abuse is, therefore, to be referred to as *tevadiyappaiyan* (son of a whore), or *appan peyarteriyadavan* (one who does not know his father).

not merely moulded a person's normative structure (*jati dharma*, *kula dharma*, and so on), but also his or her range of skills and capabilities. Caste, kinship and lineage thus became crucial quality-markers in the labour market as well as indices of human types.

The subsumption of the individual into the social history of his group was the strongest element of insurance in the fluctuating and unpredictable economic universe of agriculture. The popular distinction made between *panjattu aandi* (pauper during a famine) and *parambarai aandi* (hereditary pauper) bears this out. For the individual his or her family (*kutumbam*), lineage (*kulam*), kin group and caste provided a series of concentric circles of insurance against contingency. Depending upon age, sex and economic status, the individual's roles were clearly defined in these complex support systems—a set of entitlements from these groups and a matching set of obligations to these groups, both of which changed according to their place in the support system.

The social support system also provided for social security in the modern sense. People who were too young or too old to work, people who were disabled or otherwise incapacitated, and people with differing skills and capabilities, were assured of sustenance not merely in an economic sense, but also in a social sense. Such sustenance was varied in its form. For instance, while some villages of Tondaimandalam earmarked a portion of land as *dharmaraja manyam* for the destitute and disabled, yet others in Kongumandalam earmarked entire villages as *mudavandipalayam* (village of the disabled orphans) which catered to villages of the *nadu* region. Entitlement to social security was provided by different levels of the support system. Elderly individuals were the responsibility of the younger generations within the joint family. At the same time, elderly individuals had clearly defined roles within the family. The case of children was similar. If the parents of children were to die or become incapacitated, the *pangali* brought them up. Among certain castes, such as the Pallis, destitute individuals were allowed to seek alms as *pagal veshakkaran* (alms-seekers who came in disguise during the day).[20]

Thus far we have considered the relationship between the entitlement

[20] This was in contradistinction to *raappichchai* (the nocturnal beggar) who needed no disguise in the dark. A number of people who did not wish to be identified for a variety of reasons, resorted to nocturnal begging (for example, lepers who were destitute: Sivakumar, 1991).

structure and caste society. The hierarchy of entitlements tended to coincide with the social hierarchy. However, while hierarchy is a definition of social distances among individuals and groups, the organic solidarity (in a Durkheimian sense) of the caste society tended to counter the distancing influence of hierarchy. Since caste was an institution in a predominantly agrarian society (prior to 1900), the elements of organic solidarity in agriculture were transferred to the caste society. In particular, the indispensability of labour in food-grain cultivation created a strong bond between the holders of land entitlements and the holders of service entitlements (labourers, artisans, potters, barbers). In short, the social division of labour which the castes in the village represented made for the organic solidarity of caste society. The dual character of caste society—as one of hierarchical distances and organic proximities—was reflected in a complex web of reciprocal relations, ritual and non-ritual, that were the feature of master-servant relationships.[21]

To sum up: (1) caste society as manifested in the lineage-kinship-caste continuum of social space, constituted the symbiotic social basis of *kaniyatchi* as a structure of entitlements; (2) it also provided an extensive insurance umbrella for individuals and groups who were faced with an unpredictable state of nature as an existential reality; (3) it constituted the major source of identification of human skills, aptitudes and personality—the human type—; and (4) it provided the basis for a complex network of reciprocal inter-personal relations based on hierarchy and organic solidarity.

Kaniyatchi *as an ideological system*

During the eighteenth century an interesting copperplate inscription was unearthed by the officials of the British East India Company in Tondaimandalam. This, a lengthy legend entitled *Cholan Poorva Pattaiyam*, described the manner in which a Chola king created *kaniyatchi* settlements in Kongumandalam. The inscriptions discussed the activities of a king from Cholamandalam in Kongumandalam, and took their socio-economic paradigm from Tondaimandalam (namely *kaniyatchi*). For the present discussion the very beginning of the story is relevant:

[21] This point, of course, is the subject of that substantial body of literature on what is known as the 'jajmani system'. The desertion ritual that we discussed earlier is the best illustration of such relationships. For an extensive discussion see Sivakumar and Sivakumar (1993).

the king of the Cholas, along with the king of the Pandyas and the chief (*talaivar*) of Tondaimandalam decided to spend some time in Kanchipuram in Tondaimandalam in order to contemplate his course of action. He wanted to create human settlements. At the time of the event, the place was described as being a jungle. As the two kings and the chieftain made a small clearing in the jungle for their stay, there appeared before them the female goddess Udiramari (or Karumari) in all her terrible and fearsome aspect. She said to them, 'This place, Kanchi, is my eternal (*jenma*) *kaniyatchi*. How dare you sully this place! I shall destroy you forthwith!' At this, the three human beings fell at her feet and prayed for mercy. The goddess granted mercy on condition that they build her a shrine and offer sacrifices annually. The rest of the legend consists of the repetition of this theme as the three moved from place to place, creating human settlements in Kongumandalam. In every village, a shrine for the goddess was created and sacrifices were offered.

The repetitiveness of the theme, and the powerful description of the deity and the fear of the kings and the chieftain, was clearly designed to convey the following messages. (1) *Kaniyatchi* was the material aspect of the cosmic relationship between the earth and the state of nature as reprsented by the goddess. (2) The acquisition of *kaniyatchi* by human beings was subject to the will and pleasure of the state of nature. (3) This was to be eternally acknowledged through the enshrinement of the nature goddess in every village where *kaniyatchi* was granted by the king to Vellalas. (4) The entitlement to *kaniyatchi* was, therefore, of divine origin and had divine sanction; it was an inviolable privilege granted by the nature goddess. (5) Even the kings were mere agents of divinity and therefore had no authority to appropriate an individual's *kaniyatchi* entitlement, as long as the individual behaved accountably towards the goddess with a regular offer of sacrifice. (6) The king and *kaniyatchikkarar* alike were accountable to the divine forces (as represented by the nature goddess) for their conduct; for these entitlements were her boon to them.

To revert to the legend of *Cholan Poorva Pattaiyam*, why did the Chola king want to create *kaniyatchi*? As the legend tells it, he wanted to atone for his unaccountable exercise of power. Both this king and one of his predecessors had been tyrants, and ignored the well-being of the populace; in particular, *manyam* (revenue-free) lands had been taken over and temples had been destroyed. As a consequence, he was punished by nature with a rain of dust (*man maari*) which destroyed the crops and the economy. The establishment of *kaniyatchi* settlements was therefore an act of *parihaaram* (atonement). Not merely was *kaniyatchi* a divine gift, but its creation brought kings redemption.

Thus *kaniyatchi* and its social basis in the form of caste society, were strengthened by divine sanction; *kaniyatchi* was a highly accountable status held at the will of the nature goddess. Not only was it based on ancient traditions and customs, but these customs had a divine origin. The enshrinement of the goddess in every village was symbolic of the suggestion that people who held power at every level were under observation by the goddess who constituted the original principle of nature. This meant that the normative order was central to the social order. If appeal to tradition failed to restrain an individual in the exercise of power, threat of divine retribution certainly would.[22]

From what we have seen of the importance of organic solidarity in agriculture, the collective need for such a strong normative order is self-evident. The divine origin of *kaniyatchi* was almost always symbolised by a female goddess who represented the fearsome qualities of nature—failure of crops, disease, droughts and floods—that demanded submission by men.[23] The constant element in the various names of the goddess (*maari*) also denoted the goddess of smallpox, one of the most feared of diseases. *Maari* also meant rainfall, which was her blessing.

[22] An extensive study of the relationship between religious institutions and the agrarian society is available in Ludden's (1985) brilliant study of the Tomiraparani river basin in southern Tamil country for the period 800-1800. The pioneering studies of Stein (1980) showed the social and political role of Brahman settlement grants in the Chola country, known as *Brahmadeyas*. See also Sastri (1963) for extensive discussions on *Brahmadeyas*.

[23] Bishop Whitehead (1921) produced a fascinating study of village goddesses in South India. This ethnographically significant work, provides interesting insights into the anthropomorphism underlying the relationship between the goddess and her subjects. For instance, in many places, after a buffalo had been sacrificed, its severed right leg was clamped crosswise between the teeth, before the severed head was taken in procession round the village. As the 'natives' explained to Whitehead, this custom was analagous to the behaviour of a penitent farm servant who went to see his master with his right index finger held crosswise between his teeth as a symbol of repentance. Further, the universal practice of drinking the blood of the sacrificed animal (buffalos or goats) by the priest symbolised the incorporation of the spirit of the goddess into his self or into the collective being of the village, as represented by the priest.

Thus the religious ideology underlying *kaniyatchi* metaphorically illustrated the economic significance of the state of nature and of uncertainty in agriculture.

While most villages in eighteenth-century Tondaimandalam had a shrine for the village goddess, some in the northern parts had shrines for male equivalents like Ayyanar or Muniswaran. The major temples in most of the large villages, however, enshrined pan-Hindu gods such as Siva, Vishnuand Skanda (Kandan, Murugan). Temples were the centre of the social space of the village, and constituted the divine basis for the origin of the particular village; each of these temples (and hence the village) had a *stala puranam* (divine legend of the origin of the settlement). The complex and ever-elastic structure of Hindu mythology permitted almost limitless possibilities for the integration of thousands of gods and goddesses into a common fabric of *puranas*, and hence integration of social space through the puranic space.[24]

In terms of entitlement, the best land and the first water entitlement in the scheme of rotation of irrigation (*vatta murai*) belonged to the temples. Thus the primacy of divine forces was acknowledged at the level of the key inputs in agriculture. Management of temples and temple properties was the responsibility of *kaniyatchikkarars* who were hereditary *dharmakartas* (trustees). Aside from the smaller temples, large ones in the region, such as in Kanchipuram and Tiruvalloor, received a fraction of the output of most villages as a part of the *murai* payments from these villages.[25]

Not only *kaniyatchi* was traced to divine origin; so also were the

[24] In Whitehead (1921) is a mini-*puranam* of a local goddess, Peddamma—literally, great mother—in southern Andra Pradesh (the Telugu-speaking region to the north of Tondaimandalam) which incorporates the triumvirate of male gods Siva, Vishnu and Brahma, as manifestations of goddesses. The same message is inherent in major *puranas* of the goddess, like *Devi Mahatmiyam* or *Devi Bhagavatam.* In recent times, the millions of followers of the goddess *Adi Parasakti* of Melmaruvattur in Tondaimandalam also hold a similar belief. The widespread worship of the goddess *Sakthi* (which also means energy, capability, potentiality and skill) is said to be pre-vedic in its origin (or over 3,500 years old), and represents the ancient religious tradition of *Saaktam.*

[25] As already noted, each village also put aside a fraction of its output for *varuna japam* or *varshachalam* (prayers for rain). The importance of the temple as the foundation of the social space of *kaniyatchi* can be best understood from the fact that both the Muslim rulers or Arcot and the English (Christian) district collectors donated large sums of money to the annual festivals at Kanchipuram during the last decade of the eighteenth century.

individual castes. Caste myths that are available to us from the period 1880 to 1940 show an interesting variation. The *puranas* (myths) of the Vellala and the Karunikar—the major groups in *kaniyatchi*, one as land controller and the other as the accountant—refer to the Rg-vedic myth of Prajapati. Others, such as the Vanniyar, Mukkulattor and Nadar, refer to Indira and other gods. The myth of the Prajapati traces different *varna* (Brahman, Kshatriya, Vaisya and Sudra) to the different limbs of Prajapati, the originator of the universe. In the case of Mukkulattor (the Kallar, Maravar and Ahamudaiyar) their origin is traced to Indira, the king of gods. The Vanniyar trace their descent to a great *munivar* (sage) and *devakanyas* (divine virgins).[26]

One of the distinctive features of Tamil social history is that there had been no major claimants to the princely Kshatriya varna before the last 150 years. The Pallava, Chola and Pandya kings did have Kshatriya-like honorifics (*Verman*, *Devan*) and often traced their line of descent to the sun or the moon, as did the Rajputs of north India. Nonetheess, no single social group claimed to be Kshatriyas until the nineteenth century. Even the Vellalas, writing towards the end of the nineteenth century, claimed Vaisya (agricultural and trading) rather than Kshatriya status.[27]

This brings into discussion the varna as the paradigm for the taxonomy of castes. What little evidence we have suggests the following: first, that a high land:man ratio existed up to the end of the eighteenth century, so that it was possible for groups of people to open out fresh areas of forest land for cultivation. In fact, before the mid-nineteenth century, the history of Tamil agriculture could be viewed as a process of gradual colonisation of land. Secondly, tacit permission or support from the ruler or chieftain of the area was enough for a small group of families to move out in such a colonisation process. In addition, the tacit or covert approval of the leading *kaniyatchikkarar* of the locality was also necessary. This could be obtained if the colonising group was small enough not to pose an implicit threat to the latter's status. Thirdly, and even more slowly, over a hundred years or so, the new entitle-

[26] A large number of caste *puranas* were published during 1890-1940, few of which are available today. See, for instance, Balasubramanian (1927), Munusamy Mudaliar (1901), Servai (1940), Desigar (1930), Varadaraja Pillai (1909), Gnayiru Pareesanadar (1897), Veerapillai (1905), and Vijaya Dorasawmy Gramony (1910).

[27] See Balasubramanian (1927) and Munusamy Mudaliar (1901).

ment holders could legitimise their status within the caste taxonomy of the region by making a claim of descent from a divine source or an eminent personality of the past, and by adopting the honorific of the dominant caste. Finally, the claim was invariably granted, and a suitable adjustment made in the local caste taxonomy. However, the extent to which this process of socio-economic mobility occurred in a region depended upon a large number of factors, such as irrigation, availability of arable land, and the political complex of the rulers who supported the colonising groups.

In terms of ideology, the varna scheme constituted the basic paradigm, and new caste groups were defined as combinations of varnas (*varna samgraha*). The directions in which these combinations were considered were in terms of *anulomam*—when the male founder of the group was born from an inter-varna marriage between the male of a higher varna and a female of a lower varna, or *pratilomam*, when the female of the inter-varna marriage came from a higher varna. Thus the ground-level manifestation of the caste (namely *jati*) was linked *puranically* through myth and legend to the original paradigm of varna.[28] Caste society, at least until the nineteenth century, must have been extremely fluid. As long as there remained land to be colonised, there remained the possibility of new castes.

The expression of group identity, *engal inam* (our group or our people), could mean several levels of social apperception depending upon the context. For purposes of marriage, it came down to the levels of kinship and *nadu*. At the other extreme, for purposes of political lobbying, it could cut across a wide region involving all those who bore an honorific (for example, Mudali). At the regional level, castes which were numerically small, like potters, barbers and smiths, aligned themselves with the dominant castes (in the sense of Srinivas) to form conglomerations called *edangai* (left-hand conglomerations) and *valangai* (right-hand conglomerations).[29] However, at the regional level there

[28] See, for instance, the case of the inscription in Nagaswamy (1973) which describes how stone-inscribers were granted the *varna* status of the descendants of Viswakarma (the divine architect). Today, almost all members of the carpenter and smith caste groups refer to themselves as descendants of Viswakarma. Some even refer to themselves as Viswakarma Brahmans because of their ritual entitlement to wear the sacred thread.

[29] The categories of *edangai* and *valangai* appear to have existed in most parts of south India. Available evidence suggests that these categories have ex-

were other categories or groups (to be considered shortly) which were also important.

Finally, let us look into the role of religious ideology in the day-to-day life of the individual. Most villages boasted a *panchangan* (almanac reader) who suggested the appropriate times for sowing the crops, ploughing, marketing the crops, setting out on a venture of consequence, arranging marriages, initiating important discussions, and arriving at decisions. The appropriateness of time for various activities was decided by the planetary confluence that characterised the time of one's birth (*piranda neram*) and similar confluences at the time of the proposed activity. Aside from the almanac, innumerable beliefs about omens guided people regarding the appropriateness of the time for various activities.

The ever-present awareness that expectations and achievements did not always match, that events in life were rarely amenable to straightforward explanation, and that particular events were always a part of an incomprehensible web of events in the past and present, encouraged the common acceptance of time (*kalam*) as a basic metaphysical category. A life-span was a part of this unchanging point of reference, as was the timing of activities (*neram*).[30]

Given the universality of the idea of re-birth, the vast number of life cycles (*jenmams*) was a direct function of moral conduct. Actions could lead to the accumulation of *pavam* (retribution) or *punniyam* (merit). The relative weight of the two types of action in a life cycle decided the course of events in the succeeding life cycles. Thus the virtuous person would gradually minimise the burden of *pavam* and enhance the amount of *punniyam*. A person who had exhausted the *pavam* from all past life cycles fulfilled the conditions necessary for liberation (*mukti*) from the cycle of births and deaths. Such liberated souls would then

isted at least since the reign of the middle Cholas in Tamilnadu (about the tenth century) (Nagaswamy 1973). There is a commonly cited hypothesis that they represented the urban and non-agrarian castes (*edangai*) and rural-agrarian castes (*valangai*). From what information we have, this hypothesis is not borne out. See, for instance, Nagaswamy (1973), Hayavadana Rao (undated), and Mackenzie Manuscripts, D2751, D3284 and D3196.

[30] Hindu belief looks at cosmic time as comprising a *kalpa* consisting of four *yugas* and each *yuga* consisting of several thousand years. The present period is said to be a part of *kaliyuga*, the *yuga* of disorder and corruption (*adharma*). Further, one day in the life of the creator Brahma is said to comprise millions of human days. See Bhagavat Gita, ch.IX, 17. See also Sivakumar (1991).

reach the abode of gods and goddesses who were *anadi* (without origin) and *anantham* (without end), and hence enter the domain of timeless eternity (*sorgam*). The god of death (*yaman*), who was *dharmarajan* (the god who judged the merit or *punyam* acquired or lost during the life cycle), was also *kalan* (the god of time). Thus the moral order of the society was also the determinant of the place of individual souls in cosmic time, in the cycle of births and deaths, and in the eventual liberation of the soul.

Traditionally, formal moral education commenced with the learning of the Tamil alphabet. The twelve Tamil vowels were taught through twelve moral precepts, in the *aththichoodi* which was the common text for learning letters. Each vowel was denoted by the name of a moral norm. More generally, moral norms were a part of the socialisation process within the family, the kin group, the caste and the village. The necessary consequence of this process was that morality of conduct was perceived at several levels relating to the context of conduct. A particular action might merit *punniyam* at one level, but not at another. So *dharmam* (or morally appropriate action) could be so in terms of *kula dharmam* (appropriate to the lineage), or *grama dharmam* (appropriate to the village)—which reference points might not suggest the same type of action.[31] The resolution of this paradox lay in the individual's motivation, whether he subjectively favoured one level of *dharmam* over the others. The final test of moral appropriateness of conduct lay in the degree of detachment of the actor's ego from the action. This detachment enabled the person to evaluate, objectively, the effects of his action on the world.[32] Withdrawal of the ego was also a withdrawal from the reciprocal element of action: hence the word *dharmam* was

[31] The central and moral philosophical preoccupation of the *Bhagavat Gita* is with *dharma sankatam* (the anguish underlying the choice between 'right' courses of action that conflict with each other), and its resolution. Its narrative consists of the advice of the god Sri Krishna to the warrior Arjuna in the battlefield, wherein the latter is reluctant to offer battle against his own kinsmen who are arrayed in the military formations of the enemy. Most commentators on the *Bhagavat Gita* agreed that the battlefield *kurukshetram* is symbolic of the individual's inner (moral) universe in which Sri Krishna and Arjuna represent forces of eternal knowledge and delusion respectively.

[32] This is again brought out in the preoccupation of the *Bhagavat Gita* with *Stita Pragna* (the one with a steadfast consciousness) in the crucial chs.II and III. In particular, the following injunction in ch.II to Arjuna: 'Holding happiness and sorrow alike, / As also gain-loss and victory-defeat, / Fight thou thine battles' (translation by authors).

also synonymous with the non-reciprocal act of giving. Since the degree of detachment varied from individual to individual, the solution to *dharma sankatam* (*dharmic* anguish arising from the need to choose between two courses of action, both of which appeared to be 'right') lay in general approval of any act of non-reciprocal giving (*dharmam*, *aram*).[33] In fact, the teaching was that the first letter in Tamil, 'a', denoted *aram* in the *athichoodi*: *aram seya virumbu* (learn to like to give).[34] *Dharmam*, as an act of giving, was a central norm in *kaniyatchi* society. However, it meant much more than charity given at the will and pleasure of the giver. No doubt the *dharmavan* (the giver of *dharmam*) enhanced his *punniyam* and performed atonement (*prayachchittam*) for his past *pavam* through *dharmam*; but there were several ways in which the society too could persuade him.

First, as a part of the reciprocal relationships underlying *kaniyatchi*, the social groups surrounding a wealthy person felt that they were enti-

[33] Note another injunction to Arjuna in ch.III of the *Bhagavat Gita*: 'Know that all actions are bound by Karma / Saving those that have the quality of *Yagna* (offering)...' (translation by authors).

[34] The most exalted life style for a member of the *brahmana varna* was one of *ooncha vrdhdhi*, wherein he was to distance himself from material give-and-take. He was to live from one day to another on the basis of alms received by him while he walked the streets singing songs of worship. His life was to be spent in pursuit of *yogakshema* and *lokakshema* (activities that enhanced the spiritual and material well-being of the community). More generally, all *grhastas* (householders) were to reach liberation through a succession of *ashramas* (stages in life) which had (ideally) to culminate in *sanyasa ashrama* (the stage of mendicancy and renunciation). The contribution of Buddhism to such a value system must have been significant in Tondaimandalam, which was ruled by Buddhist chieftains up to the seventh century, and afterwards by the Pallavas (to the tenth century) who, though not Buddhist, encouraged the amalgamation of Sanskritic, Buddhist and Tamil cultural traditions (Sathyanathaier, 1987). Particularly noteworthy are the observations of the Chinese visitor Yuan Chwang, regarding the students of the Buddhist University at Kanchi: 'Though their family be in affluent circumstances, such men make up their minds to be like vagrants, and get their food by begging as they go about. With them there is honour in knowing truth (in having wisdom) and there is no disgrace in being destitute. As to those who lead dissipated idle lives, luxurious in food and extravagant in dress, as such men have no moral excellences and are without accomplishments; shame and disgrace come on them and their ill repute is spread abroad' (Sathyanathaier, 1987, p.47). The influence of Pallava Thondaimandalam as the cultural precursor to the Chola period (tenth to thirteenth centuries) is also evident from the fact that of the 14 Chola epigraphic records on the subject of education, eleven were located in Tondaimandalam (Sathyanathaier, 1987, p.68).

tled to receive help (*udavi*) from him. The *pangalis*, the *tayadis*, the caste-fellows, fellow-villagers and servants had varied entitlements in the form of acts of *udavi* from a wealthy person. For instance, the custom of bestowing gifts (*muyi*) on the bride or bridegroom in a marriage involved public announcement of the name of the giver, his relation with the bride (or groom), and the size of the gift. This custom was an explicit articulation of the principle that the size of the gift should match the status of the giver. Aside from *muyi* there was a large number of occasions where a kinsman could *ask* for the help of the wealthy person, because he felt entitled to it.[35] Likewise, an assortment of entitlements to *dharmam* was vested in the different social groups to which the wealthy person belonged. To the village as a whole, the wealthy person had the obligation to contribute to public expenditure (for example, desilting channels, repairing roads or culverts) known as *ur podhu*. An important aspect of *ur podhu* was the periodic festivals in the temples. A wealthy person was expected to make a suitable contribution on these occasions, and for these he was suitably honoured in public (*mariadai*). The refusal of a person of wealth to submit to the reciprocal entitlements of others meant his immediate alienation. Extreme non-compliance lead to the ostracising of the individual. The importance of this threat need not be overemphasised in the context of the society where the individual was socially defined. Such extremes rarely occurred.

Thus, secondly, a person might fulfil his *obligations* of *dharmam* out of fear of losing his status in the society. The fear of a fall from public esteem was a motivation strong enough to ensure compliance. Public esteem and the concomitant sense of honour (*manam*) and the loss of it (*avamanam*) constituted more important elements of status even than wealth. For the *avamanam* of an individual also meant the loss of face (*avamariadai*) of the social groups to which he belonged. Conversely, the social group enhanced its prestige and status when its members enhanced their own public esteem.

Thirdly, a person could attain a certain form of immortality through

35 Consider, for instance, the practice of *kumbidu kooli* or *kumbidu swatantram* which persists in parts of Thanjavur district even today. A poor person (usually a farm servant) would approach a well-to-do person and offer *kumbidu* (greetings) so as to be given some rice or paddy as a reward (*kooli*) for the greeting.

acts of philanthropy. He would win the goodwill of fellow-men who would remember him with respect. The desire for immortality was most natural in a society whose religious ideology constantly stressed the transitory nature of the human being and his world: *Kayame idu poyyada / Verum katradaitta paiyada* (This body is an illusion / A mere bag full of wind).

Acts of *dharmam* leading to immortality (through public goodwill) often resulted in the construction of temples and choultries (hospices), endowment of funds for public worship, and the patronage of arts and literature. The opposite of public goodwill, namely public ill-will (*vayithu erichchal*), was a crucial sanction in the face-to-face society that *kaniyatchi* entailed. To be identified as a *kanjan* (a tight-fisted individual) was tantamount to being subject to public contempt and dislike. The patronage of poets was expected on the part of rich people. The poet would compose songs in praise of the patron's charitable qualities, and in return would receive gifts. When a poet was turned away he could sing a song of dispraise (*aram pattu*) which was meant to strike shame and fear into the heart. Such songs were institutionalised ways in which loss of public goodwill was communicated to individuals.

From *kaniyatchikkarar* to the king, this emphasis on *dharmam* constituted a compulsive force which encouraged patronage. The desire for the salvation of the soul, when compounded with a strong awareness of entitlements to patronage in the society, led to a dual definition of wealth and power: they were instruments of personal enhancement and they were also instruments of benevolence (*oppuravu*). The recorded history of the Tamil country is full of references to the benevolence of men of substance (from kings to wealthy villagers) who built temples and ponds, and supported the arts. What history has not recorded is the vastly greater amount of wealth and effort which such men must have expended towards the welfare of individuals from their kin groups, castes and localities.

There are three important implications of *dharmam* as a central norm in the ideological universe: (1) moral obligations and entitlements associated with *dharmam*, constituted a powerful basis for redistribution of wealth; (2) in the process, the moral universe constituted a strong basis for the cohesiveness and solidarity of social groups; (3) the compulsive force of the norms of *dharmam*, acted as a strong counter to concentration of wealth and power; the importance of this political implication cannot be understated in the hierarchical polity that *kaniyatchi* was; *dharmam* was the basis of stability and solidarity—the ideological content of social orderliness.

The last of the above implications of the moral order had an important role to play in the definition of the political society. Let us, therefore, consider the political society under *kaniyatchi*.

Kaniyatchi *as a transaction regime*

In the first part of this essay we discussed the relationship between the economic process and the politico-legal structure. We saw that the latter was the necessary background to the legitimacy of the entitlements underlying economic processes. We also saw that the holders of economic entitlements also have a motive for influencing the politico-legal structure to their advantage. This dynamic two-way relationship defines the forms which economic and political organisations take. The specific morphology of political and economic organisation of a society that obtains at a particular time is what we have termed a transaction regime. It would also be pertinent to recall that a transaction regime will not consist of written codes or norms alone, but also of an equal number of conventions, practices and traditions which spell out the meaning of entitlements in different contexts, thereby giving rise to 'precedents' which are crucial in the process of adjudication and arbitration.

Let us now turn to *kaniyatchi* as a transaction regime. We have considered at length the structure of economic entitlements, its social dimension and their ideological foundations. We have seen that economic, social and ideological organisations were dovetailed to the imperatives of agriculture in general, and the economics of paddy in particular. While, therefore, a certain degree of uniformity is to be seen in the socio-economic organisations, necessitated, for instance, by the uniformity in cropping, a considerable flexibility is also perceived in these organisations, which may be necessitated by differences in local conditions. This suggests a decentralised political system as the logic of *kaniyatchi*. In order to illustrate this, let us revert to the legend of Atandai.

It may be recalled that when the Vellalas colonised the region under Atandai they were granted *kaniyatchi manyam* and the degree of autonomy that went with it. Equally important, there is no reference to the grant of rights of succession to any of the successors of Atandai. Even Atandai's status is not clear in the legend; was he the ruler of Tondaimandalam, or was he an agent of the Chola king? Another legend, the caste *purana* of the Vellalas, has extensive references to the happenings in Tondaimandalam, but does not refer to any rulers. This suggests that political society in Tondaimandalam was very decentralised.

For the Tamil country as a whole, the historical works of Stein and others have effectively argued the case for a decentralised political order. Stein refers to the Tamil state as a 'segmentary state'; the major locus of political authority was at the level of a locality known as *nadu*.[36] The *nadu*, which usually consisted of a cluster of villages, was represented by leaders known as *nattar*. These *nattar* were powerful and commanded the following of the people of the *nadu*. Even kings as powerful as Raja Raja I had only a small territory surrounding their capital under their direct command. The kings ruled through the *nattar*. However, while there is much epigraphic data about *nadu* and *nattar* for the Tamil country as a whole, little can be inferred from them regarding the source of the power of *nattar*. We know merely that they were powerful local leaders. There is even a thesis that when the imperial Cholas tried to reorganise *nadus* into conglomerates of *nadus* called *valanadu*, the *nattar* countered with their own conglomerates called *periyanadu*.[37]

As regards Tondaimandalam, unique historical circumstances pre-empted the possibility of dialectic between *nattar* and kings; as we have seen earlier, there were not many kings for several centuries for the region as a whole. This, and the grant of autonomy (as per the legend of Atandai), suggest a more powerful role for the *nattar*. We are also fortunate in finding a certain amount of important information regarding the *nattar* of the eighteenth century in the official records of the British East India Company. From this, and from our own field data on contemporary *nadu* organisations, we are able to make two important inferences.

Firstly, eighteenth-century Tondaimandalam *nattar* were leaders of their castes in the *nadu*. East India Company officials not only noted this, but tried to put the *nattar* to use as administrators in the service of

[36] Stein (1980).
[37] Hall (1980).

the Company. This was an acknowledgement of the power of caste in the society, and of *nattar* as the most influential men within the caste. Secondly, our field data on contemporary *nadu* organisation among Vellalas and Pallis, point to the *nadu* as the spatial limit to affinal kinship in the society until the end of the nineteenth century.[38] As we have noted earlier, the society favoured a girl's mother's brother and father's sister's son as the preferred husbands. If this were not possible for any reason, other grooms might be considered. At any rate, a husband outside the *nadu* was not considered even if he came from the same caste. In short, the *nadu* constituted the spatial domain of endogamy within each caste, be it Vellalas or Pallis.

The *nadu* in eighteenth-century Tondaimandalam can then be seen as a very special aspect of the *kaniyatchi* society. While the village constituted the domain of agnatic kinsfolk (*pangali*), the *nadu* constituted the outer limit to affinal kinship. Thus the *nadu* was the domain of kinship within the caste. This being so, the kin group constituted a powerful interest group controlling not merely entitlements to land, water and wealth, but also manpower. Needless to say, all the pre-British rulers (including the Muslim Nawabs of Arcot) found the *nadu* to be the most important economic and political category with which they had to come to terms. The *nadus* controlled the source of revenue as well as the supply of manpower to the kings. They also constituted the primordial basis of group identity in a region. While the term *engal inam* (our people) had its outer limit, as pointed out earlier, so the kinship network found its boundary at the level of the *nadu*.

The *nattar* were, thus, beyond doubt, the most influential people in pre-British Tamil society, whose word controlled the economics and politics of kingdoms.[39] The ultimate weapon in their hands against an autocratic ruler was the threat of desertion. Carefully-engineered desertion at the time of harvest or sowing could paralyse the kingdom economically, and render the king devoid of social meaning (a king with-

[38] Today *nadu* is not longer a unit of political space, yet it is a very important aspect of the social space. All our informants told us that the conception of *nadu* as a spatial limit to affinal kinship is the ancient basis of *jati* (caste).

[39] Sivakumar and Sivakumar (1993) provide a detailed description of a major dispute involving a district collector and the *nattar* of Chengalpattu district during 1795-96. During 1795-96 at least, the *nattar* were strong enough to carry their influence into the very citadel of company administration, namely the Board of Revenue.

out subjects). Desertion was, indeed, relatively easy in a region ruled for most of the time by petty chieftains and rulers; all one had to do was to leave the territory en masse and temporarily cross over into the territory of an adjoining chieftain. Thus the ancient saying: *'Kodungol mannan alum nattinum / Kadumpuli vazhum kade nanru'* (Better the jungle inhabited by the cruel tiger / Than the kingdom ruled by an unjust king).

The *nadu* was also the most important level of the judicial apparatus of the society. All inter-village issues relating to disputed entitlements, from water and land to ritual prerogatives, were arbitrated and adjudicated by the *nattar*.[40]

According to tradition, Tondaimandalam consisted of 1,900 *nattams* (settlements) in 79 *nadus*, which were clustered around 23 *kottams* (fortified settlements). By the eighteenth century, these *kottams*, which were conceived as the fortified defensive rallying-point of *nadus*, appear to have lost their significance. This was because policing and watch-and-ward duties in the region had been handed over to Telugu-speaking *palayakkarar*. These *palayakkarar* were descendants of the bands of marauders from the semi-arid region north of Tondaimandalam, and came from warlike Telugu castes of Raju and Naidu.[41] As British East India Company records suggest, the *nattar* of the region found the transaction cost of engaging them as guards (*kaval*) of *nadus* to be lower than that of engaging in constant wars with them. The erstwhile marauders now known as *palayakkarars*, undertook policing duties on contract for specific groups of villages. Except for some pockets, such as the Karunguzhi area, *palayakkarar* had become institutionalised as providers of *kaval*. Each village paid a share of the annual output (not exceeding two per cent) known as *kaval murai* (entitlement for *kaval*) or *menkaval* or *kandayam*, to the *palayakkarar*. In addition, the *palayakkar* of late eighteenth-century Tondaimandalam claimed political allegiance to the Raja of Chandaragiri, north of Tondaimandalam. The raja himself had accepted the suzerainty of the Nawab of Arcot. The civil and revenue administration of the nawabs was run by a many-tiered bureaucracy consisting of officials designated

[40] There is a sizeable volume of literature addressing this issue in the official records for the period 1780-90. See Sivakumar and Sivakumar (1993) for details.

[41] For details regarding these *palayakkarar* see ibid.

as *amildar*, *stala karnam*, *mazumdar* and *kanungo*. Most of these officials subsisted on revenue-free land grants (*manyam*, which the Muslims called *inam*). As noted already, village accounts of the period distinguish between *tarappadi manyam* (ancient grants whose size varied according to the quality of land), and *sanad manyam* (those granted by the Nawab's order or *sanad*). It would appear that the formal political apparatus of late eighteenth-century Tondaimandalam consisted of the nawab's civil administration run by his bureaucracy, and his police administration run by the Raja of Chandragiri through the *palayakkarar*. This is the structure which the British perceived when a large segment of fertile land known as the Jagir of Chengalpattu—the present-day district of Chengalpattu, and their first major territorial acquisition in South India—was granted to them in the 1760s. When the British began efforts at the revenue administration of this area in the 1780s, they discovered that the political and economic power of the region actually lay with the *nattar*. For one thing, the nawab's administration (which had an institutionalised system of bribery known as *nazar*) existed merely as a token. Secondly, the *palayakkarar* too were mainly at the mercy of the *nattar*, who could even order the non-payment of *kaval* fees if the *palayakkarar* functioned contrary to the interest of the *kaniyatchikkarar*. Most *palayakkarar* were also extremely poor, heavily in debt, and beset with internecine squabbles over rights of succession.

In short, the *nattar* had successfully pre-empted the challenge of marauding bandits, institutionalised them as their own policemen, and countered and economically overwhelmed the nawabs and their bureaucracy.[42] They had reason to believe that they could deal in a similar fashion with the new rulers of the region in the 1780s. Their ancestors were followers of Atandai, who had survived and prospered in the region for nearly a millennium; they had seen the hoards of Malik Kafoor, the raider from Delhi, pass by, and they had survived the annexation of their territory by the Vijayanagar empire.

For both the British and the *nattar*, their encounter on the land-revenue question of the 1780s was a strange experience. True the *nattar* and the British had had dealings with each other since the setting up of Fort St. George at Madras in the seventeenth century. However, the

[42] See ibid. for a castewise breakdown of the private creditors of the nawabs of Arcot in 1802.

British, who had hitherto dealt mainly with the trading Chetty caste of the region, encountered the *nattar* as a political force for the first time in the 1780s. As for the *nattar*, they saw the British as rulers rather than traders for the first time in the 1780s. The nature of this encounter and its aftermath are the subject of the ensuing discussion.

The kumbini *and the ghost of* kaniyatchi

In more senses than one, the British ruler was an enigma to the Tamil of Tondaimandalam. He spoke a strange tongue, dressed differently, and had eating-and-drinking habits and gods which were unknown. Even more important, he did not belong to any caste and arrived at Fort St. George utterly alone, devoid of any known support system such as the kin group; and this did not seem to worry the Briton either. Finally, he was backed by a corporate support system which was unheard of and enigmatic—an immense creature with its head lying across the distant seas and its limbs (employees) working to invisible commands; limbs that traded, conquered, connived, intrigued and eventually laid claim to the *melwaram* (the upper share or the ruler's share of the village output). The *kumbini*, as the Company was known, was a trader, a general, a king, but ever invisible. One could see only its myriad limbs—its functionaries in the Tamil country. Until the last decade of the eighteenth century the *nattar* reacted to the *kumbini* as they did to the nawab's administration, with a combination of appeals, bribes and intrigues, and finally, in 1795/6, with a brilliantly organised act of mass desertion. But by the end of the century the *nattar* were faced with the grim reality of failure. The *kumbini*, with which they could never deal face-to-face, was not vanquishable. It was like the mythological demons which grew back their limbs however many times they were severed. The *kumbini* was a transaction regime, an ideological system, and a social order which was completely alien and overpowering in the magnitude of its strength. Let us consider these aspects of the *kumbini* in some detail.

In the first place, the economic organisation of the *kumbini* ran counter to that of the *kaniyatchi*. The *kumbini* was a chartered company, an organisational form that arose out of the circumstances surrounding European trading with far-flung corners of the world. After the discovery of the sea route to India, it was understood that traders in Europe were in for a boom in their activity; but this possible boom also had certain riders attached to it. Given the high transportation costs, given that trading credit had to be advanced for a period of eight to twelve months on each venture, and given the high risk associated with each venture, gains could be made only if economies of scale existed in the organisation of business. These economies related to transport costs, insurance and management. Thus an organisation which could trade profitably had to own, or be able otherwise to commandeer, a sizeable fleet leading to a turnover large enough to minimise unit cost of transportation. Not merely this, there had to be as many ships returning with merchandise as setting out from the trading metropolis. Further, given the high risk, its distribution over a large number of investors became necessary. This implied an organisation that limited risk to the amount of financial participation in a venture; an investor's liabilities, in the case of the failure of a venture, could not exceed that of the capital invested by him in the venture. Finally, management had to be extensive enough in its functions to be able to deal with a number of issues, from diplomacy to the organisation of its own security—quite aside from trading.

All these led to the emergence of the chartered company, the predecessor of the modern corporation. The main feature of this organisation was centralisation of decision-making, of leadership and of managerial policy. These organisational imperatives and characteristics were the opposite of what the paddy economy of Tondaimandalam threw up. While the former stressed economies of scale, the latter featured diseconomies of scale. While the former was a mega-hierarchy dealing with large regions, the latter was an aggregation of small hierarchies across the region. While the former spelt out and modified its norms regarding its produce and input structure always with an eye to profit, the latter relied on complex sets of traditions and local norms that were oriented towards survival and subsistence. Most importantly, while the economic activity of the former stressed the individual (profits for shareholders, commissions and private trading incentives for the employees), the latter stressed the collectivity of individuals because of the organic solidarity of the agrarian universe at the village level.

Secondly, as a social group, the *kumbini* represented the trading bourgeoisie of Britain, a group which had grown strong and powerful by the mid-eighteenth century in Britain. This bourgeoisie was naturally both insensitive to and impatient with the social compulsions of agriculture even in Britain.[43] The sociology of Tamil agrarian relations was both irrelevant and alien to it, and more so the sociological compulsions of paddy cultivation. Thus all the social institutions of *kaniyatchi* were, at best, things to be tolerated as long as they could be used to serve the interests of the *kumbini.*

Thirdly, the ideology that the *kumbini* represented was a combination of capitalism and Christianity. Addressing itself to the problem of an appropriate revenue settlement in the 1790s, the Board of Revenue in Madras declared its allegiance to Smithian political economy.[44] We have noted elsewhere that this was far from the genuine intentions of the *kumbini.*[45] Be that as it may, the *kumbini* did look upon the agriculturist as a landed proprietor whose profits should be shared with the state. This conception of the *kaniyatchikkarar* was farthest from the existential reality of the Tondaimandalam agriculture. The former was not a 'landed proprietor', nor was he a profit-oriented producer.

If the *kumbini*, as a chartered company, was the secular face of the British state, the Church of England was its religious face. Not merely the Church of England, but all forms of Christianity found official or quasi-official encouragement under the rule of the *kumbini.* In many respects the church represented the ideological antithesis of Hinduism. The former was monotheist while the latter was pantheist; the former was bound by value absolutism, while the latter emphasised (equally) the context of value judgments; the former was centrally organised, whereas the latter was devoid of such a structure; the former emphasised theology; while the latter related man and god through mythology, the former was exclusive in its outlook and hence propagated itself through proselytisation, while the latter was ever inclusive and propagated itself through constant addition to its corpus of mythology; and finally, while the former had no social theory, the latter operated through the paradigm of varna. Consequently, there could never be even the slightest compatibility between the religious ideology of the conquering British and their Tamil subjects. Even in the last two decades of the eighteenth century, aggressive attempts were afoot at

[43] This was the period when the ideas of Adam Smith had gained acclaim, not only in Britain but in Europe. This was also the period of fermentation of the ideological contempt for the landed class, that found expression in the writings of Ricardo and his disciples. See also Ambirajan (1978).

[44] Board of Revenue (Consultation) Proceedings, vol.170.

[45] Sivakumar and Sivakumar (1993). See also Stein (1989).

mass conversion and the setting-up of exclusive native-Christian settlements, though they came to naught.[46] The religious factor was to remain one of the major undercurrents in the tension between the Tamil and his *dorai*, the white man. There is, for instance, even a strong argument that religion lay at the heart of the first Indian mutiny at Vellore in 1806.[47]

Finally, the *kumbini* and *kaniyatchi* represented different transaction regimes. The former was a bureaucracy representing post-feudal property rights of the West. Property and the entitlements attached to it had acquired modern forms in Britain by the end of the eighteenth century. In particular, land entitlements were considered as a right of property enjoyed at the will and pleasure of the unitary state.[48] In Tondaimandalam on the contrary, *kaniyatchi* entitlements at the village level had led to a segmentary state. More importantly, notwithstanding the late eighteenth-century acknowledgement of *nadu* and *nattar* as influential elements of the native polity, they had no place in the long-term calculations of the *kumbini*. There could be only one solution to the question of land revenue—that which suited the political-economic interests of the *kumbini*. Thus it was often mere window-dressing when officials of the *kumbini* debated the historical appropriateness of the raiyatwari land-revenue settlement, which looked upon the agriculturist as a mere *raiyat* or farmer who was entitled to till the soil as long as he paid the revenue.[49] There was no illusion about the fact that *kaniyatchi*, as a transaction regime, had to go.

The first of the blows against *kaniyatchi* was the decree in the 1790s that converted all the *melwaram* (the state's share of paddy output) into cash payments. In so doing, the *kumbini* transferred the risk of seasonal price fluctuations on to the *kaniyatchikkarar*. In the process the *kumbini* was also avoiding the cost of procuring and marketing grain. The earlier system of collecting paddy revenue in kind meant that the state

[46] Father Maneti, a missionary, sought the grant of several hundred acres of land in the region to accommodate settlements for newly-converted natives. There were vociferous protests from the *nattar*, but the debate continued for nearly a decade before it died down. See Sivakumar and Sivakumar (1993).

[47] Stein (1989).

[48] Place's *Report on the Jagir of Chengalpattu*, Chengalpattu District Records, vol.497.

[49] Even such 'authentic' historical arguments in support of the raiyatwari settlement as those of Thomas Munro appeared to have contained a well-planned element of fiction. See Stein (1989).

shared the risk of price fluctuation with the *kaniyatchikkarar* and *payirkarar*. Under the new arrangement, the *kumbini* not only took its revenue in cash, but also arrogated to itself the right to decide whether the paddy prices quoted by *kaniyatchikkarar* were 'fair' or not. This also meant increased seasonal fluctuations in paddy prices, as the influx of produce in the market after each harvest must have greatly increased soon after the decree; so that the *kaniyatchikkarar* and *payirkarar* were forced to market their produce at lower post-harvest prices than before. The overall effect of this new arrangement was, therefore, to introduce a new element of risk which was entirely borne by the subjects of the *kumbini*.

The unilateral imposition of this decree also meant the abandonment of the entitlement of the *kaniyatchikkarar* to be consulted before any change in the revenue assessment.[50] Their political status under the *kumbini* was unequivocally clarified for them: they could appeal, they could petition, and they could plead, but they could do little to influence the *kumbini*.

The raiyatwari regulations of 1802 were accompanied by a proclamation which stated that every cultivator had to apply for, and acquire, a *patta* (permit) entitling him to cultivate the lands in the village. Thus the complex of *kaniyatchi* entitlements over land were replaced at one stroke by the single entitlement of a *raiyatwari-pattadar* (the holder of a raiyat's *patta*). At the same time, the entire traditional scheme of distribution of village output was reduced to irrelevence by the inclusion of all the *swatantram* and *murai* in the state's revenue demands. The combined effect of these measures was the destruction of *kaniyatchi* as a transaction regime. More important, the economic basis of organic solidarity of the village was rudely shaken by these measures. No longer could economic risk be shared by all the members of the village economy, nor did traditional entitlements have legitimacy under the new regime.

The judgment of the Madras Supreme Court in the Tondayarpettai dispute in September 1808 was a portent of the events to follow and of the purpose of the government. This was a dispute between two groups

[50] This was the contention of the *kaniyatchikkarar* in the disputes with the *kumbini* in the last two decades of the eighteenth century. Towards the end of the century, a government proclamation accepted this in principle. See Sivakumar and Sivakumar (1993).

of Britons who had conflicting claims over parts of the land in Tondayarpettai. One group had leased its lands from the original Vellala *kaniyatchikkarar* of the village, whose *kaniyatchi* had been 'resumed' (the official term for forced dispossession of *kaniyatchikkarars*) in 1794 by the collector of the district. The other group had leased the same lands from the erstwhile *payirkarar* of the Gramini caste who now claimed *kaniyatchi*. The issue was posed in the hearing as to whether *kaniyatchi* could be 'resumed' in such a fashion, and whether it could be arbitrarily bestowed upon persons or groups by the *circar* (government). The judgment was categorically in the negative. The claim of the Vellalas to *kaniyatchi* was accepted without reservation. Yet the ruling went against them. For irrespective of their unfair and forced dispossession in 1794, they were not *in possession* of the village in 1808. It was a proclamation of the historic change that was already a *fait accompli*: 'But everything in the village is proved to have since changed hands. Its constitution appears to have changed; the *meerassee* [*kaniyatchi*] *privileges are all done away*: a new order of things, and a different set of proprietors must be taken to have succeeded.'[51]

In 1808, the Supreme Court judgment was the announcement of *kaniyatchi* as the existential reality of the time, as well as of its eventual demise in the times to come. Yet the demise did not occur till the 1860s. For, among other things, *kaniyatchi* found a strong protagonist in Ellis, the Collector of Madras during the second decade of the nineteenth century. From the 1820s, his deep and detailed understanding of the system, and his official papers on it, became standard reference points on all issues regarding *kaniyatchi*.[52] The courts almost always found Ellis' authority on these matters to be final.

There was no doubt that *kaniyatchi* as a structure of entitlements had to go, but without harm (if it were possible) to the socio-economic hierarchy of the rural areas, for the latter was a useful instrument with which the agrarian population could be controlled. Thus the power of the Vellalas could be useful, provided their power served the interests of the *kumbini*. The perspective was that of the first half of the nineteenth century. However, because of Ellis's influence, in a large number of court cases during this period, the award went in favour of the *kaniyatchikkarar*. In particular, the *kumbini* found that the courts

[51] Bailey and Hudleston (1862). Emphases added.
[52] Ibid.

repeatedly upheld the entitlement of the *kaniyatchikkarar* to veto the entry of 'outsiders' as *payirkarar*. There were two sets of transaction costs which must have dominated the consideration of revenue authorities at this time: the cost of slowly infiltrating 'outsiders' into cultivation, and the cost of liquidating *kaniyatchi* to provide free access to land.

By the 1850s, the latter proved to be lower. For one thing, revenue earnings lagged far behind the growing military and civil expenditure of the *kumbini*. For another, the cost of compensating *kaniyatchikkarar* for lands to be acquired for railways threatened to be prohibitive. Thus the 1859 land settlement was a decisive stroke against what remained of *kaniyatchi*. The Regulation which came to be known as the Darkhast Regulation (from the Persian for 'application'), scrapped the entire nomenclature of *kaniyatchi* land entitlements in the official records. All individuals requiring land for cultivation had to make a *darkhast* for the extent of land they required. This was sanctioned after scrutiny, and what was sanctioned was the only land which a person was entitled to cultivate, buy or sell. There were no more *kaniyatchikkarar* or *payirkar*; there were only raiyats. Any raiyat could purchase land from any other raiyat, but the act of purchase was legitimate only when the purchaser also transferred the *patta* (the slip of paper authorising such entitlements) to his name. The government also made it clear that these regulation were non-justiciable in any court of law. *Kaniyatchi* as a political order, as a formal transaction regime, as a structure of entitlements, was laid to rest. As a young British official put it in the mid 1860s, '...all that remained was to exorcise its ghost'.[53]

The exorcists

As an aftermath of the Darkhast Regulation the organisation of the rural economy and rural society in Tondaimandalam became increasingly schizophrenic. On the one hand, the economic compulsions of paddy and of irrigation continued to generate the decentralising imperatives of *kaniyatchi*. On the other, the new transaction regime was based on centralised and bureaucratic management of the economy, society and polity. From the ground level the tendency was to subvert and break down the centralised political order to the advantage of the villagers, while from the top of the bureaucratic-political pyramid the tendency

[53] Crole (1879).

was to manipulate and compel the society and economy to dovetail themselves to the objectives of the centralised regime. This was a destructive dichotomy: the village did not have the power effectively to subvert the state, but nor could the state prevent a rapid escalation in the transaction cost of manipulating and forcing the village to meet its demands. By the end of World War One, there was no doubt that the raj had become prohibitively costly to the British.[54] There was also no doubt that the agrarian economy would stagnate, as it actually did in the last 75 years of the raj.

Of the *kaniyatchi* institutions that ceased to exist politically, the most important was the *nadu*. With the consolidation of the centralised regime, *nadu* and its political imperative the segmentary state, became defunct. At the same time, as mentioned earlier, the organisational compulsions of agriculture remained. Kinship and caste continued to be the key organisational forms to ensure (a) access to information about the bureaucracy; (b) access to positions of power within the bureaucracy; (c) an insurance against risk in agriculture, and (d) factor rewards in agriculture. These institutions continued to be important, even if they had lost political legitimacy within the new entitlement structures. Likewise, at the village level rules of reciprocity and the ideological universe appear to have persisted well into the present century, even if custom and tradition, on which the normative order was based, had little legal meaning.

By 1860, however, the *kumbini* had ceased to exist, and so also *kaniyatchi* as a transaction regime. Further, Queen Victoria's imperial government (which replaced the *kumbini* with its bureaucracy) had a vast territory to administer, and required networks of contact and communication with the native populace, both for the sake of administration and for keeping itself informed about its subjects. The government came to rely on the networks of kinship and caste for these purposes, for there were no other networks available.

Kinship and caste, therefore, became instruments used by both the ruler and the ruled to maximise their respective advantages in the cen-

[54] The relationship between the growth in the size of the colonies and transaction costs has not been the subject of any historical work in Britain known to us. However, Morris (1979) has an interesting description of the growing expense of the 'raj' on which the sun never set. By 1900, according to Morris, the 'raj' had become an economic burden on London.

tralised colonial regime. This process became informally routinised by the end of the nineteenth century. In short, caste had effectively penetrated the imperial administration. As we shall see shortly, there was a price to be paid in the form of far-reaching changes in the institutional character of caste.

At the same time by the end of the first half of the twentieth century, the imperial government had also built up a vast legal, administrative and political apparatus, which rejected the ideological principles underlying caste society. Equality of opportunity, equality of all natives before the law, and equal franchise entitlements for all native adults, constituted the elements of the liberal democratic British legacy to Indian political society. While *kaniyatchi* perceived equality and justice as mutually exclusive normative categories, the liberal democratic tradition viewed equality as a precondition for social justice. Formally, ascriptive categories of status like caste acquired a subversive character within this emergent political ideology. Traditional avenues of social mobility, such as sanskritisation, had little legitimacy. Yet sanskritisation was more pronounced in this period as newer social groups gained access to power and resources within the political order. This is evidenced in the spate of caste *purana* (legends) which came into existence during 1870-1920, using the varna paradigm for socially and ideologically legitimising these mobile castes.[55]

The Tamil country was also subject to considerable Westernisation during the first half of the twentieth century. Formal Westernisation was increasingly found to be necessary in the process of gaining or claiming status within the urban milieu, where English education had been in vogue since the 1850s. The urban centres, with their growing white-collared middle-class populations, became places where it was fashionable to articulate Western liberal values. The urban middle-class Tamil of the 1940s (such as those in Madras) presented a picture of contradiction: a dual conception of social status, emerging from a dual social background—caste society and urban society. In the reconciliation of the two, caste underwent morphological changes; but there were other non-ideological influences too. Let us consider the changes.

One of the most important sources of change was the way caste was used by the state and the agrarian population. As in the case of all locality-based groups, so also in the case of castes: economies of scale

[55] See note 26.

were to be had in the formation of larger regional bodies when dealing with a centralised bureaucracy. Thus by the end of the nineteenth century, the Tamil country had witnessed the emergence of caste conglomerates that covered many districts: the *Mukkulattor Sangam* (covering the castes of Maravar, Kallar and Ahamudaiyar), and the *Vanniya Kula Kshatriya Sangam* (covering the castes of Pallis, Nayakars, Gounders and Padaiyatchis). These were newly-emergent dominant castes which formed new regional associations. Regional Vellala lobbies also came into existence, sometimes involving relaxation of the *nadu* rules of endogamy. In Tondaimandalam, the Reddys, the Naidus and the *Vanniya Kula Kshatriyas* emerged as rivals to the Vellalas, both as dominant castes and as conglomerates. The result was the gradual replacement of the village by the region as the main concern of caste organisations, which became informal regional loci of political power. Increasingly, intra-caste relationships among individuals became subjects of political patronage rather than of reciprocal relationships. The extent of wealth which commanded political patronage within the caste came to replace moral accountability to fellow-castemen in importance. In short, while castes retained many of their sociological features such as endogamy, nevertheless they took the form of sources of regional power for individual leaders who were increasingly in conflict with each other; each had his own group of clients, his own strategies to countermand the moves of his opponents, and his own contacts with the bureaucratic seats of power. Castes became informal fiefdoms within the colonial power hierarchy. In this milieu, the gradual decline of the older moral order was inevitable.

This tendency was further accentuated by the fact that the bureaucracy, both at the regional capital of Madras and at the district capitals, preferred to deal with the population through men of 'influence' and 'standing'. Caste politics therefore became important avenues through which the colonial regime exercised its power over the population. Encouraging and patronising particular members of the leadership of a particular caste became the easy way to manipulate the caste. Likewise, encouraging or discouraging particular castes with patronage became an easy way to maintain a favourable balance of power. During the twentieth century, the Tamil country witnessed the rise of two powerful political movements which were born in this process—the Justice Party, which supported the continuation of British rule in India, and the

Suya Mariyadai Iyakkam (Self-Respect Movement), which opposed the Congress party's attempt to mobilise support for its national struggle for independence. It is important to note that both movements formally opposed the caste ideology based on varna—the former from a liberal platform and the latter from a rationalist perspective.

After independence, the Congress Party assumed power in Tamilnadu, as it did in the rest of the country. The leadership of the party at the centre consisted of strong liberal democrats. Nehru, whose charismatic qualities put him in a position of unassailable power, was also a highly Westernised liberal socialist. The power of the centralised state apparatus increased several-fold in the decades after independence. This happened in two ways: (a) the economic philosophy of socialism strongly emphasised state control of the economy and of major industries through direct ownership; and (b) the onset of economic development meant vastly increased resource control (both direct and indirect) in the hands of the state.

These developments further weakened the role of the village in caste society. Older traditions and customary relationships in the village between employer and employed were also weakened by the formal extension to the village of the political ideology and institutions of liberal democracy through the ægis of *panchayati* raj. From the 1950s, village *panchayats* (local-government committees) were micro-level representations of the concept of equal opportunity for all, and had quotas of seats reserved for castes which were considered to constitute the weaker sections of the society. Resource entitlements were thus separated formally from entitlements towards membership of local government bodies—the very opposite of *kaniyatchi*—wherein the *ur* was a representative body of people with land entitlements.

The relationship between employee and employer also underwent a dramatic transformation from the 1940s onwards. Until then, a majority of employees were *padiyals*, employed on annual contracts of *padi* or salaries, as were their eighteenth-century ancestors.[56] However, the substantial increase in population, and the consequent decline in the land:man ratio, had put this system under considerable pressure at village level. Data for the region suggest a gradual decline in the *padiyal* system in favour of the *coolie* (casual labour) system during the first

[56] S.S. Sivakumar (1978) and R. Arunachalam (1985).

half of the twentieth century.[57] The 1960s constituted the most dramatic turning-point in this respect. From this period onwards the *padiyal* system declined perceptibly.

Let us consider the implication of this dramatic change in some detail. We had seen earlier that *padi*, *swatantram* and personal incentives constituted three distinctive elements of the eighteenth-century labour market. The first guaranteed subsistence to the labourers, while the other two involved risk-sharing and incentive principles. The replacement of *padi* with daily wages therefore threatened the subsistence of labourers who had at most one hundred days of work available to them in agriculture during the year.[58] From the point of view of the employers, the availability of a larger number of labourers made it cheaper to employ them on a casual basis; there was no need to hold them as *padiyals* in anticipation of a peak season with excess demand for labour.

The massive economic encouragement given by the government to farmers who adopted modernised agricultural practices constituted the final blow to the *padiyal* system. This programme, known euphemistically as the Green Revolution, involved provision to farmers of a technology package consisting of high-yielding varieties of seeds, fertilisers and irrigation (wells and pumpsets), backed by elaborate input subsidies, credit subsidies and price support. The availability of subsidised credit for the installation of privately-owned wells and pumpsets, freed individual farmers from the need to subscribe to the community-based irrigation-cum-cropping regimen which was centuries old.[59] Thus the labour demand of individual farmers could be staggered at times of their own choosing, resulting in the collapse of the organisational integration of land use, labour market and irrigation management in the village. The institutionalised basis of spreading risk across the different participants in agriculture collapsed at the same time. Risk became an

[57] Ibid.

[58] Sivakumar (1977) and Arunachalam (1985) estimate that there were 70 to 80 days of work available in agriculture in the villages studied by them during 1975-76 and 1982-85 respectively. These villages had a cropping intensity of 1.4 and 1.3 respectively.

[59] For instance, from 1959-60 to 1984-85, the number of diesel pumpsets used for agriculture increased from 18,654 to 1,731,016 in Tamilnadu. The area under pumpsets as a supplementary source irrigation rose by 241 per cent during the period 1959-75 (Government of Tamilnadu Season and Crop Reports).

individual undertaking for employer as well as employee. The impact of the growth of private well-irrigation on the older communal system of crops, and water and land management is best illustrated in the northern part of Tondaimandalam (North Arcot district) where hundreds of irrigation tanks have been abandoned, and then been encroached upon by individual farmers. In many villages the erstwhile tank can be seen as a part of the cultivated area. It is also interesting to recall that by 1973 this small district had one-fifth of all the irrigation pumpsets used in India. Thus caste and kinship ceased to be meaningful as institutionalised distributors of risk and as determinants of factor rewards at the village level. Caste and kinship thus became divorced from their agrarian origins once and for all.

At the same time, the political role of caste (and within it kinship) as a power lobby intensified considerably. This was natural considering the vast resources at the command of the state, resources which every caste sought to manipulate and commandeer to its advantage. The 1970s also saw the emergence of the *Vivasayigal Sangam* (Farmers' Association) as a powerful multi-caste political lobby. But within the next ten years, this organisation split into factions representing leadership from different castes. Thus the macro-level political reality of the Tamil country of the 1980s was a heightened continuum of that of the first half of the century. Politics was still caste politics. At the same time, until the 1980s, political parties in Tamilnadu articulated varied ideologies—from rationalism, through Tamil nationalism, to Marxism and Nehruvian socialism. The post-colonial transaction regime extended and deepened the colonial ideological dichotomy. In addition, by the 1980s the village had been completely subsumed and rendered devoid of significance within the regional control of development programmes, and within the ambit of regional caste politics. As noted earlier, the rapid eclipse of the older agro-irrigation imperatives for organic solidarity, and its replacement with the atomistic farmer and farm-worker, speeded up this process of subsumption. During the 1980s, the government decided to dispense with hereditary *karnams* (village accountants) and replace them with itinerant graduates designated as Village Administrative Officers (V.A.O.s). While this reflected the concern of the government to provide employment to a few thousand unemployed graduates, it was also symbolic of the final political reduction of the village. The V.A.O.s were not obliged to live in the villages under their care, which meant that villagers had now either to await their passing visits, or meet them in the nearest town where they lived, if they wished to conclude any business relating to their lands, crops, bank loans and agricultural subsidies. Village accounts had no longer to be placed in the villages, a far cry from the detailed variety of village accounts in the eighteenth century. The colonial transaction regime had nearly attained its full potential, four decades after the end of colonial rule.

Conclusion

In Tondaimandalam, rain-fed tanks, which irrigated the lands in most villages, created organisational imperatives in the eighteenth century. The need to follow uniform cropping patterns for efficient water use, and the consequent seasonality of paddy cultivation, led to strong inter-linkages between the use of and entitlements over land, labour and water. Centralised management of thousands of such tank, paddy and labour systems was riven with diseconomies of scale, given the heterogeneity in soil types, tank sizes and water-holding capacities, and the command and catchment areas of these tanks. The transaction regime, therefore, reflected both the uniformity of entitlements over land, water and labour, and the heterogeneity of localities in their inter-linkages. The transaction regime was a decentralised one, with the village as the monad and the *nadu* as the most significant and powerful socio-political level of aggregation.

Sociologically speaking, kinship and caste were the crucial organisational institutions. They were the categories around which hierarchy, the social division of labour, a reciprocity of inter-personal relationships, and norms of conduct, had evolved over the centuries. They ensured the identification of differences in skills and human types; they helped risk to be shared by groups; they constituted the basic sources of information on all matters; and they enabled the determination of factor rewards. Tradition and custom, which were immensely powerful, constituted the guidelines of human conduct. At the same time, traditions and customs were flexible, and permitted contextual interpretation. The ideology underlying the varied taxonomy of castes derived from the varna paradigm, which was attributed to a divine origin.

The political order and the transaction regime were expressed in terms of divine ordering of entitlements to land, water and political power. These entitlements, therefore, were to be held under a condition of strict accountability on the part of the landed to the rest of society, and of the ruler to the ruled. This led to the evolution of the *nadu* as the most powerful locus of political authority. The *nadu* itself derived its power as the spatial boundary to kinship in a group of villages which were the former's *raison d'être.* The political order was therefore a highly decentralised one.

The advent of the British raj witnessed the accession to power of a people who shared none of the elements of the indigenous social order, its agrarian imperatives and political compulsions. They were the product of an entirely different history, and constituted a transaction regime that stood opposed to *kaniyatchi* at every level. The history of *kumbini* rule over the region was one of an inconclusive struggle between the two transaction regimes. But the end of *kumbini* rule and its replacement by the imperial government signalled the formal demise of *kaniyatchi* as a transaction regime. In particular the *nadu* as a locus of formal political authority ceased to exist.

However, the organisational compulsions of paddy and of community irrigation continued to create a need for the caste, kinship and custom-based social order at ground level. Thus the last ninety years of colonial rule was a period of pre-emptive struggle: the bureaucracy sought to use these traditional institutions to meet the objectives of the imperial government; the native population sought to penetrate the imperial bureaucracy in order to meet the needs of different caste and kinship-based power groups. The effect was the emergence of mega-power groups based on caste and kinship which replaced the *nadu*. The decentralised segmentary state gave way to a centralised political authority. The transaction regime thus acquired characteristics of a schizophrenic political order: the agrarian economy continued to necessitate a decentralised polity, while imperial politics necessitated centralised management of the agrarian population.

This was also the character of the political order that came into being at the time of independence; for independence came as a takeover of the then existing bureaucracy and judiciary by Indian leadership. Significantly, there was very little attempt to reconsider the historical processes of the previous two centuries. The new leadership, which was mostly trained in Western intellectual tools, naturally did not feel the need to do so.[60] The result was the advent of a transaction regime that emphasised every one of the socio-political tenets of the colonial order in the name of independent self-government. The colonial dichotomy persisted and intensified. Caste as the basic source of regional power grew in strength, while formal institutions of the state used a rhetoric that condemned caste (in much stronger language) as a subversive relic of history. The village, which had been the source of power of the erstwhile *nadu*, continued to be subsumed with greater speed within a bureaucratic structure which greatly expanded after independence. The emergence of development planning created unprecedented resources in the hands of the state, and thus expedited the process. The 'Green Revolution' finally broke down the village social order. It enabled individual farmers to acquire private irrigation, thus 'liberating' them from the community-based irrigation and cropping regimen. With this the traditional labour market collapsed. The farmer who could free himself from the traditional cropping pattern could now regulate his demand for labour to his advantage by staggering his cultivation and crops, thus capitalising on off-season labour surplus. For labour this meant destitution, since the subsistence component of its reward (*padi*) ceased to exist while risk persisted. The organic solidarity of the village had become a matter of history by the 1980s.[61]

60 The sole exception was Gandhi's (1910) (1978) *Hind Swaraj*. This 60-page, imaginary dialogue between a moderate nationalist and an extreme nationalist begins with a debate on the meaning of *swaraj* (synonymous with 'home rule' at that time). Gandhi's contention in the debate was that he objected to colonial rule, not because the rulers were British by race, but because they represented the superimposition of a culture and a value system that was destructive of the older Indian culture and values. In fact, Gandhi made it clear in *Hind Swaraj* that he would find no reason to remove the British from power if they could ensure the survival and development of Indian civilisation. Conversely, he categorically refused to acknowledge that *swaraj* or 'home rule' simply meant the occupation of the white man's chair by a brown man. Not merely in the *Hind Swaraj*, but also in his political actions, Gandhi seems to have functioned with an 'ideal type' of pre-British-ruled India that resembled *kaniyatchi* (Sivakumar and Sivakumar 1993). Without distracting the reader unduly, we might point out that recent research shows that *kaniyatchi*-type entitlements existed in Royalaseema of Andhrapradesh as well as in the coastal tracts of Goa, in the name of Gankare. Likewise, the Patidars of Gujarat and the Chowbeesee of the Haryanvi Jats evoke memories of *kaniyatchi*.

61 The spate of popular Tamil films since the early 1980s romanticising the village *kattuppaadu* (order or discipline) and the wealthy dominant-caste benefactors are, in our opinion, gestures of a collective nostalgia over an irrevocably

This essay questions the existing paradigms of the agrarian social order, most of which, among economists, have relied upon a belief that it is necessary only to establish an analytically consistent set of abstract propositions of human behaviour, relaxation of the assumptions of which leads to completeness.[62] The major thrust of social anthropology and ethnography is in the opposite direction: preoccupation with the details of a given empirical situation lends substance to theoretical propositions which are, at best, analytical similes.[63] The present work rests on the latter belief, or praxis. We have used some of the recent developments in the theory of economic organisations as the initial points of reference and have focused on the social history of Tondaimandalam during the last two centuries. This exercise has brought us to an important conceptual category—the transaction regime.

This refers to a set of entitlements over resources, its concomitant politico-legal norms, both explicit and implicit, and the dynamics of interaction among individuals and groups to gain advantage through these norms. Economic processes always function on the foundations of transaction regimes. The costs involved are transaction costs. Our study of Tondaimandalam leads us to infer that political processes are ways in which different groups of individuals try to distribute transaction costs among themselves. The concept of a transaction regime can also relate systems of religious ideology, social stratification and social organisation, in a common frame of analysis, without any deterministic compulsions.[64]

In the pre-British transaction regime, the village was the main locus of the incidence of transaction costs, and hence was also the main source of political power in the form of the *nadu*. Conversely, the king,

---

extinct past. In fact, the most telling gesture symbolising the political demise of the village was the replacement of the post of hereditary *karnam* (accountant) with that of an itinerant Village Administration Officer (V.A.O.). Many of these V.A.O.s today visit the village only if called upon to do so by urgent and unavoidable business. We are informed that thousands of unemployed graduates who applied for the posts of V.A.O.s were interviewed by members of a committee headed by a retired major-general of the Indian army.

[62] For a detailed argument see Sivakumar (1986, 1992).

[63] Georgescu-Roegen (1971).

[64] It would, for instance, be possible to study the social and economic organisations without wishing to establish a deterministic marriage of Marxian economics and the so-called CDAWN school (Bardhan 1989, introduction). At the same time, one need not presuppose these organisations to be subject to purely stochastic growth processes (Ijiri and Simon 1977).

or the ruler who was the head of the state apparatus, bore a much smaller part of the transaction costs and derived his power from the political support of the *nadus*. The *nadu* thus constituted the crucial interface between formal political authority—the king or the ruler—and the actual sources of power—the agrarian populace of the village. Needless to say, the *nadu* was the crucial entity of economic, social and political organisation.

The colonial transaction regime and its successor depended on the appropriation of power within the fold of a centralised bureaucracy, which had no formal relationship with the organisational imperatives of either agriculture or the society. These regimes could achieve their objectives only by systematic reduction of the power of the villages through a legal apparatus that reduced the entitlement structures of the villages into derivatives of the state's blue-print. In this process, the burden of transaction costs shifted inexorably to the state and its bureaucracy. Increasing state power meant increasing transaction costs to the state. If there is any substance in the speculation that the possession of India as a colony became increasingly uneconomical to Britain, the speculation would also hold true for the independent Indian state. Centralised economic planning through state control of finance, prices, wages, industrial investment and agricultural inputs could be seen to enhance transaction costs significantly.

The effect of mounting transaction costs for the state would eventually affect the economic process itself. The stagnation of agriculture during the colonial era and the disproportionately sluggish response of the economy to the massive resource outlay in development by the state, during the first four decades of planning, might have some common causes in terms of transaction regimes. This paper is, however, not the place to go into the details of recent Indian economic problems. Suffice it to suggest the above line of enquiry as an interesting area for further research.

References

Ackerloff, G.A., 1984. *An Economic Theorist's Book of Tales* (Cambridge).

Alchian, A.A. and H. Demsetz, 1972. 'Production, information costs and economic organization', *American Economic Review* (December).

Ambirajan, S., 1978. *Classical Political Economy and the British Policy in India* (Cambridge).

Arrow, K., 1970. 'The organization of economic activity: issues pertinent to the choice of market versus non-market allocation', in R. Haverman and J. Margolis (eds.), *Public Expenditure and Policy Analysis* (Chicago).

—, 1985. 'The economics of agency', in J. Pratt and R. Zeckhanser (eds.), *Principals and Agents: the structure of business* (Cambridge, Mass.).

Arunachalam, R., 1985. 'Poverty, affluence and agrarian structure', unpublished PhD thesis, University of Madras.

Bailey, W.H. and W. Hudleston, 1862. *Papers on Mirasi Rights* (Madras).

Balusubramanian Vallai, 1927. *Velalarodu Totramum Avardam Varalarum* (Tamil: The Origin and the History of the Vellalas) (Madras).

Bardhan, P.K. (ed.), 1989. *The Economic Theory of Agrarian Institutions* (Oxford).

Barnam, H.N. and L. Squire, 1979. *A Model of an Agricultural Household: theory and evidence* (Baltimore and London).

Bell, C. and T.N. Srinivasan, 1989. 'Some aspects of linked product and credit market contracts among risk-neutral agents', in Bardhan (1989).

Bhagavad Gita, 1983. Sanskrit text with commentaries by Swami Chitbhavananda (Madras).

Chayanov, A.V., 1966. *Theory of Peasant Economy*, trans. and ed. by D. Thorner, B. Kerbley, and R.E.F. Smith (Homewood Cliff, Ill.).

Cholan Purva Pattaiyam 1957. (Tamil), edited by Chandrasekharan (Madras).

Coase, R., 1937. 'The nature of the firm', *Economica* 4.

—, 1960. 'The problem of social cost', *Journal of Law and Economics* 3.

Commons, J.R., 1957. *Legal Foundations of Capitalism* (Madison).

Crole, C.S., 1879. *Chingleput District Manual* (Madras).

Demsetz, H., 1988. *The Organisation of Economic Activity* (Oxford).

Desigar Thiruvambala, 1930. *Varnasrama Chandrika* (Tamil-Grantam; Madras, 1930).

Domar, E., 1968. 'Review of A.V. Chayanov's *Theory of Peasant Economy*', *American Economic Review* (June).

Gandhi, M.K., 1978. *Collected Works*, vol.X (New Delhi).

Georgescu-Roegen, N., 1960. 'Economic theory and agrarian economics', *Oxford Economic Papers*, (February).

—, 1971. *The Entropy Law and the Economic Process* (Cambridge, Mass.).

Gnayiru Parisanadar, 1897. *Karunikar Puranam* (Tamil: The Legend of the Karunikar Caste) (Madras).

Grossman, S. and O. Hart, 1982. 'The costs and benefits of ownership: a theory of vertical and lateral integration', *Journal of Political Economy* (August).

Hall, K.R., 1980. *Trade and Statecraft in the Age of the Cholas* (New Delhi).

Harriss, J., 1982. *Capitalism and Peasant Farming: structure and ideology in Northern Tamilnadu* (Bombay).

Hart, O. and B. Holmstrom, 1987. 'The theory of contracts', in T. Bewley (ed.), *Advances in Economic Theory: Fifth World Congress* (Cambridge).
Hayavadana Rao, n.d. *A Study of Caste* (Madras).
Heady, O.E., 1952. *Economics of Agricultural Production and Resource Use* (New York).
Hirschleifer, J. and J.G. Riley, 1979. 'The analytics of uncertainty and information: an expository survey', *Journal of Economic Literature* (December).
Ijiri, Y. and H.A. Simon, 1977. *Skew Distribution and the Sizes of Business Firms* (Amsterdam).
Iyengar, S. Sundararaja, 1921. *Land Tenure in Madras Presidency* (Madras).
Janakiraman, T., 1981. *Uyir Thaen* (Tamil: The Honey of Life) (Madurai).
Knight, F.H., 1951. *The Economic Organization* (New York).
Ludden, D., 1985. *Peasant History in South India* (Princeton).
March, J.G. and H.A. Simon, 1958. *Organizations* (New York).
McCloskey, D.N., 1976. 'English open fields as behavior towards risk', in P. Uselding (ed.), *Research in Economic History: an annual compilation of research* , vol.I (Greenwich, Conn.).
Meade, J.E., 1952. 'External economies', *Economic Journal* (March).
Menard, C., 1989. 'Why are there organisations?', *Working Paper* no.89/1 (Paris).
Morris, J., 1979. *Farewell the Trumpets: an imperial retreat* (Harmondsworth).
Munusamy Mudaliar, M., 1901. *Velan marabu Vilakkam* (Tamil: An Explanation of the Vellala Tradition) (Madras).
Nagaswamy, R., 1973. *Yavarum Kelir* (Tamil: Ancient Tamil Life: a reconstruction) (Madras).
Perumal Pillai, Tirisirapuram, A., 1922. *Adi Dravidar Varalaru* (Tamil: History of the Adi Dravidar Caste) (Madras).
Ross, S., 1973. 'The economic theory of agency: the principal's problem', *American Economic Review* 63.
Saminatha Iyer, 1950. *En Sarithiram* (Tamil: My History) (Madras).
Sankar, U. and G. Mythili, 1991. *Uncertainty and Farm Production Decisions* (Bombay).
Sastri, K.A.N., 1955a. *The Colas* (Madras).
—, 1955b. *A History of South India from Pre-historic Times to the Fall of Vijayanagar* (Madras).
—, 1963. *Development of Religion in South India* (Madras).
Sathyanathaier, 1987. *Studies in the Ancient History of Tondaimandalam* (Madras).
Savage, L., 1972. *The Foundations of Statistics* (New York).
Scherer, F.M., 1988. 'Corporate take-overs: the efficiency argument', *Journal of Economic Perspectives* (Winter)
Sen, A.K., 1981. *Poverty and Famines: an essay on entitlement and deprivation* (Delhi).
Sengunda Manadu, 1926. *Proceedings of the Provincial Conference of the Sengundar Caste* (Madras).
Servai, S., 1940. *Muvendar Kula Saridai* (Tamil: A History of the Kallar, Maravar and Agambadiar Castes) (Madras).
Shackle, G.L.S., 1955. *Uncertainty in Economics* (Cambridge).
Simon, H.A., 1957. *Models of Man* (New York).

—, 1991. 'Organizations and markets', *Journal of Economic Perspectives* (Spring).

Sivakumar, C., 1989. 'The lord with a cockerel pennant: power, violence, hierarchy and multi-dimensionality in Skanda Puranam', *Archives de Sciences Sociales des Religions* 61, 1 (Paris).

Sivakumar, C. and S.S. Sivakumar, 1979. 'Jati and class at Asthapuram-Kanthapuram: some comments towards a structure of interests', *Economic and Political Weekly* (annual number, February).

—, 1993. *Peasants and Nabobs: agrarian radicalism in late eighteenth-century Tamil country* (Delhi).

—, 1994. 'Rural change, social history and participant observation' in A.M. Shah *et al.*, *Essays in Honour of Professor M.N. Srinivas*, vol.I (New Delhi).

Sivakumar, S.S., n.d. 'The Theory of Peasant Economy: a study in neo-populism', unpublished PhD thesis, University of Delhi.

—, 1978a. 'Aspects of agriculture in Tamilnadu' (serialised in three parts), *Economic and Political Weekly* (May).

—, 1978b. 'Transformation of traditional economy in Tondaimandalam (1760 AD-1900 AD)', *Social Scientist* (May).

—, 1980. 'Disguised unemployment technology and private property', in V. Vyasulu (ed.), *Technology in the Indian Environment* (New Delhi).

—, 1986. 'Is the method all madness? Comments from a participant observer economist', *Social Research* (Winter).

—, 1991. 'Perception of leprosy in agriculture', in H. Srinivasan (ed.), *Social Science Research and Social Action for Leprosy Control* (Madras).

—, 1992. 'Field work: comments towards a praxis', in *Proceedings of the University Grants Commission Conference on Research Methodology* (New Delhi).

Sivakumar, S.S. and B. Batavia, 1992a. 'Induced innovations, transaction costs and economic organisations: some theoretical considerations' (mimeo., Madras).

—, 1992b. 'Transaction costs and allocative efficiency: some theoretical comments relating to India's economic liberalisation' (mimeo., Madras).

Stein, B., 1980. *Peasant State and Society in Medieval South India* (Oxford).

—, 1989. *Thomas Munro: the origins of the colonial state and his vision of the empire* (New Delhi).

Stigler, G., 1961. 'The economics of information', *Journal of Political Economy* 69.

Stiglitz, J.E., 1975. 'Incentive, risk and information: notes towards a theory of hierarchy', *Bell Journal of Economics* (Autumn).

—, 1991. 'Symposium on organisations and economics—introduction', *Journal of Economic Perspectives* (Spring).

Thirumalai, R., 1981. *Studies in the History of Ancient Townships in Pudukkottai* (Madras).

Ulaganatha Pandithar, n.d. *Jati Beda Villakkam* (Tamil: An Explanation of Differentiation of Castes) (Madras).

Varadaraja Pillai, A., 1909. *Sree Karunika Puranam* (Tamil: The Legend of the Karunikar Caste) (Salem).

Veerappillai, Ki, 1905. *Vanniyar Puranam* (Tamil: The Legend of the Vanniyar) (Madras).

Vijaya Doraswamy Gramony, 1910. *Arya Kshatriya Vilakkam* (Tamil: A Treatise on the Gramany Caste) (Madras).
Whitehead, 1988. *The Village Gods of South India* (New Delhi, 1921, reprint).
Williamson, O.E., 1975. *Markets and Hierarchies* (New York).
—, 1979. 'Transaction-cost economics: the governance of contractual relations', *Journal of Law and Economics* 22.
—, 1985. *The Economic Institutions of Capitalism* (New York).
—, 1986. *Economic Organisation: firms, markets and policy control* (Brighton).

Archival Records

Board of Revenue Proceedings (volumes relating to 1780-1800), Tamilnadu Archives, Madras.
Chingleput District Records (vols.440-450, 480-498), Tamilnadu Archives, Madras.
Madras Collectorate Records (selected volumes for the period 1800-1825), Tamilnadu Archives, Madras.
Proceedings of the Sudder Adalat Courts and Munsif's Courts for Madras Presidency (Judicial Department) (selected cases for 1800-1860).
Mackenzie Manuscripts (Tamil and Grantam) (selected manuscripts: legends and *kaifieths* pertaining to villages, left- and right-hand castes, and *palayakkarar*), Government Oriental Manuscripts Library, Madras.